Polymer Processing and Characterization

Polymer Processing and Characterization

Edited by
Bethany Ferguson

WILLFORD PRESS
www.willfordpress.com

Published by Willford Press,
118-35 Queens Blvd., Suite 400,
Forest Hills, NY 11375, USA

ISBN: 978-1-68285-781-6

Cataloging-in-Publication Data

Polymer processing and characterization / edited by Bethany Ferguson.
    p. cm.
Includes bibliographical references and index.
ISBN 978-1-68285-781-6
1. Polymers. 2. Polymer engineering. 3. Polymer melting. 4. Polymers--Analysis. I. Ferguson, Bethany.
TA455.P58 P65 2020
620.192--dc23

For information on all Willford Press publications
visit our website at www.willfordpress.com

WILLFORD PRESS

Contents

Preface...VII

Chapter 1   **An Assessment of Reuse of Light Ash from Bayer Process Fluidized Bed Boilers in Geopolymer Synthesis at Ambient Temperature** 1
Woshington S. Brito, André L. Mileo Ferraioli Silva, Rozineide A. A. Boca Santa, Kristoff Svensson, José Antônio da Silva Souza, Herbert Pöllmann and Humberto Gracher Riella

Chapter 2   **Cadmium Sulfide Thin Films Deposited onto MWCNT/Polysulfone Substrates by Chemical Bath Deposition** 7
M. Moreno, G. M. Alonzo-Medina, A. I. Oliva and A. I. Oliva-Avilés

Chapter 3   **Two-Dimensional Modeling of Thermomechanical Responses of Rectangular GFRP Profiles Exposed to Fire** 17
Lingfeng Zhang, Weiqing Liu, Guoqing Sun, Lu Wang and Lingzhi Li

Chapter 4   **Synthesis and Characterization of Porous Fly Ash-Based Geopolymers using Si as Foaming Agent** 34
D. Kioupis, Ch. Kavakakis, S. Tsivilis and G. Kakali

Chapter 5   **Role of Dielectric Constant on Ion Transport: Reformulated Arrhenius Equation** 45
Shujahadeen B. Aziz

Chapter 6   **Feasibility of Reprocessing Natural Fiber Filled Poly(lactic acid) Composites: An In-Depth Investigation** 56
Sujal Bhattacharjee and Dilpreet S. Bajwa

Chapter 7   **Evaluating Mechanical Properties of Few Layers MoS_2 Nanosheets-Polymer Composites** 66
Muhammad Bilal Khan, Rahim Jan, Amir Habib and Ahmad Nawaz Khan

Chapter 8   **Preparation and Characterization of Nano-Dy_2O_3-Doped PVA + $Na_3C_6H_5O_7$ Polymer Electrolyte Films for Battery Applications** 72
J. Ramesh Babu, K. Ravindhranath and K. Vijaya Kumar

Chapter 9   **Rake Angle Effect on a Machined Surface in Orthogonal Cutting of Graphite/Polymer Composites** 81
Dayong Yang, Zhenping Wan, Peijie Xu and Longsheng Lu

Chapter 10   **Molecular Association Studies on Polyvinyl Alcohol at Different Concentrations** 89
Richa Saxena and S. C. Bhatt

Chapter 11   **Effect of Molecular Chain Structure on Fracture Mechanical Properties of Aeronautical Polymethyl Methacrylate using Extended Digital Image Correlation Method** 94
Wei Shang, Xiaojing Yang, Xinhua Ji and Zhongxian Liu

Chapter 12   **Polymer-Cement Mortar with Quarry Waste as Sand Replacement** 102
D. N. Gómez-Balbuena, T. López-Lara, J. B. Hernandez-Zaragoza, R. G. Ortiz-Mena,
M. G. Navarro-Rojero, J. Horta-Rangel, R. Salgado-Delgado, V. M. Castano and
E. Rojas-Gonzalez

Chapter 13   **Needleless Electrospinning and Electrospraying of Mixture of Polymer and
Aerogel Particles on Textile** 112
Abu Shaid, Lijing Wang, Rajiv Padhye and Amit Jadhav

Chapter 14   **Deproteinization of Nonammonia and Ammonia Natural Rubber Latices by
Ethylenediaminetetraacetic Acid** 119
N. Moonprasith, A. Poonsrisawat, V. Champreda, C. Kongkaew, S. Loykulnant
and K. Suchiva

Chapter 15   **Experimental and Numerical Study of Texture Evolution and Anisotropic
Plastic Deformation of Pure Magnesium under Various Strain Paths** 126
Hamad F. Alharbi, Monis Luqman, Ehab El-Danaf and Nabeel H. Alharthi

Chapter 16   **Combined Effects of Sustained Loads and Wet-Dry Cycles on Durability of
Glass Fiber Reinforced Polymer Composites** 138
Mengting Li, Jun Wang, Weiqing Liu, Ruifeng Liang, Hota GangaRao and Yang Li

Chapter 17   **Functional Modification Effect of Epoxy Oligomers on the Structure and
Properties of Epoxy Hydroxyurethane Polymers** 152
Victor Stroganov, Oleg Stoyanov, Ilya Stroganov and Eduard Kraus

Chapter 18   **Microstructure and Damping Property of Polyurethane Composites Hybridized
with Ultraviolet Absorbents** 168
Jiang Chang, Bing Tian, Li Li and Yufeng Zheng

Chapter 19   **Preparation and Sealing Performance of a New Coal Dust Polymer Composite
Sealing Material** 177
Bo Li, Junxiang Zhang, Jianping Wei, and Qiang Zhang

Chapter 20   **Preparation of Poly(L,L-Lactide) Microparticles via Pickering Emulsions using
Chitin Nanocrystals** 187
T. S. Demina, Yu. S. Sotnikova, A. V. Istomin, Ch. Grandfils, T. A. Akopova and
A. N. Zelenetskii

Chapter 21   **The Synthesis and Characterization of Poly (methyl methacrylate-tourmaline acrylate)** 195
Yingmo Hu, Xubo Chen and Yunhua Li

Permissions

List of Contributors

Index

Preface

I am honored to present to you this unique book which encompasses the most up-to-date data in the field. I was extremely pleased to get this opportunity of editing the work of experts from across the globe. I have also written papers in this field and researched the various aspects revolving around the progress of the discipline. I have tried to unify my knowledge along with that of stalwarts from every corner of the world, to produce a text which not only benefits the readers but also facilitates the growth of the field.

Polymer consists of various repeated sub-units which together make up a large molecule. There are two types of polymers – natural and synthetic or man-made. Natural polymeric materials include amber, wool, shellac, silk and natural rubber. Synthetic polymers include polystyrene, polyethylene, polyvinyl chloride, synthetic rubber, neoprene, silicon, Bakelite and PVB. The process of polymerization includes the combining of various monomers into a network that is covalently bonded. Both natural and synthetic polymers are created through polymerization. The identity of the constituent monomers of a polymer and its microstructure are the most basic properties of polymers. This book includes some of the vital pieces of work being conducted across the world, on various topics related to polymers. It presents researches and studies performed by experts across the globe. The extensive content of this book provides the readers with a thorough understanding of the subject.

Finally, I would like to thank all the contributing authors for their valuable time and contributions. This book would not have been possible without their efforts. I would also like to thank my friends and family for their constant support.

Editor

An Assessment of Reuse of Light Ash from Bayer Process Fluidized Bed Boilers in Geopolymer Synthesis at Ambient Temperature

Woshington S. Brito ⓘ,[1] André L. Mileo Ferraioli Silva ⓘ,[1] Rozineide A. A. Boca Santa,[2] Kristoff Svensson,[3] José Antônio da Silva Souza,[3] Herbert Pöllmann,[3] and Humberto Gracher Riella[2]

[1]Pará Federal University, 66075-110 Belém, PA, Brazil
[2]Chemical Engineering Department, Santa Catarina Federal University, 88040-900 Florianópolis, SC, Brazil
[3]Department of Geosciences and Geography, Martin Luther University of Halle-Wittenberg, Halle 06108, Germany

Correspondence should be addressed to Woshington S. Brito; wsbrito3@gmail.com

Academic Editor: Munir Nazzal

Sustainable civil construction in the future, besides having low energy consumption and greenhouse gas emissions, must also adopt the principle of reusing wastes generated in the production chain that impact the environment. The aluminum production chain includes refining using the Bayer process. One of the main wastes produced by the Bayer process that has an impact on the environment is fly ash. Geopolymers are cementitious materials with a three-dimensional structure formed by the chemical activation of aluminosilicates. According to studies, some are proving to be appropriate sources of Al and Si in the geopolymerization reaction. The research reported here sought to assess the possibility of reusing fly ash characteristic of the operational temperature and pressure conditions of Bayer process boilers in geopolymer synthesis. Geopolymerization reaction was conducted at an ambient temperature of 30°C, and the activator used was sodium hydroxide (NaOH) 15 molar and sodium silicate (Na_2SiO_3) alkaline 10 molar. Fly ash and metakaolin were used as sources of Al and Si. XRD, XRF, and SEM Techniques were used for characterizing the raw materials and geopolymers. As a study parameter, the mole ratios utilized followed data from the literature described by Davidovits (year), so that the best results of the geopolymer samples were obtained in the 2.5 to 3.23 range. Resistance to mechanical compression reached 25 MPa in 24 hours of curing and 44 MPa after 28 days of curing at ambient temperature.

1. Introduction

The chain for producing aluminum from bauxite to primary aluminum includes refining using the Bayer process. One of the main waste products of the Bayer process is the fly ash collected from the cyclone overflow of the fluidized bed boilers at operational conditions of 900°C of temperature and 120 kPa of pressure. The pressure and temperature conditions influence the specification and morphology of the ash waste generated [1].

In the Bayer process, the circulating-type fluidized bed boilers operate at specific conditions of 900°C and 120 kPa as a means of guaranteeing maintenance of the bed and generate a specific fly ash characteristic of that process. Fly ashes are classified into two types. These are C type class with a higher amount of calcium oxide and F type class with a lower amount of calcium oxide [2].

When the sum of the levels of silica, alumina, and iron oxide is greater than 70%, the fly ash is classified as class F [3].

Geopolymers are synthesized at a different temperature by the alkaline activation of aluminosilicates derived from natural minerals, calcined clay or industrial by-products 2008. This activation is generally done with metakaolin silicates of sodium or potassium. Geopolymers are inorganic ligands with good resistance to high temperatures and degradation by acid, as well as good mechanical properties. They are thus an attractive alternative to Portland-type cement, and their use makes it possible to recycle large quantities of industrial wastes. The mechanical properties of geopolymeric materials depend upon the alkaline cation (Na$^+$), the SiO$_2$/Al$_2$O$_3$ molar relation as used by Davidovits and the conditions under which the reaction occurs [4, 5].

The term "geopolymer" describes the chemical properties of aluminosilicate-based inorganic polymers. Geopolymers present properties of cement and therefore have great potential for use in the civil construction industry [6].

The water trapped in a network of geopolymers generates porosity, which results in a reduction of mechanical properties. Their three-dimensional structure is made up of SiO$_2$ and MAlO$_{4t}$ tetrahedrons, where M is a monovalent cation, typically Na$^+$. This network is comparable to some zeolites but differs in its amorphous character. The polymeric character of those materials increases with the SiO$_2$/Al$_2$O$_3$ relation, as the aluminum atoms cross the SiO$_2$ tetrahedron chains [7].

Changing the SiO$_2$/Al$_2$O$_3$ relation in geopolymers thus enables the synthesis of materials with different structures. The geopolymerization mechanism is particularly difficult due to the kinetics of the reaction. However, the majority of authors agree that the mechanism involves dissolution, followed by gel polycondensation [8, 9].

Studies show that the mechanical forces of geopolymers increase with elevation of the temperature of kaolin calcination in order to generate metakaolin; however, the ideal temperature for calcination of kaolin is around 700°C for 2 hours [10, 11].

The development of ash-based geopolymer derived from burning coal shows promise as a means of reusing that waste. The worldwide demand for coal will continue to grow up to 2030, achieving a duplication in relation to current demand. With the increasing consumption of coal for generating energy, there will be an increased production of ashes as a byproduct of burning coal [12].

The cements produced by alkaline activation of aluminosilicates were studied in a search for materials with bonding properties more resistant than those of current Portland cements, as well as materials that can be produced with low cost raw materials, little expenditure of energy, and especially with low levels of toxic gas emissions into the atmosphere [13].

Those studies point to excellent possibilities for being implanted around the world and such materials can be produced on a large scale, suppressing the demand for cement in a market that is growing every year. In that context, this study was performed with the objective of assessing the microstructural properties and resistance to compression of geopolymeric materials produced using the light ash characteristic of the Bayer process.

2. Materials and Methods

Fly ash waste was collected from the cyclone overflow of the fluidized bed circulating-type boilers of the Bayer process at conditions of 900°C of temperature and 120 kPa of pressure. That waste was utilized as a source of Si and Al in

TABLE 1: Compositions formulated using fly ashes and metakaolin for obtaining geopolymers.

Geopolymers	Fly ash (mass %)	Metakaolin (mass %)	SiO$_2$/Al$_2$O$_3$
1	100	0	4.4
2	94.24	5.76	4.11
3	90.39	9.61	3.94
4	84.62	15.38	3.72
5	75	25	3.4
6	69.2	30.8	3.23
7	61.54	38.46	3.041
8	53.85	46.15	2.87
9	42.31	57.69	2.65
10	34.62	65.38	2.52

TABLE 2: Molar concentration and proportions of the activating solution.

Activator	NaOH	Na$_2$SiO$_3$
Molar concentration	15	10
Proportion (NaOH : Na$_2$SiO$_3$)	1 : 3 in mass of solution	

geopolymer synthesis. Besides the fly ash, another source of Si and Al was metakaolin produced by the calcination of kaolin at a temperature of 800°C for two hours.

To analyze the crystalline structure of ash, kaolin, and metakaolin, X-ray diffractometry (XRD–Bruker LyuxEye) was utilized. For chemical analyses, X-ray fluorescence was employed using equipment from AxiosMinerals (PANalytical), with a ceramic X-ray tube and rhodium anode. For the morphology study, we made use of the scanning electron microscopy (SEM) technique using Zeiss MEO 1430 equipment. Ash and metakaolin were submitted to granulometric analysis of their particles, in a Fritsch Analysette 22 Micro tec plus laser granulometer.

The mole ratios utilized by Davidovits (SiO$_2$/Al$_2$O$_3$) are an important parameter for orienting the best compositions for geopolymer synthesis [14]. Ten compositions of geopolymers were formulated using fly ash and metakaolin as a base, with different mole ratios among SiO$_2$/Al$_2$O$_3$. Table 1 presents the geopolymers studied with their respective SiO$_2$/Al$_2$O$_3$ ratios.

The alkaline activator used for synthesizing the geopolymer was composed of a solution of sodium hydroxide (NaOH) micro pearl (neon, 97% purity). A solution of alkaline sodium silicate (Na$_2$SiO$_3$) (Manchester Química do Brasil S.A., SiO$_2$/Na$_2$O = 3.2) was also used. Table 2 shows the composition of the activator.

The compositions were prepared in a mechanical mixer and placed in cylindrical molds with 100 mm height and diameter of 50 mm. After molding, the molds were submitted to curing at an ambient temperature of 30°C. Geopolymers at 24 hours and 7 and 28 days of curing were submitted to compression resistance testing in an Emic SSII300 press. Geopolymers before and after synthesis were submitted to XRD analysis in order to assess their degree of polymerization. SEM tests were performed after the geopolymer synthesis.

TABLE 3: Chemical composition in % of ash and of metakaolin.

Material	SiO$_2$	Al$_2$O$_3$	CaO	Fe$_2$O$_3$	Na$_2$O	MgO	TiO$_2$	P$_2$O$_5$	Loi
Ash	42.531	16.399	19.005	7.081	0.941	0.264	0.897	0	12.882
Metakaolin	53.36	43.58	0	0.6	0.33	0	1.51	0.13	0.49

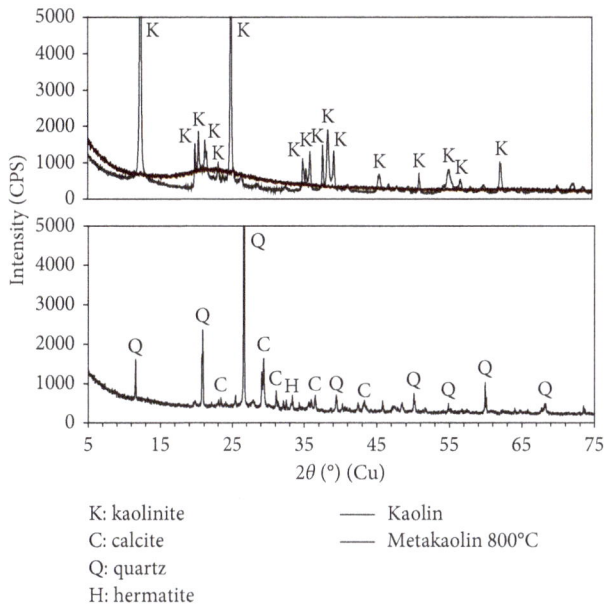

K: kaolinite   —— Kaolin
C: calcite   —— Metakaolin 800°C
Q: quartz
H: hermatite

FIGURE 1: XRD of kaolin, of metakaolin, and of fly ash.

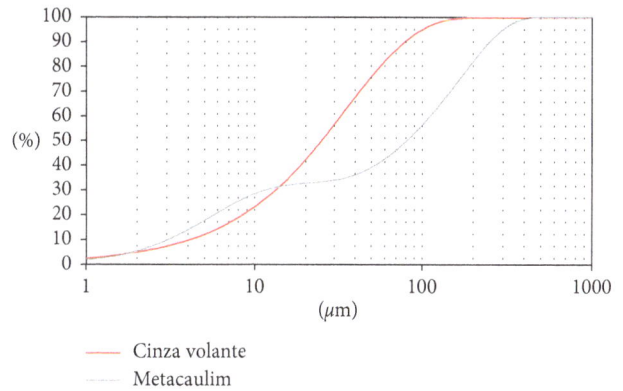

FIGURE 2: Distribution of the size particles for fly ash and metakaolin.

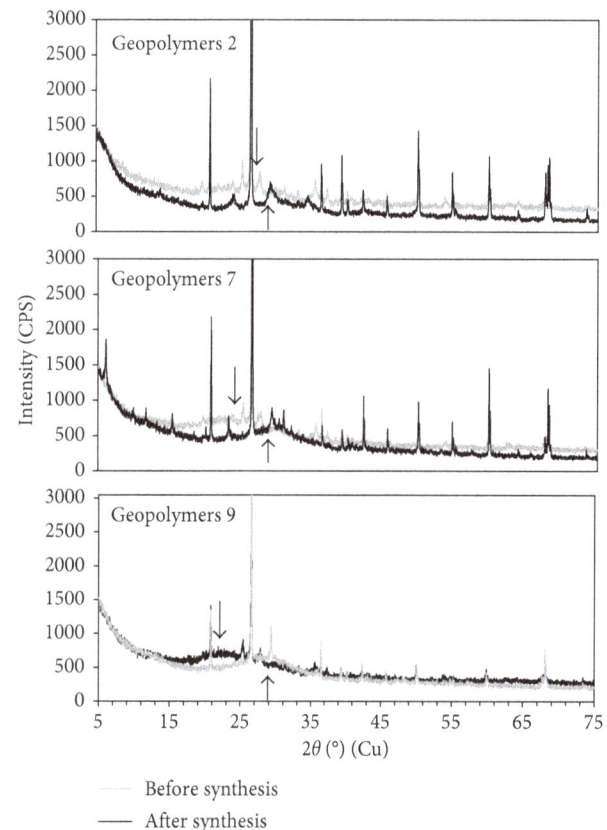

FIGURE 3: XRD of geopolymers 2, 7, and 9 before and after synthesis.

3. Results

Table 3 presents the results of the XRF analyses of the materials used in geopolymer synthesis. One may observe (Table 3) the 42.531% level of SiO$_2$ and 16.399% level of Al$_2$O$_3$ in the fly ash used, which indicates that ash is a rich source of Al and Si. Metakaolin (Table 3) is composed of 43.58% Al$_2$O$_3$ and 63.36% SiO$_2$ which characterizes it as a predominant source of aluminum in proportion [15–18].

The use of fly ash waste from burning charcoal as a source of Si and Al favors the geopolymerization reaction and guarantees good durability for the geopolymer [19–23].

According to the result from XRF, the fly ash generated in the Bayer process under the operational temperature and pressure conditions may be classified as C class type fly ash, due to the sum of the levels of silica, alumina, and iron oxide being lower than 70%.

Figure 1 presents the result of the XRD test of fly ash, of kaolin, and of metakaolin calcined at 800°C for 2 hours.

As shown in Figure 1, the calcination of kaolin at 800°C for two hours was satisfactory because it generated a change in the crystalline phase, transforming into an amorphous phase, which characterizes metakaolin.

The change occurring in the material after thermal treatment frees Al and Si for the geopolymerization reaction. Kaolin calcined at a range from 600°C to 800°C for a time interval from 2 to 4 hours, shows itself to be adequate for obtaining metakaolin applied to the development of geopolymer [24–26].

The fly ash generated in aluminum refineries presents quartz and calcite in its composition as shown in the diffractogram for ash in Figure 1.

The distribution of the particle size for the raw materials used in synthesizing the geopolymer resulted in an average

(a)

(b)

FIGURE 5: Scanning electron microscopy (SEM): (a) Geopolymer 9 increase 500x; (b) Geopolymer 4 increase 500x.

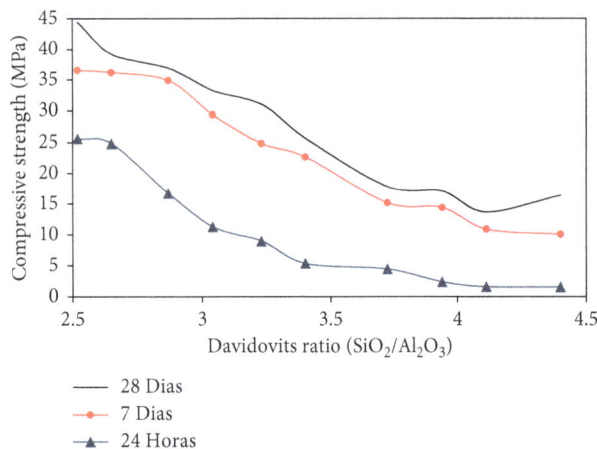

FIGURE 4: Compressive Strength (MPa) of the geopolymers.

diameter (d_{50}) of 24 μm for the fly ash and 88 μm for the metakaolin. Figure 2 presents the results of the laser granulometric analysis for fly ash and metakaolin.

Nath et al. [27] worked on a synthesized geopolymer with fly ash at a granulometry of 4.7 μm and studied the geopolymerization reaction using metakaolin with an average diameter of 18 μm.

Figure 3 presents the result of the XRD analysis for geopolymers 2, 7, and 9 with the arrow showing the halo before and after synthesis. According to Salih et al. [5] the geopolymer with the best formed and defined halo indicates larger amorphous zones and thus greater regions of reaction occurrence. In his studies, Salih et al. [5] showed that the greater displacement of the halo after synthesis indicates geopolymers with a greater degree of geopolymerization and thus greater mechanical resistance.

According to the data presented in Figure 3, geopolymer 2 presented a low halo formation, indicating a lower degree of geopolymerization in relation to geopolymer 7. Geopolymer 9 presented a good definition and displacement of the halo before and after the synthesis. That agrees with Figure 4 that presents the results of mechanical compressive strength in MPa, indicating that geopolymer 9 has greater mechanical compressive strength to geopolymers 7 and 2.

Geopolymers with a ratio in the 2.5 to 3.23 range, as indicated by Davidovits, presented mechanical resistance to geopolymers with a greater ratio. Geopolymer 9 with a ratio of 2.65 presented a high compressive strength, reaching values of 25 MPa, 36.22 MPa, and 44 MPa in 24 hours, 7 days, and 28 days of cure, respectively, at ambient temperature. From the results of Figure 4, the geopolymers from samples 8, 9, and 10 in 24 hours achieved the resistance of conventional Portland-type concrete at 28 days of curing, but compressive strength continues to increase in the tested range and is thus higher [28].

Figure 5 presents the result of scanning electron microscopy (SEM) with 500x expansion of geopolymer 9 in (a) and of geopolymer 4 in (b).

In Figure 5, one may observe that geopolymer 9 presented a denser and more homogenous morphology when compared to geopolymer 4. The more uniform morphology

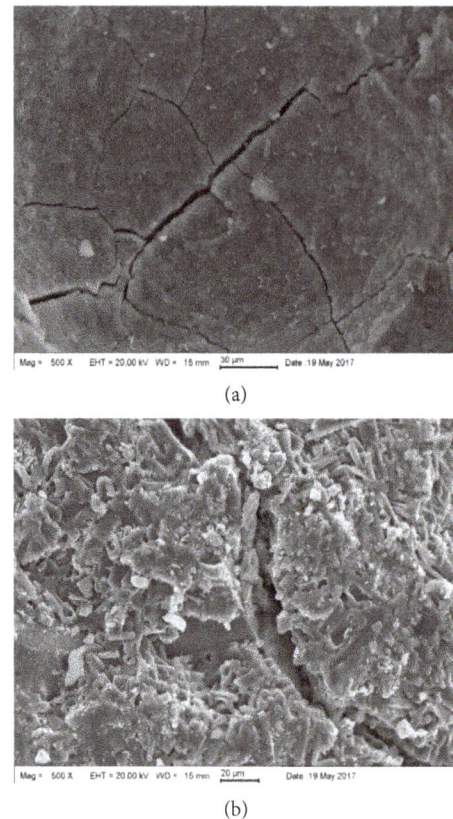

of geopolymer 9 is in accord with the results obtained in the compression resistance tests and also indicated a greater degree of geopolymerization after the reaction. Salehi et al. [29] found in his studies that the denser and more uniform morphological aspects provide greater mechanical performance to the geopolymers and thus greater advance in the geopolymerization reaction.

4. Conclusion

This research made it possible to develop a geopolymer from metakaolin and industrial waste from the Bayer process that contained amorphous aluminosilicates in their composition. The waste utilized was fly ash generated in the operational conditions of the Bayer process. As seen in the procedures, the synthesis of the geopolymer according to the 2.5 to 3.3 mole ratios utilized by the researcher Davidovits was favorable to the process, and geopolymers with greater mechanical compressive strength were obtained. Geopolymers synthesized at ambient temperature with an activator having a composition of 15 molar of sodium hydroxide and 10 molar of alkaline sodium silicate appear to be a viable alternative for supplying demands for cement, achieving resistance superior to that of conventional concretes derived from Portland-type cement. Furthermore, the geopolymers were produced with low emissions of CO_2, when compared with conventional cements, which is environmentally favorable. With the characterization of the synthesized geopolymers, it was concluded that the waste from the Bayer process presents a characteristic favorable to the synthesis of geopolymeric materials, classified by the study as class C ash. Our research concludes that synthesized geopolymers have great potential for production of geopolymeric materials.

Conflicts of Interest

The authors declare that they have no conflicts of interest.

Acknowledgments

The authors would like to thank CNPq, CAPES, UFPA, PRODERNA, UFSC, Martin Luther University Halle-Wittenberg, and the Hydro Alunorte company.

References

[1] M. Ahmaruzzaman, "A review on the utilization of fly ash," *Progress in Energy and Combustion Science*, vol. 36, pp. 327–363, 2010.

[2] A. Narayanan and P. Shanmugasundaram, "An experimental investigation on flyash-based geopolymer mortar under different curing regime for thermal analysis," *Energy and Buildings*, vol. 138, pp. 539–545, 2017.

[3] J. B. M. Dassekpo, X. Zha, and J. Zhan, "Synthesis reaction and compressive strength behavior of loess-fly ash based geopolymers for the development of sustainable green materials," *Construction and Building Materials*, vol. 141, pp. 491–500, 2017.

[4] P. Duxson, J. L. Provis, G. C. Lukey, S. W. Mallicoat, W. M. Kriven, and J. S. J. Deventer, "Understanding the relationship between geopolymer composition, microstructure and mechanical properties," *Colloids and Surfaces A: Physicochemical and Engineering Aspects*, vol. 269, pp. 47–58, 2005.

[5] M. A. Salih, A. A. A. Abang, and N. Farzadnia, "Characterization of mechanical and microstructural properties of palm oil fuel ash geopolymer cement paste," *Construction and Building Materials*, vol. 65, pp. 592–603, 2014.

[6] J. Davidovits, *Geopolymer: Chemistry and Applications*, Institut Géopolymère, Saint-Quentin, France, 3rd edition, 2011.

[7] J. Davidovits, *Geopolymer: Chemistry and Applications*, Institut Geopolymere, St-Quentin, France, 2nd edition, 2008.

[8] P. Duxson, J. A. Fernandez, J. L. Provis, G. C. Lukey, A. Palomo, and J. S. J. Van Deventer, "Geopolymer technology: the current state of the art," *Journal of Materials Science*, vol. 42, no. 9, pp. 2917–2933, 2007.

[9] P. Duxson, J. A. Fernandez, J. L. Provis, G. C. Lukey, A. Palomo, and J. S. J. Van Deventer, "The role of inorganic polymer technology in the development of 'green concrete'," *Journal of Materials Science*, vol. 42, pp. 2917–2933, 2007.

[10] M. R. Wang, D. C. Jia, P. G. He, Y. Zhou, and M. Lett, "Influence of calcination temperature of kaolin on the structure and properties of final geopolymer," *Materials Letters*, vol. 64, no. 22, pp. 2551–2554, 2010.

[11] R. Cioffi, L. Maffucci, and L. Santoro, "Optimization of geopolymer synthesis by calcination and polycondensation of a kaolinitic residue," *Resources, Conservation and Recycling*, vol. 40, pp. 27–38, 2003.

[12] J. P. Castro-Gomes, A. P. Silva, R. P. Cano, J. Durán Suarez, and A. Albuquerque, "Potential for reuse of tungsten mining waste-rock in technical-artistic value added products," *Journal of Cleaner Production*, vol. 25, pp. 34–41, 2012.

[13] J. D. Gavronski, *Mineral Coal and Renewable Energies in Brazil*, Federal University of Rio Grande do Sul, Porto Alegre, Brazil, 2007.

[14] R. A. A. B. Santa, A. M. Bernardin, H. G. Riella et al., "Geopolymer synthetized from bottom coal ash and calcined paper sludge," *Journal of Cleaner Production*, vol. 57, pp. 302–307, 2013.

[15] H. K. Tchakoute, C. H. Rüscher, J. N. Y. Djobo, B. B. D. Kenne, and D. Njopwouo, "Influence of gibbsite and quartz in kaolin on the properties of metakaolin-based geopolymer cements," *Applied Clay Science*, vol. 107, pp. 188–194, 2015.

[16] H. Y. Zhang, V. Kodur, S. L. Qi, L. Cao, and B. Wu, "Development of metakaolin-fly ash based geopolymers for fire resistance applications," *Construction and Building Materials*, vol. 55, pp. 38–45, 2014.

[17] P. Duan, C. J. Yan, W. J. Luo, and W. Zhou, "A novel surface waterproof geopolymer derived from metakaolin by hydrophobic modification," *Materials Letters*, vol. 164, pp. 172–175, 2016.

[18] M. Ardanuy, J. Claramunt, and R. D. T. Filho, "Cellulosic fiber reinforced cement-based composites: a review of recent research," *Construction and Building Materials*, vol. 79, pp. 115–128, 2015.

[19] M. Marcin, M. Sisol, and I. Brezani, "Effect of slag addition on mechanical properties of fly ash based geopolymers," *Procedia Engineering*, vol. 151, pp. 191–197, 2016.

[20] M. Soutsos, A. P. Boyle, R. Vinai, A. Hadjierakleous, and S. J. Barnett, "Factors influencing the compressive strength of fly ash based geopolymers," *Construction and Building Materials*, vol. 110, pp. 355–368, 2016.

[21] X. Ma, Z. H. Zhang, and A. G. Wang, "The transition of fly ash-based geopolymer gels into ordered structures and the effect on the compressive strength," *Construction and Building Materials*, vol. 104, pp. 25–33, 2016.

[22] C. D. Atis, E. B. Gorur, T. Karahan, C. Bilim, S. Likentapar, and E. Luga, "Very high strength (120 MPa) class F fly ash geopolymer mortar activated at different NaOH amount, heat curing temperature and heat curing duration," *Construction and Building Materials*, vol. 96, pp. 673–678, 2015.

[23] F. N. Okoye, J. Durgaprasad, and N. B. Singh, "Effect of silica fume on the mechanical properties of fly ash based-geopolymer concrete," *Ceramics International*, vol. 42, no. 2, pp. 3000–3006, 2016.

[24] B. B. K. Diffoa, A. Elimbi, C. J. D. Manga, and H. T. Kouamo, "Effect of the rate of calcination of kaolin on the properties of metakaolin-based geopolymers," *Journal of Asian Ceramic Societies*, vol. 3, no. 1, pp. 130–138, 2015.

[25] J. Davidovits, *Geopolymers Based on Natural and Synthetic Metakaolin–A Critical Review*, Elsevier, New York, NY, USA, 2015.

[26] V. Medri, S. Fabbri, J. Dedecek, Z. Sobalik, Z. Tvaruzkova, and A. Vaccari, "Role of the morphology and the dehydroxylation of metakaolins on geopolymerization," *Applied Clay Science*, vol. 50, no. 4, pp. 538–545, 2010.

[27] S. K. Nath, S. Maitra, S. Mukherjee, and K. Sanjay, "Microstructural and morphological evolution of fly ash based geopolymers," *Construction and Building Materials*, vol. 111, pp. 758–765, 2016.

[28] A. M. Rajabi and F. O. Moaf, "Simple empirical formula to estimate the main geomechanical parameters of preplaced aggregate concrete and conventional concrete," *Construction and Building Materials*, vol. 146, pp. 485–492, 2017.

[29] S. Salehi, M. J. Khattak, A. H. Bwala, and F. S. Karbalaei, "Characterization, morphology and shear bond strength analysis of geopolymers: implications for oil and gas well cementing applications," *Journal of Natural Gas Science and Engineering*, vol. 38, pp. 323–332, 2017.

Cadmium Sulfide Thin Films Deposited onto MWCNT/Polysulfone Substrates by Chemical Bath Deposition

M. Moreno,[1] G. M. Alonzo-Medina,[1] A. I. Oliva,[2] and A. I. Oliva-Avilés[1]

[1]*División de Ingeniería y Ciencias Exactas, Universidad Anáhuac Mayab, Carretera Mérida-Progreso Km. 15.5 AP 96 Cordemex, 97310 Mérida, YUC, Mexico*
[2]*Centro de Investigación y de Estudios Avanzados del IPN, Unidad Mérida, Departamento de Física Aplicada, AP 73 Cordemex, 97310 Mérida, YUC, Mexico*

Correspondence should be addressed to A. I. Oliva-Avilés; andres.oliva@anahuac.mx

Academic Editor: Kaveh Edalati

Cadmium sulfide (CdS) thin films were deposited by chemical bath deposition (CBD) onto polymeric composites with electric field-aligned multiwall carbon nanotubes (MWCNTs). MWCNT/polysulfone composites were prepared by dispersing low concentrations of MWCNTs within dissolved polysulfone (PSF). An alternating current electric field was "in situ" applied to align the MWCNTs within the dissolved polymer along the field direction until the solvent was evaporated. $80 \, \mu m$ thick solid MWCNT/PSF composites with an electrical conductivity 13 orders of magnitude higher than the conductivity of the neat PSF were obtained. The MWCNT/PSF composites were subsequently used as flexible substrates for the deposition of CdS thin films by CBD. Transparent and adherent CdS thin films with an average thickness of 475 nm were obtained. The values of the energy band gap, average grain size, rms roughness, crystalline structure, and preferential orientation of the CdS films deposited onto the polymeric substrate were very similar to the corresponding values of the CdS deposited onto glass (conventional substrate). These results show that the MWCNT/PSF composites with electric field-tailored MWCNTs represent a suitable option to be used as flexible conducting substrate for CdS thin films, which represents an important step towards the developing of flexible systems for photovoltaic applications.

1. Introduction

Cadmium sulfide (CdS) has been widely and successfully used as a window layer for optical and photovoltaic applications [1–3]. CdS grows naturally as n-type semiconductor with an energy band gap of 2.42 eV at room temperature [4], resulting in a desirable material for solar applications due to its optical and electrical properties. For solar cell applications, CdS thin films have been used as a window material together with several semiconductors such as CdTe, Cu_2S, $CuInSe_2$, InP, and $Cu(In,Ga)Se_2$, and solar cell efficiencies between 14% and 19% have been reported [5–10]. Several techniques have been extensively used to prepare CdS in a thin film configuration onto different substrates, such as electrodeposition [11], chemical bath deposition [12], spray deposition [13], femtosecond pulsed laser deposition [14], molecular beam epitaxy [15], successive ionic layer adsorption and reaction

[16], screen printing [17], close spaced sublimation [18], RF sputtering [19], metal organic chemical vapor deposition [20], pulsed laser deposition [21], and physical vapor deposition [22], among the most relevant ones. Among these methods, the chemical bath deposition (CBD) has been demonstrated to be a suitable technique to deposit CdS thin films for large area industrial processes [23, 24]. CBD is a simple and low cost technique which does not require high vacuum conditions and, since it is a low-temperature technique ($<90°C$), does not limit the choice of the substrate material. It is well known that the physical and chemical properties of semiconducting thin films are highly dependent on the early stages of growth. In the same level of importance, the choice of the substrate also defines the film quality and properties. In the field of II–VI compound semiconductor deposition, Pyrex® [25], glass [26], quartz [14], and silicon [27, 28] are

among the most used materials due to their availability, cost-effectiveness, and inert character. Such substrates do not react with the deposited semiconductors and the surface promotes the formation of optically smooth thin films [29]. Despite the important results and presence of the semiconducting thin films deposited onto the mentioned substrates, their weight, rigidity, and fragility are the main drawbacks for more versatile and far-reaching applications in several technological fields. As an example, one of the major obstacles in the development of low cost and high efficiency solar cells is the use of glass substrates. Ordinary glass substrates have a poor thermal conductivity, making it extremely difficult to maintain a constant annealing temperature across a large area panel and to avoid thermal stresses which may cause breakage during the fabrication [30]. As a response for the mentioned challenges, the usage of flexible substrates, mainly based on polymers and metal foils, has emerged as a convenient option to deposit semiconducting thin films, offering several advantages for many technological and scientific applications [4, 29–35]. In particular, the growth of CdS onto different polymers such as polyethylene terephthalate [4, 32–35], flexible polymer-based plastic [29], polyimide [31], and polycarbonate [32–34] has been reported and the resulting films exhibit adequate physical properties, similar to those of the CdS films deposited onto glass. In the configuration of a thin film-based solar cell, the electrical characteristics of the substrate are of paramount importance. For insulating substrates (such as glass) a transparent conducting layer is generally deposited onto the surface of the substrate in order to generate an electrode that, together with the back conducting contact, closes up the "electrical circuit" across the solar cell allowing the usage of the induced photovoltage. However, the high temperatures required to deposit such conducting layers onto polymeric (flexible) substrates may induce damage and/or melting of the substrates as they generally have low glass transition temperatures. In recent years, polymers reinforced with carbon nanotubes (CNTs) have gained attention due to the novel and versatile properties that CNTs offer to the polymeric matrix [36–38]. It has been extensively reported that CNTs importantly modify the physical properties of the polymers when the CNTs are incorporated into the matrix to conform the polymer composites. Electrical, mechanical, and thermal properties of the composites are notoriously enhanced due to presence of CNTs, as CNTs partially transfer their physical properties to the polymeric matrix. Regarding the electrical properties, the inclusion of a very low content of CNTs within a polymer matrix (<1 wt%) promotes an increase of several orders of magnitude of the electrical conductivity compared to the conductivity of the pure polymer [39–41]. The combination of CNTs with polymers offers an attractive route not only to the development of reinforced polymer films but also to introducing new electronic properties based on morphological and electrical modifications. This represents a very convenient feature in the photovoltaic area, where the flexibility and enhanced electrical properties of the substrates become relevant requirements for an improved performance.

Here we present a study of the optical, morphological, and structural properties of CdS thin films deposited by CBD onto flexible composite substrates. Low concentrations of multiwall carbon nanotubes (MWCNTs) were incorporated into dissolved polysulfone and an alternating current (AC) electric field was "in situ" applied during the fabrication process to align the MWCNTs along the electric field direction. Polysulfone was selected as the (thermoplastic) polymeric matrix due to the combination of several high-performance properties such as high thermal stability, high toughness and strength, good environmental stress cracking resistance, high heat deflection temperature (~174°C), and high transparency and combustion resistance [42]. These adequate characteristics for solar applications along with the capability of being dissolved in chloroform (solvent used to disperse the MWCNTs) supported the election of the mentioned polymer for this work. Solid MWCNT/PSF composites with an increased electrical conductivity were obtained by the solution casting technique and used as flexible substrates for the CdS deposition. The average values of thickness, energy band gap, grain size, rms roughness, and crystalline structure of the CdS thin films deposited onto the polymeric composites were obtained and compared to the well-known values of CdS deposited onto glass. The aim of this investigation is to demonstrate that it is possible to fabricate CdS thin films onto a CNT-based polymeric substrate by the CBD technique. The capability to deposit CdS thin films with adequate optical, morphological, and structural properties onto a flexible and electrically enhanced composite represents an important step towards the development of flexible materials for photovoltaic applications.

2. Materials and Methods

2.1. Fabrication and Electrical Characterization of MWCNT/Polysulfone Composite Substrates. MWCNTs grown by chemical vapor deposition ("Baytubes C150P") were supplied by Bayer Material Science [43]. The MWCNTs have inner and outer diameters of approximately 4 nm and 13 nm, respectively, and length in the range of 1–4 μm, yielding a distribution of CNT aspect ratio ranging from 80 to 300 [44]. The polymer matrix used in this work was polysulfone (PSF, "UDEL P1700") supplied by Solvay Advanced Polymers [42]. PSF is an engineering thermoplastic with high thermal stability, transparency, and low moisture absorption characteristics among the most relevant characteristics. MWCNT/PSF composite films were fabricated by solution casting using 0.5 weight percent (wt%). To fabricate the composite films, 2.5 mg of MWCNTs was dispersed in chloroform ($CHCl_3$) using an ultrasonic bath (Cole-Parmer 08895-59, 40 kHz) for 120 min (Figure 1(a)). In a parallel process, 500 mg of PSF pellets was dissolved in 2 mL of $CHCl_3$ forming a viscous PSF/$CHCl_3$ solution (Figure 1(b)). The PSF/$CHCl_3$ and MWCNT/$CHCl_3$ solutions were then mixed in a common beaker for 15 min (Figure 1(c)) and the MWCNT/PSF/$CHCl_3$ final solution was cast onto a custom-made glass mold with aluminum electrodes for the application of an AC electric field (Figure 1(d)).

The AC voltage applied to the solution was 120 V_{rms} at 60 Hz and the electrode separation was 20 mm, for an electric field of 6 kV_{rms}/m. After the solution casting, $CHCl_3$ was

MWCNT/CHCl$_3$

(a)

PSF/CHCl$_3$

(b)

MWCNT/PSF/CHCl$_3$

(c)

MWCNT/PSF/CHCl$_3$

AC source

(d)

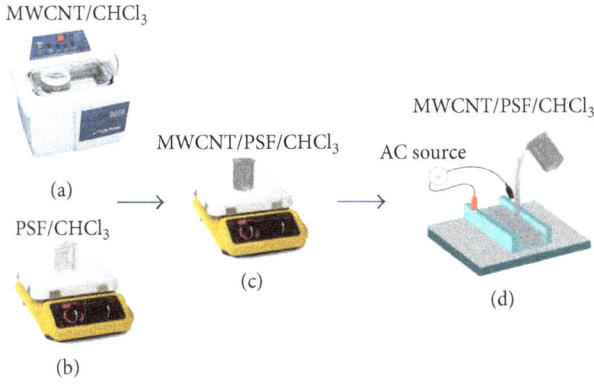

FIGURE 1: Schematic of the solution casting process for the preparation of MWCNT/PSF composites with electric field-aligned MWCNTs.

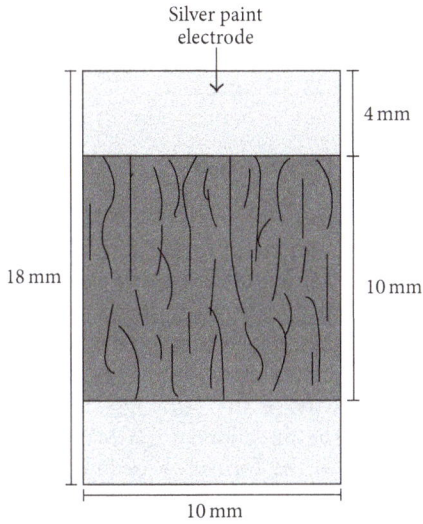

Silver paint electrode

4 mm

18 mm

10 mm

10 mm

FIGURE 2: MWCNT/PSF specimen for electrical characterization.

evaporated for 24 h and the viscous solution turned into solid MWCNT/PSF films. Five samples were obtained from each batch. The direct current (DC) electrical conductivity of the samples in the direction of the MWCNT alignment was measured. DC electrical resistance of MWCNT/PSF films was measured at room temperature (25°C) with a digital multimeter (Agilent 34401A). For the measurement of the electrical resistance, two silver-painted electrodes located at the film edges of the specimen were used. The specimen length was 18 mm with 4 mm long electrodes, leaving an effective span (L) of ~10 mm between the silver electrodes; see Figure 2. The specimen width was 10 mm and its nominal thickness was ~80 μm. In order to minimize surface effects in the electrical measurements, silver paint electrodes were painted completely covering the ends of the specimens. A DC voltage of 40 V was applied between the electrodes and the electrical conductivity (σ_e) was calculated using the measured electrical resistance (R) and the specimen dimensions as

$$\sigma_e = \frac{L}{AR}, \quad (1)$$

where A is the cross-sectional area of the specimen.

2.2. Deposition of Cadmium Sulfide by Chemical Bath Deposition.

MWCNT/PSF specimens with a surface area of 1 cm × 1 cm were used as substrates for the CdS deposition by CBD. As a first step, the substrates were cleaned following a protocol which includes washing steps with liquid soap and isopropyl alcohol, ultrasonic agitation, distilled water for rinsing, and drying with compressed air. The cleaning process of the substrates was carried out right before the CdS deposition. For the chemical bath preparation, four aqueous solutions were prepared as follows: 200 mL of 0.50 M potassium hydroxide (KOH), 80 mL of 1.5 M ammonium nitrate (NH$_4$NO$_3$), 80 mL of 0.025 M cadmium chloride (CdCl$_2$), and 80 mL of 0.20 M thiourea (CS(NH$_2$)$_2$). Distilled water was used to dissolve all the reagents. The chemical bath, with a final volume of 440 mL, was prepared by mixing the mentioned solutions. Five substrates were vertically placed in Teflon substrate holders and immersed within the chemical bath. During the magnetic stirring, the bath was heated using a hot plate until reaching 80°C, which was the temperature selected for the deposition process [45]. When the solution with the potassium hydroxide, cadmium chloride, and ammonium nitrate reaches 80°C, the thiourea was added as a final step, initiating the process of CdS formation. Once the thiourea was added to the solution, the chemical bath changed its coloration from transparent to a yellowish tone, indicating the formation of the CdS. The yellowish coloration reflects the optical property of the material and resembles the correspondent wavelength (512 nm) of the CdS energy band gap (2.42 eV). The growth of CdS by CBD is given by the decomposition of the thiourea (CS(NH$_2$)$_2$) in presence of a cadmium salt in a basic solution with ammonia as complexing agent. The importance of the formation of the tetra-amino-cadmium complex ions, $[Cd(NH_3)_4]^{2+}$, is determinant for the liberation of the Cd^{2+} ions and a further recombination with S^{2-} ions to produce CdS [46]. The chemical process for the formation of the CdS can be described through the following chemical reactions [46]:

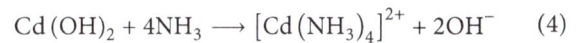

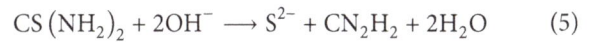

$$Cd^{2+} + 2OH^- \longrightarrow Cd(OH)_2 \quad (2)$$

$$NH_4^+ + OH^- \longrightarrow NH_3 + H_2O \quad (3)$$

$$Cd(OH)_2 + 4NH_3 \longrightarrow [Cd(NH_3)_4]^{2+} + 2OH^- \quad (4)$$

$$CS(NH_2)_2 + 2OH^- \longrightarrow S^{2-} + CN_2H_2 + 2H_2O \quad (5)$$

$$[Cd(NH_3)_4]^{2+} + S^{2-} \longrightarrow CdS(s) + 4NH_3 \quad (6)$$

In the CBD technique, deposition of the CdS (metal-chalcogenide) occurs in the surface of the substrates immersed in aqueous solution containing both the metal and the chalcogenide ions. The deposition occurs when the ionic product of the metallic and chalcogenide ions exceeds the solubility product Ksp (for CdS at 25°C, Ksp = 10^{-28}) [47]. The ions are chemically attached in the aqueous medium and precipitate onto the surface of the substrate producing the metal-chalcogenide film. For the formation of the CdS, Ortega-Borges and Lincot [48] proposed an atom-by-atom growth mechanism in an ammonia-thiourea system through the decomposition of adsorbed thiourea molecules due to the formation of an intermediate complex with cadmium hydroxide. In another report, Doña and Herrero proposed a modification of the mechanism of CdS growth through the kinetic study of another intermediate species, the hydroxy-amino complexes [49]. Through the study of the CdS microstructure, films can grow by heterogeneous and/or homogenous precipitation. The CdS deposition in the chemical bath takes place onto the surface of the substrates and at the inner reactor walls (heterogeneous reaction) as well as in the chemical bath (homogeneous reaction). The homogeneous reaction is not desirable due to the formation of colloids, which precipitate as powders and promote the formation of films with low adherence [48]. The substrates were finally taken out after 60 minutes. The CdS/composite samples were rinsed with distilled water and the deposited CdS of one surface of the substrate was removed using isopropyl alcohol. As a final step, the thickness of the CdS films was measured by a high-resolution profilometer (Dektak 8) through a step on the corner of the substrate where the material was carefully removed using isopropyl alcohol. CdS thin films with an average thickness of 475 nm were obtained from the CBD technique.

2.3. Characterization of CdS Thin Films

2.3.1. Optical Properties. It is well known that CdS is a semiconductor with a direct energy band gap (E_g). The absorption coefficient (α) obeys the equation [50]

$$\alpha = \frac{k}{h\nu} \left(h\nu - E_g \right)^{1/2}, \qquad (7)$$

where k is constant, h is the Planck constant, and ν is the light frequency. E_g value of the CdS film was estimated from the straight line fitted at the absorption edge of the $(\alpha h\nu)^2$ versus $h\nu$ plot by extrapolating the linear portion of the graph of the energy axis at $\alpha = 0$. Spectrophotometry UV-Vis (Thermo Scientific Evolution 300) was used to evaluate the absorption spectrum of the CdS films in the wavelength range 300–650 nm and E_g value of the films was estimated. Although the MWCNT/PSF composites were fabricated following the same procedure and the morphology of all the samples was similar, it would not be correct to use a single MWCNT/PSF substrate as the one reference (to be subtracted) for the spectrophotometer during all the optical characterization. Thus, in order to adequately capture the optical properties of the CdS films, MWCNT/PSF substrates were labeled and the absorption spectrum of each one was individually measured and registered before the CdS deposition. Once the CdS was deposited by CBD, the absorption spectra of the system CdS/substrate were registered and the absorption spectrum of the corresponding labeled substrate was subtracted as a final step, therefore eliminating the optical contribution of the composite substrate for each individual sample.

2.3.2. Morphology. The morphology of the CdS films was analyzed by AFM and SEM. AFM images of $1\,\mu m \times 1\,\mu m$ (SPM AMBIOS Universal) were obtained from the surface of the CdS films and the average grain size and rms roughness

were estimated. The rms roughness was calculated using the WSxM software of "Nanotec Electronica" Company [51]. To calculate the mean grain size, a home-built algorithm implemented in the software "Mathematica" was used. The algorithm transforms the digital image of local heights into a continuous function which represents a "mathematical mold" (a contour map) of the film surface. The lateral grain size is defined as the larger distance between the intersections of the contour map (grain boundaries) and the average grain size of the image is estimated using a maximum and minimum criterion. A detailed explanation of the algorithm can be found elsewhere [52, 53]. Additionally, SEM images (JEOL JSM-6010LA) were obtained from the samples.

2.3.3. Crystalline Structure. To investigate the crystalline structure and preferential orientation of the CdS films, X-ray diffraction (XRD) profiles were obtained from the samples by two different configurations, the grazing incident angle configuration (Siemens D5000) and the Bragg-Brentano configuration (Bruker D8 Advance).

3. Results and Discussion

3.1. Electrical Conductivity of the MWCNT/PSF Substrates. The DC electrical conductivity (σ_e) of the MWCNT/PSF samples (>20) obtained from different batches was estimated by measuring the electrical resistance of the samples and using (1). The average σ_e value was measured as 7.44×10^{-2} S/m, being such a conductivity value 13 orders of magnitude higher than the conductivity of the pure PSF ($\sim 10^{-15}$ S/m). The increase of σ_e in the direction of the MWCNT alignment can be correlated to the formation of interconnected conducting paths spanning the electrodes. As a consequence, the electrons can easily travel through the conducting paths and the electrical conductivity increases in such a direction. In previous reports by our group, the application of electric fields to fabricate aligned MWCNT/PSF composites (technique here used) and the correlation between the morphologies observed at different length scales, from macro- to nanoscale, have been reported and the electrical properties of the composites have been explored [41, 54]. From these reports, the electric field alignment of MWCNTs can be appreciated at the macroscale by optical images [41, 54], at the microscale by SEM images of aligned MWCNT bundles [54], and at the nanoscale, where the alignment of individual MWCNTs was appreciated by TEM images [41]. The results observed here are thus in good agreement with such previous works that report that the orientation of the CNTs within the polymer significantly enhances the value of electrical conductivity [41, 54, 55]. Although the electrical conductivity of the MWCNT/polymer composite is still significantly lower than the conductivity of a metallic layer ($\sim 10^4$ S/m), the notorious increase of such a value due to the presence of the tailored MWCNTs represents an important result towards the fabrication of electrically enhanced polymer composites, and optimized procedures and strategies to reach higher electrical conductivities for composites are object of current investigation.

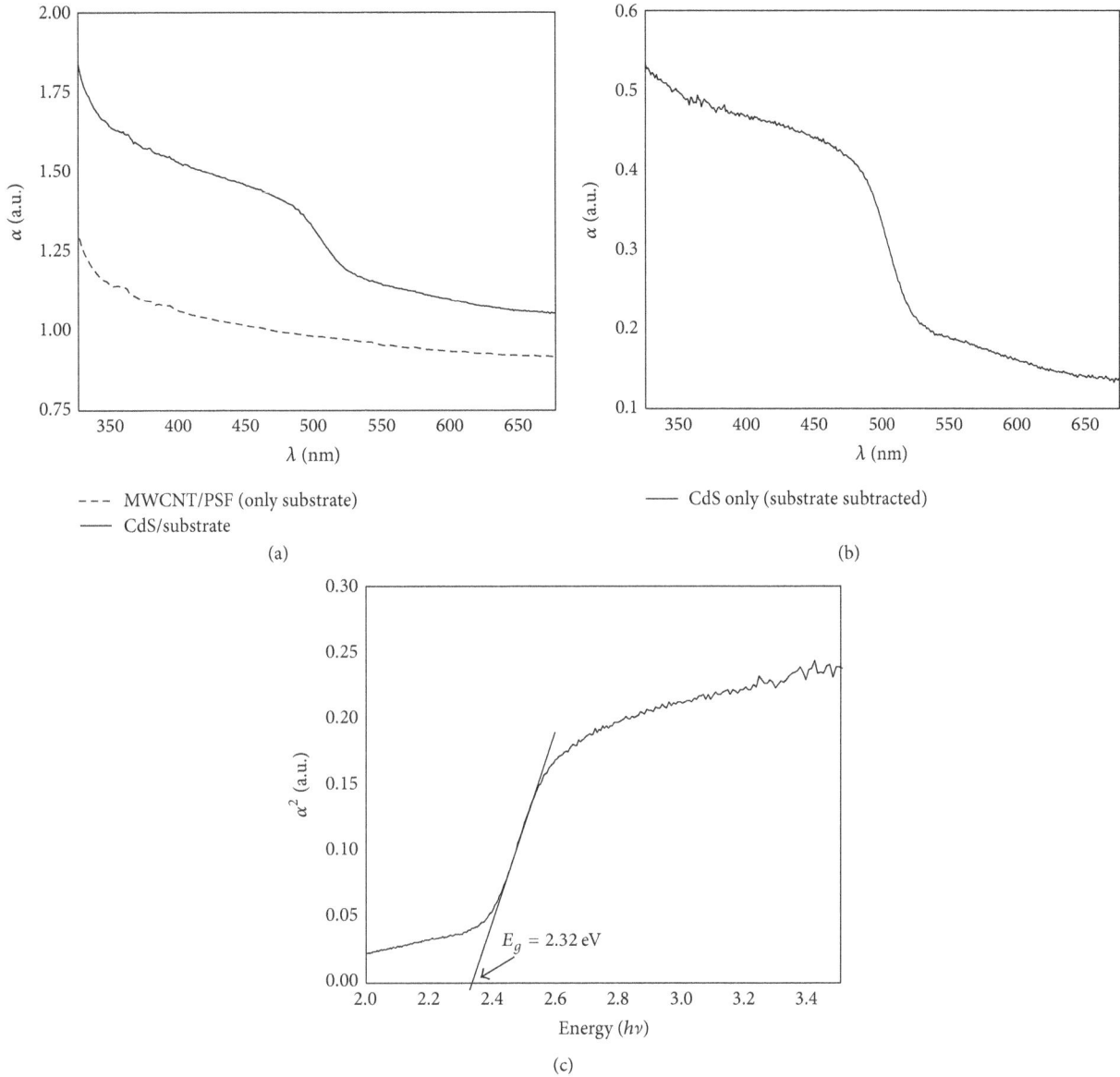

--- MWCNT/PSF (only substrate)
— CdS/substrate

(a)

— CdS only (substrate subtracted)

(b)

(c)

FIGURE 3: Optical characterization of a representative CdS thin film deposited by CBD. (a) Absorbance spectra of the MWCNT/PSF substrate (dashed line) and of the CdS/substrate system (solid line) as a function of the wavelength, (b) absorbance spectrum of the CdS/film obtained by subtracting the reference spectrum (substrate), and (c) α^2 versus energy ($h\nu$) plot obtained from data of Figure 3(b). The straight line in the linear portion of the plot is included for clarification purposes.

3.2. Absorbance Spectrum and Energy Band Gap of the CdS Films. The absorbance spectrum of the CdS films was obtained by UV-Vis spectrophotometry and E_g was estimated using (7). Figure 3(a) shows representative plots of the absorption spectra registered from the MWCNT/PSF substrate before the CdS deposition (dashed line) and from the system CdS/substrate (solid line). It can be observed that the intensity of the absorption spectrum of the substrate solely (without CdS film) is lower than the spectrum of the system CdS/substrate, thus allowing the optical characterization of the CdS film. The absorption of the spectrum (reference) was then subtracted from the spectrum of the whole CdS/substrate system and the resulting spectrum,

correspondent to the CdS thin film, is shown in Figure 3(b). It can be observed that a notorious decrease on the absorbance occurs in the wavelength range from 450 nm to 550 nm. Such a decreasing behavior within the mentioned wavelength range was consistently observed for all the CdS samples. A typical α^2 versus energy ($h\nu$) plot is shown in Figure 3(c). The plot in Figure 3(c) was obtained from the data of the absorbance spectrum in Figure 3(b) and by means of (7), and E_g of 2.32 eV was estimated (value of energy when the straight line crosses $\alpha = 0$). The solid line extrapolating the linear portion of the graph is presented for clarification purposes. Similar process for the optical characterization was applied to all the CdS/composite samples and the average E_g

was calculated as 2.33 ± 0.04 eV. The average value of E_g is very similar to E_g reported for CdS thin films deposited onto glass at room temperature (2.31–2.34 eV) under similar experimental conditions [56, 57]. Faraj and Ibrahim carried out a comparative study of CdS deposited by thermal evaporation onto a polymeric substrate (polyethylene terephthalate, PET) and onto glass [4]. They found that E_g of CdS deposited onto PET was slightly lower (2.41 eV) than E_g of the CdS deposited onto glass (2.43 eV). In another report, Lee evaluated the influence of the substrate (different polymers and glass) on the optical properties of CdS films deposited by CBD [32]. He found that the energy band gaps of the samples deposited onto polycarbonate and PET substrates were 2.32 and 2.31 eV, respectively, and such a value increased to 2.37 for the CdS deposited onto glass. Author also suggested that the deviation of E_g values from the standard bulk value (2.42 eV) can be explained on the basis on the small grain size of the substrates, which could explain the lower average value of E_g obtained in this investigation. Moon et al. reported a comparative study of the CdS film properties deposited onto different substrates by RF sputtering [34]. Authors reported that E_g values of CdS films deposited onto polycarbonate and PET were not consistent with E_g of the CdS deposited onto glass and suggest that such an inconsistency may be associated with the optical properties of the polymer substrates, which partly absorb the light in the high energy range of the visible spectrum. The results obtained here show that the selection of MWCNT/PSF polymeric substrate with a low concentration of MWCNTs does not importantly affect the optical properties of the deposited CdS thin films as compared to the CdS films deposited onto glass.

3.3. Morphology of CdS Films.

The rms roughness of the polymeric substrate and the mean grain size and rms roughness of the CdS films were estimated from AFM images by means of the procedure previously described. Figure 4(a) shows a two-dimensional micrograph of a scanned region of $1\,\mu m \times 1\,\mu m$ of the MWCNT/PSF polymeric substrate. As can be appreciated, the roughness is very low and a mean rms roughness value of 0.285 nm is estimated, being this value in the order of the values typically observed on glass substrates (~0.35 nm) [58]. Figure 4(b) presents a two-dimensional micrograph of a scanned region of $1\,\mu m \times 1\,\mu m$ where the grains and their boundaries can be appreciated. A markedly different morphology with respect to the morphology of the substrate in Figure 4(a) is observed in Figure 4(b), due to the formation of the CdS thin film.

As previously mentioned, a home-built algorithm implemented in the commercial software "Mathematica" was used to calculate the average grain size of the sample. The algorithm uses the difference of intensities of the image to assign the grain limits and then to calculate the required parameters. Figure 4(c) shows an example of the grain boundaries of the AFM micrograph in Figure 4(b) traced by the algorithm. The mean grain size and rms roughness of the CdS films deposited onto the polymeric substrate were calculated as 271 nm and 11 nm, respectively. For CdS film deposited onto a glass substrate by similar experimental conditions, the mean grain

size and rms roughness were estimated as 269 nm and 21 nm, respectively [56]. SEM image of the CdS film is presented in Figure 5. Some grains with size larger than the mean size can be appreciated but in general a uniform morphology of the surface can be observed, corresponding to the surface morphology observed by AFM on a smaller region; see Figure 4(b). The uniform morphology of the surface can be correlated with a good formation and adhesion of the CdS thin film onto the polymeric substrate. These results confirm that the morphological characteristics of the CdS films are very similar to those of the CdS deposited onto glass, suggesting that the proposed polymeric composite represents a suitable option to be used as a substrate for CdS films.

3.4. Crystalline Structure.

XRD profiles of CdS films deposited onto MWCNT/PSF substrates in different experiments were obtained by the grazing incident angle and Bragg-Brentano configurations in order to evaluate the crystalline structure and preferential orientation of the CdS films and the reproducibility of the deposition process. The XRD profiles of the composite substrates solely (without CdS film) were also obtained from the two configurations. Figure 6(a) shows a representative XRD profile obtained by the grazing angle configuration. The XRD spectra confirm the formation of CdS in a face-centered cubic (fcc) crystalline structure "Hawleyite" with (111) as the preferential orientation, as can be noticed from the presence of the main peak at $2\theta = 26.54°$. Secondary peaks of the cubic structure at 44.02° and 52.14° were also detected. A wide peak is observed around $2\theta = 18°$ which corresponds to the composition of the polymeric substrate. A peak at 29.4° (marked with "∗" in Figure 6(a)) was also noticed; such a peak corresponds to the "fixing putty" used to fix the sample to the sample-holder on the diffractometer. Figure 6(b) shows a representative XRD profile by the Bragg-Brentano configuration. The main peak at 26.54° was again detected, confirming the cubic structure and preferential orientation of the CdS films. No peak was observed at 29.4° as no "fixing putty" was required for the Bragg-Brentano configuration. The observed CdS crystalline structure and preferential orientation are consistent with previous reports of CdS deposited onto glass [46, 56–59]. Also, it has been reported that fcc structure is the most stable structure for CdS films obtained by means of low-temperature techniques (such as the CBD) while the hexagonal structure (Greenockite) is the most stable one for high-temperature techniques (e.g., sputtering or closed space sublimation) [29, 34, 60].

4. Conclusions

CdS thin films were deposited by CBD onto flexible polymeric substrates. AC electric field was applied during the fabrication of the polymeric composite substrates to induce preferential alignment of the MWCNTs within the polymer during the fabrication process. After the solvent evaporation, solid MWCNT/polysulfone composites with MWCNTs aligned in the electric field direction were obtained, with an electrical conductivity 13 orders of magnitude higher than the conductivity of the pure polymer. CdS thin films

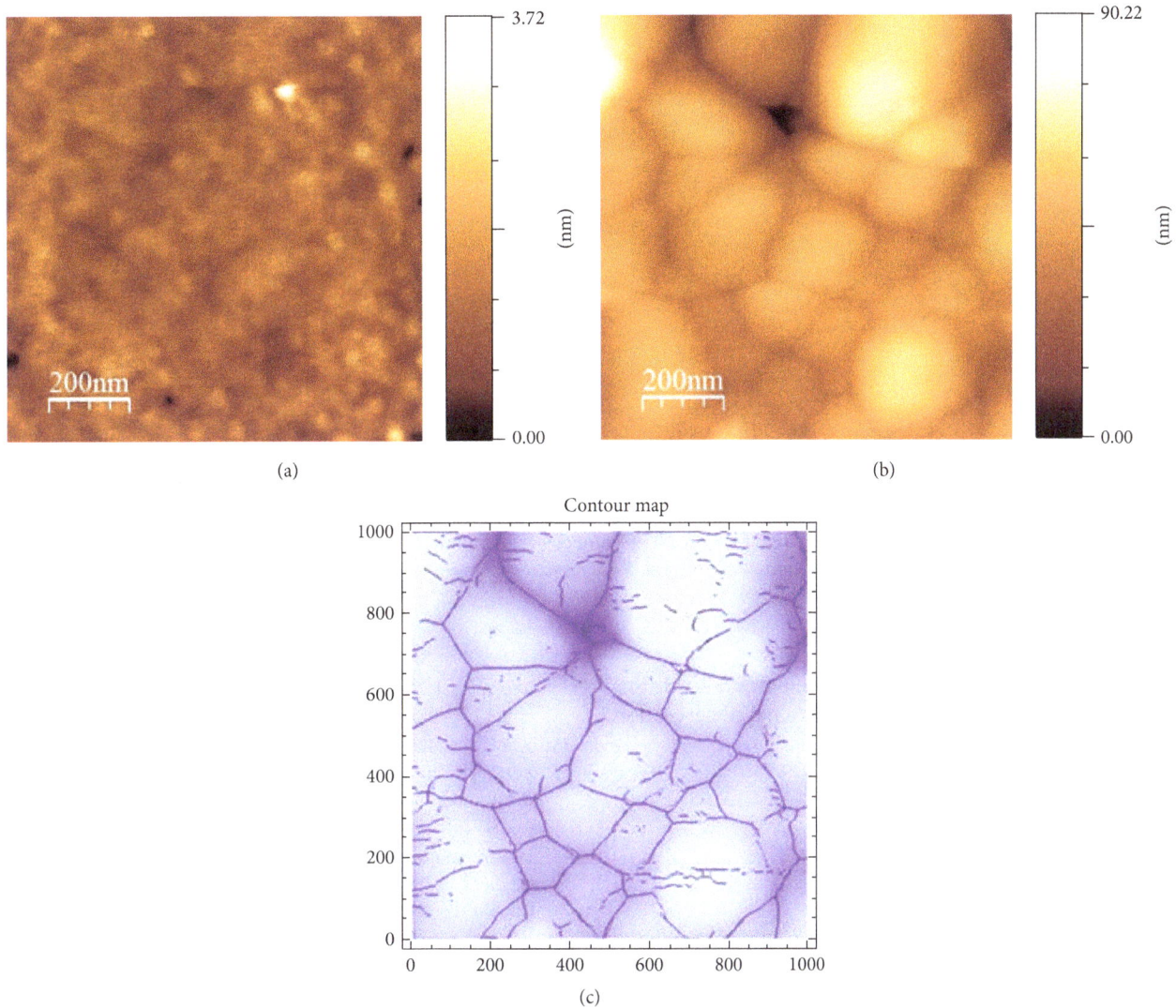

Figure 4: AFM images of the CdS film deposited onto MWCNT/PSF substrate. (a) Two-dimensional image of $1\,\mu m \times 1\,\mu m$ of the MWCNT/PSF polymeric substrate (without film), (b) two-dimensional image of $1\,\mu m \times 1\,\mu m$ of the CdS film deposited by CBD, and (c) contour map with the grain boundaries traced by the home-built algorithm implemented in "Mathematica" to calculate the mean grain size of the CdS.

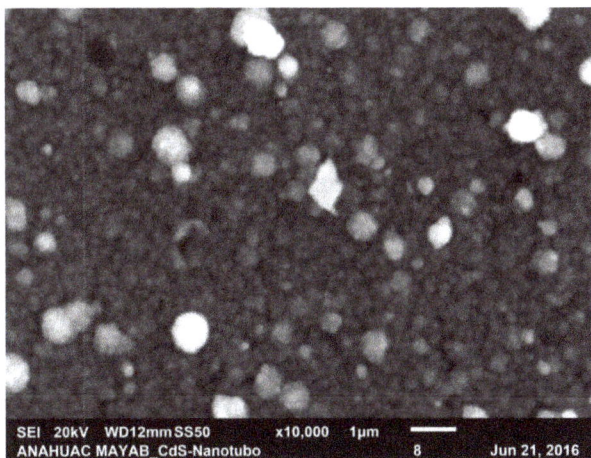

Figure 5: SEM image of the CdS film deposited by CBD onto MWCNT/PSF polymeric substrate.

were deposited onto the MWCNT/PSF substrates by CBD and the optical, morphological, and structural properties were evaluated. CdS thin films with an average thickness of 475 nm, good adherence, and transparency were successfully obtained by the proposed procedure. The average energy band gap was measured as 2.33 eV, a similar value to that of the CdS deposited onto glass. The mean grain size (271 nm) and rms roughness (11 nm) were found to be very similar to the values reported for CdS deposited onto glass by similar experimental conditions. XRD patterns measured by two different configurations confirm the formation of CdS in Hawleyite (fcc) crystalline structure with (111) as preferential orientation, which is in agreement with the most stable structure for CdS grown by low-temperature techniques, such as the CBD. From the observed results it can be concluded that the electrically enhanced MWCNT/PSF flexible composites are suitable substrates for the deposition

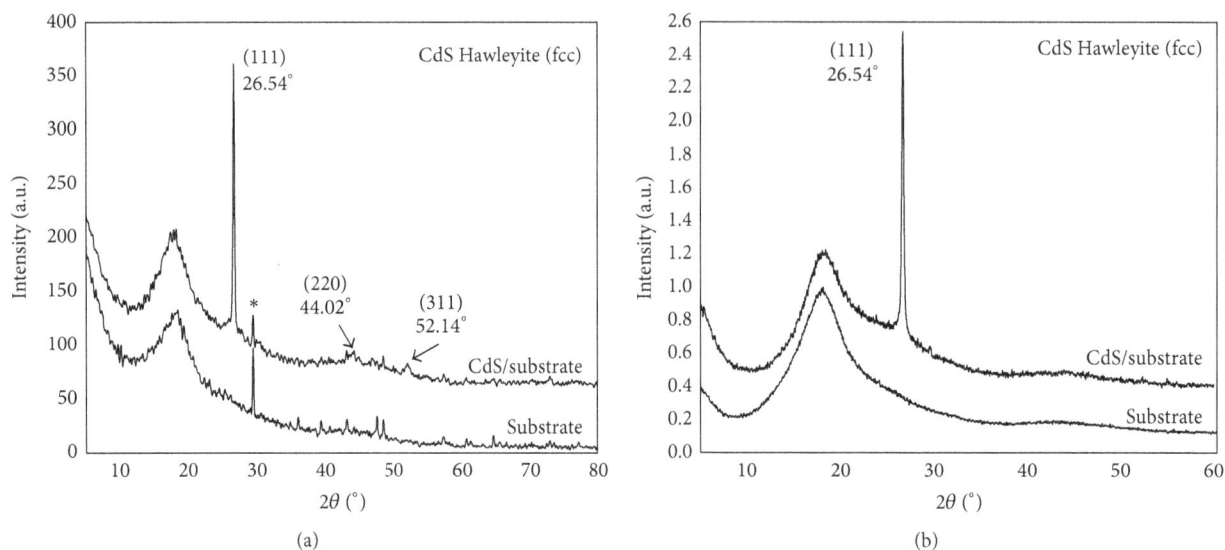

FIGURE 6: XRD profiles of CdS films deposited onto MWCNT/PSF composite substrates by two different configurations: (a) grazing incident angle configuration and (b) Bragg-Brentano configuration. The XRD profile of the composite substrate solely (without CdS film) is also presented on each plot.

of CdS by CBD, representing an important step towards the development of more versatile and functional materials for photovoltaic applications.

Competing Interests

The authors declare that there are no competing interests regarding the publication of this paper.

Acknowledgments

This work was supported by the "Fondo Sectorial de Investigación para la Educación" through the SEP-CONACYT Grant no. 235905 (A. I. Oliva-Avilés) and by the Projects no. 251932 and no. 235637 (CONACYT). The authors also thank the technical assistance of Daniel Aguilar, Emilio Corona, Mauricio Romero, Wilian Cauich (CINVESTAV), Rubén Domínguez, Yazmín Sabido (Universidad Anáhuac Mayab), and Francis Avilés (CICY).

References

[1] R. Tenne, V. M. Nabutovsky, E. Lifshitz, and A. F. Francis, "Unusual photoluminescence of porous CdS (CdSe) crystals," *Solid State Communications*, vol. 82, no. 9, pp. 651–654, 1992.

[2] V. Ruxandra and S. Antohe, "The effect of the electron irradiation on the electrical properties of thin polycrystalline CdS layers," *Journal of Applied Physics*, vol. 84, no. 2, pp. 727–733, 1998.

[3] B. Su and K. L. Choy, "Electrostatic assisted aerosol jet deposition of CdS, CdSe and ZnS thin films," *Thin Solid Films*, vol. 361, pp. 102–106, 2000.

[4] M. G. Faraj and K. Ibrahim, "Comparison of cadmium sulfide thin films deposited on glass and polyethylene terephthalate substrates with thermal evaporation for solar cell applications,"

Journal of Materials Science: Materials in Electronics, vol. 23, no. 6, pp. 1219–1223, 2012.

[5] X. Wu, J. C. Keane, R. G. Dhere et al., "16.5%-Efficient CdS/CdTe polycrystalline thin-film solar cells," in *Proceedings of the 17th European Photovoltaic Solar Energy Conference (EU PVSEC '01)*, p. 995, 2001.

[6] K. Ramanathan, M. A. Contreras, C. L. Perkins et al., "Properties of 19.2% efficiency ZnO/CdS/CuInGaSe$_2$ thin-film solar cells," *Progress in Photovoltaics: Research and Applications*, vol. 11, no. 4, pp. 225–230, 2003.

[7] K. D. Dobson, I. Visoly-Fisher, G. Hodes, and D. Cahen, "Stability of CdTe/CdS thin-film solar cells," *Solar Energy Materials and Solar Cells*, vol. 62, no. 3, pp. 295–325, 2000.

[8] R. L. N. Chandrakanthi and M. A. Careem, "Preparation and characterization of CdS and Cu2S nanoparticle/polyaniline composite films," *Thin Solid Films*, vol. 417, no. 1-2, pp. 51–56, 2002.

[9] V. K. Kapur, B. M. Basol, and E. S. Tseng, "Low cost methods for the production of semiconductor films for CuInSe$_2$/CdS solar cells," *Solar Cells*, vol. 21, no. 1–4, pp. 65–72, 1987.

[10] M. Purica, E. Budianu, E. Rusu, and P. Arabadji, "Electrical properties of the CdS/InP heterostructures for photovoltaic applications," *Thin Solid Films*, vol. 511-512, pp. 468–472, 2006.

[11] O. A. Ileperuma, C. Vithana, K. Premaratne, S. N. Akuranthilaka, S. M. McGregor, and I. M. Dharmadasa, "Comparison of CdS thin films prepared by different techniques for applications in solar cells as window materials," *Journal of Materials Science: Materials in Electronics*, vol. 9, no. 5, pp. 367–372, 1998.

[12] A. E. Rakhshani and A. S. Al-Azab, "Characterization of CdS films prepared by chemical-bath deposition," *Journal of Physics Condensed Matter*, vol. 12, no. 40, pp. 8745–8755, 2000.

[13] S. Mathew, P. S. Mukerjee, and K. P. Vijayakumar, "Optical and surface properties of spray-pyrolysed CdS thin films," *Thin Solid Films*, vol. 254, no. 1-2, pp. 278–284, 1995.

[14] X. L. Tong, D. S. Jiang, Y. Li, Z. M. Liu, and M. Z. Luo, "Femtosecond pulsed laser deposition of CdS thin films onto quartz

substrates," *Physica Status Solidi (A)*, vol. 203, no. 8, pp. 1992–1998, 2006.

[15] P. Boieriu, R. Sporken, Y. Xin, N. D. Browning, and S. Sivananthan, "Wurtzite CdS on CdTe grown by molecular beam epitaxy," *Journal of Electronic Materials*, vol. 29, no. 6, pp. 718–722, 2000.

[16] B. R. Sankapal, R. S. Mane, and C. D. Lokhande, "Deposition of CdS thin films by the Successive Ionic Layer Adsorption and Reaction (SILAR) method," *Materials Research Bulletin*, vol. 35, no. 2, pp. 177–184, 2000.

[17] D. Patidar, R. Sharma, N. Jain, T. P. Sharma, and N. S. Saxena, "Optical properties of CdS sintered film," *Bulletin of Materials Science*, vol. 29, no. 1, pp. 21–24, 2006.

[18] Muneeb-ur-Rehman, A. K. S. Aqili, M. Shafique, Z. Ali, A. Maqsood, and A. Kazmi, "Properties of silver-doped CdS thin films prepared by closed space sublimation (CSS) techniques," *Journal of Materials Science Letters*, vol. 22, no. 2, pp. 127–129, 2003.

[19] I. Mártil de la Plaza, G. González-Díaz, F. Sánchez-Quesada, and M. Rodríguez-Vidal, "Structural and optical properties of r.f.-sputtered CdS thin films," *Thin Solid Films*, vol. 120, no. 1, pp. 31–36, 1984.

[20] M. Tsuji, T. Aramoto, H. Ohyama, T. Hibino, and K. Omura, "Characterization of CdS thin-film in high efficient CdS/CdTe solar cells," *Japanese Journal of Applied Physics*, vol. 39, no. 7, pp. 3902–3906, 2000.

[21] O. Vigil-Galán, J. Vidal-Larramendi, A. Escamilla-Esquivel et al., "Physical properties of CdS thin films grown by pulsed laser ablation on conducting substrates: effect of the thermal treatment," *Physica Status Solidi A: Applications and Materials Science*, vol. 203, no. 8, pp. 2018–2023, 2006.

[22] S. A. Mahmoud, A. A. Ibrahim, and A. S. Riad, "Physical properties of thermal coating CdS thin films using a modified evaporation source," *Thin Solid Films*, vol. 372, no. 1, pp. 144–148, 2000.

[23] J. Herrero, M. T. Gutiérrez, C. Guillén et al., "Photovoltaic windows by chemical bath deposition," *Thin Solid Films*, vol. 361, pp. 28–33, 2000.

[24] J. Britt and C. Ferekides, "Thin-film CdS/CdTe solar cell with 15.8% efficiency," *Applied Physics Letters*, vol. 62, no. 22, pp. 2851–2852, 1993.

[25] B. Ullrich and R. Schroeder, "Green emission and bandgap narrowing due to two-photon excitation in thin film CdS formed by spray pyrolysis," *Semiconductor Science and Technology*, vol. 16, no. 8, pp. L37–L39, 2001.

[26] N. Romeo, A. Bosio, R. Tedeschi, and V. Caneveri, "Growth of polycrystalline CdS and CdTe thin layers for high efficiency thin film solar cells," *Materials Chemistry and Physics*, vol. 66, no. 2, pp. 201–206, 2000.

[27] X. L. Tong, D. S. Jiang, Y. Li, Z. M. Liu, and M. Z. Luo, "The influence of the silicon substrate temperature on structural and optical properties of thin-film cadmium sulfide formed with femtosecond laser deposition," *Physica B: Condensed Matter*, vol. 382, no. 1-2, pp. 105–109, 2006.

[28] K. P. Acharya, A. Erlacher, and B. Ullrich, "Optoelectronic properties of ZnTe/Si heterostructures formed by nanosecond laser deposition at different Nd:YAG laser lines," *Thin Solid Films*, vol. 515, no. 7-8, pp. 4066–4069, 2007.

[29] K. P. Acharya, J. R. Skuza, R. A. Lukaszew, C. Llyanage, and B. Ullrich, "CdS thin films formed on flexible plastic substrates by pulsed-laser deposition," *Journal of Physics Condensed Matter*, vol. 19, no. 19, Article ID 196221, 2007.

[30] X. Mathew, G. W. Thompson, V. P. Singh et al., "Development of CdTe thin films on flexible substrates—a review," *Solar Energy Materials and Solar Cells*, vol. 76, no. 3, pp. 293–303, 2003.

[31] X. Mathew, J. Pantoja Enriquez, A. Romeo, and A. N. Tiwari, "CdTe/CdS solar cells on flexible substrates," *Solar Energy*, vol. 77, no. 6, pp. 831–838, 2004.

[32] J.-H. Lee, "Influence of substrates on the structural and optical properties of chemically deposited CdS films," *Thin Solid Films*, vol. 515, no. 15, pp. 6089–6093, 2007.

[33] J.-H. Lee, "Structural and optical properties of CdS thin films on organic substrates for flexible solar cell applications," *Journal of Electroceramics*, vol. 17, no. 2–4, pp. 1103–1108, 2006.

[34] B.-S. Moon, J.-H. Lee, and H. Jung, "Comparative studies of the properties of CdS films deposited on different substrates by R.F. sputtering," *Thin Solid Films*, vol. 511-512, pp. 299–303, 2006.

[35] D. Lim, J.-H. Lee, and W. Song, "Improvement of the characteristics of chemical bath deposition-cadmium sulfide films deposited on an O_2 plasma-treated polyethylene terephthalate substrate," *Thin Solid Films*, vol. 546, pp. 317–320, 2013.

[36] E. T. Thostenson, Z. Ren, and T.-W. Chou, "Advances in the science and technology of carbon nanotubes and their composites: a review," *Composites Science and Technology*, vol. 61, no. 13, pp. 1899–1912, 2001.

[37] T.-W. Chow, L. Gao, E. T. Thostenson, Z. Zhang, and J.-H. Byun, "An assessment of the science and technology of carbon nanotube-based fibers and composites," *Composites Science and Technology*, vol. 70, no. 1, pp. 1–19, 2010.

[38] L. Dai, D. W. Chang, J.-B. Baek, and W. Lu, "Carbon nanomaterials for advanced energy conversion and storage," *Small*, vol. 8, no. 8, pp. 1130–1166, 2012.

[39] C. Park, J. Wilkinson, S. Banda et al., "Aligned single-wall carbon nanotube polymer composites using an electric field," *Journal of Polymer Science, Part B: Polymer Physics*, vol. 44, no. 12, pp. 1751–1762, 2006.

[40] Y. Zhu, C. Ma, W. Zhang, R. Zhang, N. Koratkar, and J. Liang, "Alignment of multiwalled carbon nanotubes in bulk epoxy composites via electric field," *Journal of Applied Physics*, vol. 105, no. 5, p. 054319, 2009.

[41] A. I. Oliva-Avilés, F. Avilés, and V. Sosa, "Electrical and piezoresistive properties of multi-walled carbon nanotube/polymer composite films aligned by an electric field," *Carbon*, vol. 49, no. 9, pp. 2989–2997, 2011.

[42] Solvay Advanced Polymers, Alpharetta, Ga, USA, http://www.solvay.com/.

[43] Bayer Material Science, Leverkusen, Germany, http://www.covestro.com/.

[44] F. Avilés, A. Ponce, J. V. Cauich-Rodríguez, and G. T. Martínez, "TEM examination of MWCNTs oxidized by mild experimental conditions," *Fullerenes, Nanotubes and Carbon Nanostructures*, vol. 20, no. 1, pp. 49–55, 2012.

[45] A. I. Oliva, R. Castro-Rodríguez, O. Solís-Canto, V. Sosa, P. Quintana, and J. L. Peña, "Comparison of properties of CdS thin films grown by two techniques," *Applied Surface Science*, vol. 205, no. 1–4, pp. 56–64, 2003.

[46] A. I. Oliva-Avilés, R. Patiño, and A. I. Oliva, "CdS films deposited by chemical bath under rotation," *Applied Surface Science*, vol. 256, no. 20, pp. 6090–6095, 2010.

[47] G. Hodes, *Chemical Solution Deposition of Semiconductor Films*, Marcel Dekker, New York, NY, USA, 2002.

[48] R. Ortega-Borges and D. Lincot, "Mechanism of chemical bath deposition of cadmium sulfide thin films in the ammonia-thiourea system," *Journal of the Electrochemical Society*, vol. 140, no. 12, pp. 3464–3473, 1993.

[49] J. M. Doña and J. Herrero, "Chemical bath deposition of CdS thin films: an approach to the chemical mechanism through study of the film microstructure," *Journal of the Electrochemical Society*, vol. 144, no. 11, pp. 4081–4091, 1997.

[50] J. Tauc, *Amorphous and Liquid Semiconductors*, Plenum Press, New York, NY, USA, 1974.

[51] I. Horcas, R. Fernandez, J. M. Gomez-Rodriguez, J. Colchero, J. Gomez-Herrero, and A. M. Baro, "WSXM: a software for scanning probe microscopy and a tool for nanotechnology," *Review of Scientific Instruments*, vol. 78, Article ID 013705, 2007.

[52] G. M. Alonzo-Medina, A. González-González, J. L. Sacedón, A. I. Oliva, and E. Vasco, "Local slope evolution during thermal annealing of polycrystalline Au films," *Journal of Physics D: Applied Physics,* vol. 45, no. 43, Article ID 435301, 2012.

[53] A. González-González, J. L. Sacedón, C. Polop, E. Rodríguez-Caas, J. A. Aznárez, and E. Vasco, "Surface slope distribution with mathematical molding on Au(111) thin film growth," *Journal of Vacuum Science and Technology A: Vacuum, Surfaces and Films*, vol. 27, no. 4, pp. 1012–1016, 2009.

[54] A. I. Oliva-Avilés, F. Avilés, V. Sosa, A. I. Oliva, and F. Gamboa, "Dynamics of carbon nanotube alignment by electric fields," *Nanotechnology*, vol. 23, no. 46, Article ID 465710, 2012.

[55] T. C. Theodosiou and D. A. Saravanos, "Numerical investigation of mechanisms affecting the piezoresistive properties of CNT-doped polymers using multi-scale models," *Composites Science and Technology*, vol. 70, no. 9, pp. 1312–1320, 2010.

[56] S. Tec-Yam, R. Patiño, and A. I. Oliva, "Chemical bath deposition of CdS films on different substrate orientations," *Current Applied Physics*, vol. 11, no. 3, pp. 914–920, 2011.

[57] M. Aguilar, A. I. Oliva, P. Quintana, and J. L. Peña, "Dynamic phenomena in the surface of gold thin films: macroscopic surface rearrangements," *Surface Science*, vol. 380, no. 1, pp. 91–99, 1997.

[58] A. I. Oliva, R. Castro-Rodríguez, O. Ceh, P. Bartolo-Pérez, F. Caballero-Briones, and V. Sosa, "First stages of growth of CdS films on different substrates," *Applied Surface Science*, vol. 148, no. 1, pp. 42–49, 1999.

[59] R. Castro-Rodríguez, A. I. Oliva, V. Sosa, F. Caballero-Briones, and J. L. Peña, "Effect of indium tin oxide substrate roughness on the morphology, structural and optical properties of CdS thin films," *Applied Surface Science*, vol. 161, no. 3, pp. 340–346, 2000.

[60] A. I. Oliva, O. Solís-Canto, R. Castro-Rodríguez, and P. Quintana, "Formation of the band gap energy on CdS thin films growth by two different techniques," *Thin Solid Films*, vol. 391, no. 1, pp. 28–35, 2001.

Two-Dimensional Modeling of Thermomechanical Responses of Rectangular GFRP Profiles Exposed to Fire

Lingfeng Zhang,[1] Weiqing Liu,[1,2] Guoqing Sun,[3] Lu Wang,[3] and Lingzhi Li[4]

[1]School of Civil Engineering, Southeast University, Nanjing, China
[2]Advanced Engineering Composites Research Center, Nanjing Tech University, Nanjing, China
[3]College of Civil Engineering, Nanjing Tech University, Nanjing, China
[4]College of Civil Engineering, Tongji University, Shanghai, China

Correspondence should be addressed to Weiqing Liu; wqliu@njtech.edu.cn and Lu Wang; kevinlwang@hotmail.com

Academic Editor: Fabrizio Sarasini

In the past three decades, one-dimensional (1D) thermal model was usually used to estimate the thermal responses of glass fiber-reinforced polymer (GFRP) materials and structures. However, the temperature gradient and mechanical degradation of whole cross sections cannot be accurately evaluated. To address this issue, a two-dimensional (2D) thermomechanical model was developed to predict the thermal and mechanical responses of rectangular GFRP tubes subjected to one-side ISO-834 fire exposure in this paper. The 2D governing heat transfer equations with thermal boundary conditions, discretized by alternating direction implicit (ADI) method, were solved by Gauss-Seidel iterative approach. Then the temperature-dependent mechanical responses were obtained by considering the elastic modulus degradation from glass transition and decomposition of resin. The temperatures and midspan deflections of available experimental results can be reasonably predicted. The overestimation of deflections could be attributed to the underestimation of bending stiffness. This model can also be extended to simulate the thermomechanical responses of beams and columns subjected to multiside fire loading, which may occur in real fire scenarios.

1. Introduction

Compared with traditional building materials, fiber-reinforced polymer (FRP) materials have many obvious advantages such as high strength-to-weight ratio, superior durability, good fatigue endurance, and rapid installation. Hence, the FRP composites have increasingly been used in civil engineering due to these excellent properties. However, the resin in the FRP composites undergoes a glass transition stage when subjected to the high temperatures around the glass transition temperatures (T_g). The resin matrix changes from a rigid, glassy solid to a softer and viscoelastic material during the glass transition stage. The mechanical properties of FRP exhibit a significantly reduction at T_g, which determines the limited use temperature for FRP composites [1]. The matrix transfers from compact viscous polymer to porous solid char and volatiles over the decomposition temperature (T_d) range. The mechanical properties of FRP composites dropped once

again. Therefore, the usage of FRP composites in engineering can be hindered by the high degradation of mechanical performance at elevated and high temperatures.

In the past two decades, a large number of experimental studies have been conducted to evaluate the thermal and thermomechanical performances of FRP materials. In 1999, the fire resistance of GFRP sprinkler pipes with empty cavity, stagnant water, and flowing water was investigated by Davies and Dewhurst [2]. The failure time of the pipes with empty cavity and stagnant water was 1.5 min and 8.5 min, respectively, while the flowing water-cooled pipe remained structural integrity subjected to 120 min fire exposure. Inspired by the water-cooling concept of filament wound GFRP pipe used in [2], this concept, developed by Keller et al., was applied in the fire endurance of pultruded GFRP multicellular panels [3, 4]. The fire performance of GFRP panels was investigated in three different experimental parts: charring of GFRP laminates, fire endurance of liquid cooling moderately sized GFRP panels, and fire resistance of structural liquid cooling full-scale GFRP panels, respectively. In the first part, the pultruded GFRP laminates started burning at roughly 6 min, and the temperature of cold face reached T_g at approximately 10 min [5]. In the second part, the temperature profiles through the thickness of lower face sheet were much lower than those of charring experiments [3]. In the third part, for the noncooled specimens, the cold face of lower face sheet reached T_g and T_d in 10 min and 57 min, respectively. The test results demonstrated that liquid cooling was an effective way to improve the fire resistance of pultruded GFRP components. Correia et al. [6] conducted structural fire endurance experiments on full-scale cellular GFRP columns subjected to one-side fire exposure. The noncooled column failed at 49 min due to global buckling, while structural function of the water-cooled columns could be maintained for two hours. In 2010, the fire resistances of noncooled and water-cooled pultruded GFRP beams with square hollow section were investigated by Bai et al. [7]. The bottom flange of the GFRP beams was subjected to ISO-834 fire. The fire resistance was over 120 min by using the water-cooled approach with a 72 mm/s flowing rate. But the noncooled specimen failed abruptly after about 38 min due to the kinking and buckling of top flange. In 2015, the fire performance of pultruded GFRP columns was investigated by Morgado et al. [8]. The columns with passive (calcium silicate boards) and active (water-cooling) fire protection systems were exposed to one-side (bottom flange) and three-side (bottom flange and two webs) fire, respectively. The test results showed that the efficiency of fire protection systems depended on the fire exposure sides. For one-side fire exposure, the water-cooling with a 72 mm/s flowing rate exhibited the best fire protection, which provided more than 120 min of fire resistance, while for three-side fire exposure, the column with calcium silicate boards showed the most effective protection, which provided roughly 40 min of fire resistance.

The mathematical modeling of the thermal response of FRP composite under elevated and high temperatures can be traced back to the 1980s. In 1981, Griffis et al. [9] proposed a 1D finite difference model to simulate the temperature profiles of graphite epoxy coupons subjected to intense surface heating. The analytical thermal responses agreed well with the test results. A thermomechanical model was developed to predict the thermal response of graphite epoxy composites and wood [10]. The proposed model can reasonably predict the temperature profiles and strength degradations. A 1D transient thermal model based on chemical kinetics, developed by Henderson et al., was applied to predict the temperature responses of fiber-reinforced phenol-formaldehyde polymer composite [11]. Good agreement was found between the analytical and experimental temperature profiles. In 1995, a thermochemical model of Thick composite laminates subjected to hydrocarbon fire was presented by Gibson et al. [12]. The 1D finite difference method was adopted to solve the proposed model. In 1996, this model was improved by Looyeh et al. [13]. The finite element method was introduced to predict the thermal performance of polymer composite materials. In 2000, Dodds et al. [14] updated the model for simulating the composite laminates. A straightforward explicit finite

difference format was applied to solve the model. In 2006, the apparent specific capacity, developed by Lattimer and Ouellette [15], was input into a heat transfer model to predict the temperature profile through E-glass/vinyl ester composite laminates. Gibson et al. [16, 17] investigated the postfire mechanical properties of polymer composites by combing the thermal model with Mouritz's two-layer mechanical model [18]. In 2012, the thermal responses of FRP laminates subjected to three-point bending and one-side heat flux were experimental and numerical investigated by Gibson et al. [19]. More recently, Miano and Gibson [20, 21] introduced a simplified thermal model by using the apparent diffusivity (ATD) method. The temperature-dependent ATD can significantly simplify the computational procedures and improve the stability of the numerical solutions. Feih et al. [22, 23] presented the thermomechanical model to predict the tension and compression properties of FRP laminates in fire. A 1D model was further developed to simulate the thermomechanical responses of sandwich composites [24]. This model also validated and applied by Anjang et al. [25, 26] to investigate the in-fire and postfire mechanical properties of FRP sandwich composite structures with balsa wood core. In 2006, a thermochemical and thermomechanical models were introduced by Keller et al. [27]. The models were used to predict the structural response of water-cooled multicellular GFRP slabs. In 2007, Bai et al. [28] proposed the chemical kinetics-based thermophysical model by considering the decomposition of matrix resin. Based on the temperature-dependent thermophysical model, 1D thermal model with the effective specific capacity was developed [29]. The predicted temperature responses agreed well with the experimental results. In 2008, the mechanism-based models [30–32] were developed to investigate the mechanical behavior of FRP composites in fire. This model was further validated on the predicting of the time-dependent temperature responses, elastic modulus degradation, and time to failure of water-cooled full-scale GFRP cellular columns [6]. In 2011, Miano [21] developed a three-dimensional (3D) finite element (FE) thermal model with ATD to estimate the temperature profiles of carbon fiber-reinforced polymer (CFRP) wing box laminates. More recently, a 3D FE model was developed by Shi et al. [33] to investigate the coupled temperature-diffusion-deformation problem of silica/phenolic composite materials. The accuracy of this model was validated by comparing the measured temperatures and displacements with numerical results.

Among the various existing thermal models, the 1D finite difference method was most frequently used to predict the thermal response of FRP composites. However, the FRP structural components applied in building construction were subjected to multidimensional fire exposures. The temperature profile of heated zone of FRP composites cannot be predicted by the use of 1D model. Hence, to address this issue, a 2D thermal model was proposed in this study. The accuracy of the proposed model was validated by the existing experimental data of pultruded E-glass/polyester rectangular tube under four-point bending and fire from one side. The comparison indicated that the developed two-dimensional thermomechanical model can be reasonable to predict the thermomechanical responses of GFRP composites.

2. Mathematical Model

2.1. Modeling Assumptions. When subjected to fire, FRP composites undergo many complex thermal, physical, chemical, and structural failure processes [35]. The challenge to accurately modeling the temperature responses is to consider complex interaction of degradation processes. Previous studies [5, 28] showed that the thermal and mechanical processes were dominated by glass transition and thermal decomposition. The convective heat transfer of volatiles flow up through the decomposition front to the surface may have a small effect on the temperature profile in the FRP material [15, 21]. Hence, in this study, the 2D thermomechanical model was simplified by considering heat conduction, physical process (glass transition), chemical (decomposition), and mechanical degradations.

2.2. Thermophysical Properties Model. During fire processes, thermophysical properties include density, specific capacity, and thermal conductivity of virgin (nonchar) and char material.

A lot of works were made to investigate these properties. The temperature-dependent density can be described by using an Arrhenius equation based on chemical kinetics [11]:

$$\frac{\partial \rho}{\partial t} = -A\rho \exp\left(\frac{-E_A}{RT}\right), \tag{1}$$

where ρ is the density; A represents the preexponential factor; E_A is the activation energy; R is the universal gas constant; and T is temperature.

The change of density can be described by (1). But in the existing literatures, specific capacity and thermal conductivity of virgin and char were usually expressed by polynomial fitting instead of analytical from. Hence, the parameters used in fitting have no clear physical meaning. Bai et al. [28] developed a model for predicting temperature-dependent thermophysical properties. In this model, a conversion degree of decomposition was introduced to characterize the pyrolysis process of polymer resin, as indicated in

$$\frac{d\alpha_d}{dt} = A_d \exp\left(\frac{-E_{A,d}}{RT}\right)(1-\alpha_d)^{nr_d}, \tag{2}$$

where α_d denotes the decomposition degree and $A_d, E_{A,d}$, and nr_d are kinetic parameters, which can be derived from thermogravimetric analysis.

Based on decomposition degree α_d, the temperature-dependent density can be obtained

$$\rho = (1-\alpha_d)\rho_b + \alpha_d\rho_a, \tag{3}$$

where ρ_b is the density of virgin composite and ρ_a is the density of char material.

The apparent specific capacity can be given as [28]

$$C_p = f_b C_{p,b} + (1-f_b)C_{p,a} + \frac{d\alpha_d}{dT}C_d, \tag{4}$$

where C_p is the apparent specific capacity, which increases due to endothermic phenomenon during decomposition process [35]; $C_{p,b}$ and $C_{p,a}$ are apparent specific capacity before and after decomposition respectively; C_d is decomposition heat; and f_b is mass fraction of undecomposed FRP materials, which can be calculated by [28]

$$f_b = \frac{M_{a,v}(1-\alpha_d)}{M_{a,v}(1-\alpha_d) + M_{a,e}\alpha_d}, \tag{5}$$

where $M_{a,v}$ ($M_{a,e}$) is the initial (final) mass of FRP material.

In this paper the main research object is the cross section of rectangular GFRP tubes; only the cross-sectional properties should be considered. Previous study [36] showed that thermal conductivity parallel to the axis of fiber (0°) was much higher than in the transverse (90°) and through thickness direction, while the thermal conductivity of transverse and through thickness direction were similar for glass fiber/polyester composites. Hence, the transverse thermal conductivity can be simplified by using the through thickness thermal conductivity. The thermal conductivity through the thickness of FRP composites can be expressed as [28]

$$\frac{1}{k_\perp} = \frac{(1-\alpha_d)}{k_{\perp b}} + \frac{\alpha_d}{k_{\perp a}}, \tag{6}$$

where $k_\perp$ denotes the through thickness thermal conductivity; the subscripts b and a represent the thermal conductivity before and after decomposition, respectively; and $k_{\perp b}$ and $k_{\perp a}$ can be obtained by using the inverse rule of mixtures:

$$\frac{1}{k_{\perp b}} = \frac{V_f}{k_f} + \frac{V_m}{k_m},$$
$$\frac{1}{k_{\perp a}} = \frac{V_f}{k_f} + \frac{V_{ga}}{k_{ga}}, \tag{7}$$

where k_f, k_m, and k_{ga} are the thermal conductivities of fiber, matrix, and volatile gas, respectively; V_f, V_m, and V_{ga} are the volume fractions of fiber, matrix, and volatile gas, respectively.

2.3. 2D Heat Transfer Model. 2D transient heat transfer equation can be expressed as

$$\rho C_p \frac{\partial T}{\partial t} = \frac{\partial}{\partial x}\left(k_x\frac{\partial T}{\partial x}\right) + \frac{\partial}{\partial y}\left(k_y\frac{\partial T}{\partial y}\right), \tag{8}$$

where k_x and k_y are the thermal conductivity in x and y directions, respectively. The term on the left side of (9) refers to the rate of change of internal energy, and the other two terms on the right side denote the net heat flux.

The boundary conditions applied in thermal analysis can be divided into three types. The first one is the Dirichlet boundary condition, which can be expressed as

$$T(x,y,t)\big|_{bd} = T(t), \tag{9}$$

where x and y represent the 2D spatial Descartes coordinates; subscript bd denotes the boundaries of rectangular tube;

and $T(t)$ is the time-dependent specific temperature on the boundary. The second one is the prescribed heat flux, which can be expressed as

$$-k_x \frac{\partial T(x,y,t)}{\partial x}\bigg|_{\text{bd}} = q(t),$$

$$-k_y \frac{\partial T(x,y,t)}{\partial y}\bigg|_{\text{bd}} = q(t), \qquad (10)$$

where $q(t)$ is the time-dependent heat flux at the boundaries. The last one is the convection between the boundary and external environment, which can be expressed as

$$-k_x \frac{\partial T(x,y,t)}{\partial x}\bigg|_{\text{bd}} = h\left(T_\infty(t) - T(x,y,t)|_{\text{bd}}\right),$$

$$-k_y \frac{\partial T(x,y,t)}{\partial y}\bigg|_{\text{bd}} = h\left(T_\infty(t) - T(x,y,t)|_{\text{bd}}\right), \qquad (11)$$

where h denotes the convective coefficients and $T_\infty(t)$ represents the time-dependent temperature of external environment.

In this study, convection and radiation between the boundaries and external/internal environment can be obtained from

$$q(t) = h_x\left[T_\infty(t) - T(x,y,t)|_{\text{bd}}\right]$$
$$+ \varepsilon_{x,r}\sigma_{x,r}\left[T_\infty(t)^4 - T(x,y,t)^4|_{\text{bd}}\right],$$
$$q(t) = h_y\left[T_\infty(t) - T(x,y,t)|_{\text{bd}}\right]$$
$$+ \varepsilon_{y,r}\sigma_{y,r}\left[T_\infty(t)^4 - T(x,y,t)^4|_{\text{bd}}\right]. \qquad (12)$$

2.4. Thermomechanical Model. Young's modulus, E, can be expressed as [31]

$$E = E_g\left(1 - \alpha_g\right) + E_r \cdot \alpha_g\left(1 - \alpha_d\right), \qquad (13)$$

where E_g is Young's modulus in glassy state and E_r is Young's modulus in leathery and rubbery states; and α_g is the conversion degree of glass transition and can be expressed by [31]

$$\frac{d\alpha_g}{dt} = A_g \exp\left(\frac{-E_{A,g}}{RT}\right)\left(1 - \alpha_g\right)^{nr_g}, \qquad (14)$$

where A_g, $E_{A,g}$, and nr_g are kinetic parameters of glass transition for FRP composites. Based on beam theory, the midspan deflection of GFRP rectangular beam under four-point bending can be calculated by

$$\delta = \frac{PL_s^3}{24EI}\left(\frac{3a}{L_s} - \frac{4a^3}{L_s^3}\right), \qquad (15)$$

where δ denotes the time-dependent midspan deflection; P is half of the total load; a is the distance between the support and one load; L_s is the span; and I is the second moment of area of the section. The total bending stiffness of the GFRP beam, EI, can be calculated as the sum of the stiffness contribution of each elements [31]:

$$EI = E_w I_w + E_{\text{tf}} I_{\text{tf}} + E_{\text{bf}} I_{\text{bf}}, \qquad (16)$$

where the subscripts w, tf, and bf denote the web, top flange, and bottom flange, respectively.

3. Solutions of Governing Equation

3.1. ADI Techniques. Equation (8) is a 2D nonlinear parabolic partial differential equation, which can be solved directly by explicit difference scheme. However, the numerical stability of this scheme is rather poor when solving the governing equation with time-dependent thermophysical properties. When heat transfer switches from 1D to 2D problem, the coefficient matrix of governing equation transfers from tridiagonal to nontridiagonal matrix by using the traditional implicit difference scheme. The formats of results are in a very complicated set of algebraic equations in two dimensions, which is difficult to solve [37]. Alternating direction implicit (ADI) format was an effective method for solving multidimensional heat transfer problems, which was first proposed by Peaceman and Rachford [38] in 1955 and developed by Douglas and Rachford [39]. By using ADI method, the 2D equation can be splitting into the same 1D implicit format with simpler tridiagonal matrix algorithm at each time step. In this study, ADI method was introduced and applied in this simulation of 2D thermal and mechanical responses of GFRP rectangular profiles.

3.2. Mesh. Figure 1 presents the "C type" profiles by considering the symmetry of shape and boundary conditions of a rectangular GFRP tube. The lower-left corner was set as the coordinate origin. In general, the solution domain was meshed into $L \times M$ elements, where M (M) is the number of elements in x (y) direction. T_s represents number of elements through the thickness. l (m) denotes the total width (height) of the GFRP tube; then the space interval can be written as $\Delta x = l/L$ or $\Delta y = m/M$. Similarly, the time step $\Delta t = t/N$ was used for simulating program, where t represents the total fire exposure time, and N is the number of time step.

3.3. Finite Difference Scheme. The ADI difference method splits the time step from n to $n + 1$ into two substeps: Step I: $n \rightarrow n + 1/2$ and Step II: $n + 1/2 \rightarrow n + 1$. In Step I, implicit scheme was used in x direction while explicit scheme was applied in y direction. In Step II, implicit scheme was then applied in y direction while explicit scheme was applied in

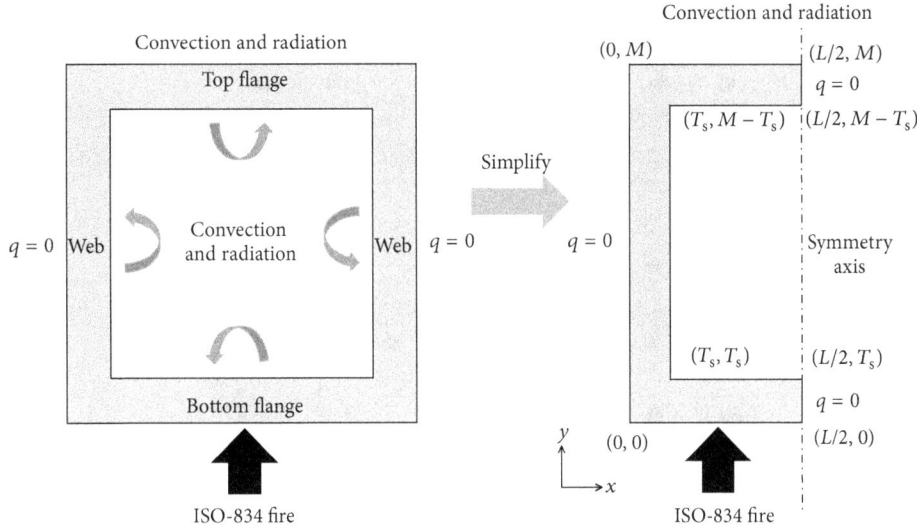

FIGURE 1: Thermal boundary conditions of GFRP rectangular tube.

x direction. For example, the difference schemes of all notes inside the domain were derived as follows:

Step I: $\dfrac{k_{x,i-1,j}^{n} + k_{x,i,j}^{n}}{2} \cdot \dfrac{T_{i-1,j}^{n+1/2} - T_{i,j}^{n+1/2}}{\Delta x} \Delta y$

$+ \dfrac{k_{x,i,j}^{n} + k_{x,i+1,j}^{n}}{2} \cdot \dfrac{T_{i+1,j}^{n+1/2} - T_{i,j}^{n+1/2}}{\Delta x} \Delta y$

$+ \dfrac{k_{y,i,j-1}^{n} + k_{y,i,j}^{n}}{2} \cdot \dfrac{T_{i,j-1}^{n} - T_{i,j}^{n}}{\Delta y} \Delta x + \dfrac{k_{y,i,j}^{n} + k_{y,i,j+1}^{n}}{2}$

$\cdot \dfrac{T_{i,j+1}^{n} - T_{i,j}^{n}}{\Delta y} \Delta x = C_{p,i,j}^{n} \rho_{i,j}^{n} \dfrac{T_{i,j}^{n+1/2} - T_{i,j}^{n}}{\Delta t/2} \Delta x \Delta y$

Step II: $\dfrac{k_{x,i-1,j}^{n+1/2} + k_{x,i,j}^{n+1/2}}{2} \cdot \dfrac{T_{i-1,j}^{n+1/2} - T_{i,j}^{n+1/2}}{\Delta x} \Delta y$ (17)

$+ \dfrac{k_{x,i,j}^{n+1/2} + k_{x,i+1,j}^{n+1/2}}{2} \cdot \dfrac{T_{i+1,j}^{n+1/2} - T_{i,j}^{n+1/2}}{\Delta x} \Delta y$

$+ \dfrac{k_{y,i,j-1}^{n+1/2} + k_{y,i,j}^{n+1/2}}{2} \cdot \dfrac{T_{i,j-1}^{n+1} - T_{i,j}^{n+1}}{\Delta y} \Delta x$

$+ \dfrac{k_{y,i,j}^{n+1/2} + k_{y,i,j+1}^{n+1/2}}{2} \cdot \dfrac{T_{i,j+1}^{n+1} - T_{i,j}^{n+1}}{\Delta y} \Delta x$

$= C_{p,i,j}^{n+1/2} \rho_{i,j}^{n+1/2} \dfrac{T_{i,j}^{n+1} - T_{i,j}^{n+1/2}}{\Delta t/2} \Delta x \Delta y,$

where the subscripts i, j denote the coordinates; Δx, Δy represent the space interval in x and y direction, respectively. As indicated in (17), based on thermophysical properties and temperatures of the last time step n ($n+1/2$), the temperature in time step $n+1/2$ ($n+1$) can be calculated by using of Gauss-Seidel iterative approach.

The above only presents the notes inside the calculation domain without any external heat source. Therefore, the differential equations related to various boundary conditions must be derived. There are a total of 16 types (8 points and 8 sides) of boundary conditions in this model. The detailed expressions of ADI difference format were given in Appendices A~O, and the corresponding schematic diagram was given in Figure 2.

The difference format of decomposition degree can be expressed as

$$\dfrac{\alpha_{d,i,j}^{n} - \alpha_{d,i,j}^{n-1/2}}{\Delta t/2} = A_d \exp\left(\dfrac{-E_{A,d}}{RT_{i,j}^{n-1/2}}\right)\left(1 - \alpha_{d,i,j}^{n-1/2}\right)^{n_d},$$

$$\dfrac{\alpha_{g,i,j}^{n} - \alpha_{g,i,j}^{n-1/2}}{\Delta t/2} = A_g \exp\left(\dfrac{-E_{A,g}}{RT_{i,j}^{n-1/2}}\right)\left(1 - \alpha_{g,i,j}^{n-1/2}\right)^{n_g},$$

(18)

where n is the current time step and $n - 1/2$ denotes the previous time step.

Based on the decomposition degree at the last time step $n - 1/2 (\alpha_{i,j}^{n-1/2})$, the difference schemes of thermophysical properties at time step n can be derived as

$$\rho_{i,j}^{n} = \left(1 - \alpha_{i,j}^{n}\right)\rho_b + \alpha_{i,j}^{n}\rho_a,$$

$$C_{p,i,j}^{n} = f_{b,i,j}^{n} C_{p,b} + \left(1 - f_{b,i,j}^{n}\right)C_{p,a} + \dfrac{\alpha_{i,j}^{n} - \alpha_{i,j}^{n-1/2}}{T_{i,j}^{n} - T_{i,j}^{n-1/2}}C_d,$$

$$f_{b,i,j}^{n} = \dfrac{M_{a,v}\left(1 - \alpha_{i,j}^{n}\right)}{M_{a,v}\left(1 - \alpha_{i,j}^{n}\right) + M_{a,e}\alpha_{i,j}^{n}},$$

(19)

$$\dfrac{1}{k_{x,i,j}^{n}} = \dfrac{\left(1 - \alpha_{i,j}^{n}\right)}{k_{\perp b}} + \dfrac{\alpha_{i,j}^{n}}{k_{\perp a}},$$

$$\dfrac{1}{k_{y,i,j}^{n}} = \dfrac{\left(1 - \alpha_{i,j}^{n}\right)}{k_{\perp b}} + \dfrac{\alpha_{i,j}^{n}}{k_{\perp a}}.$$

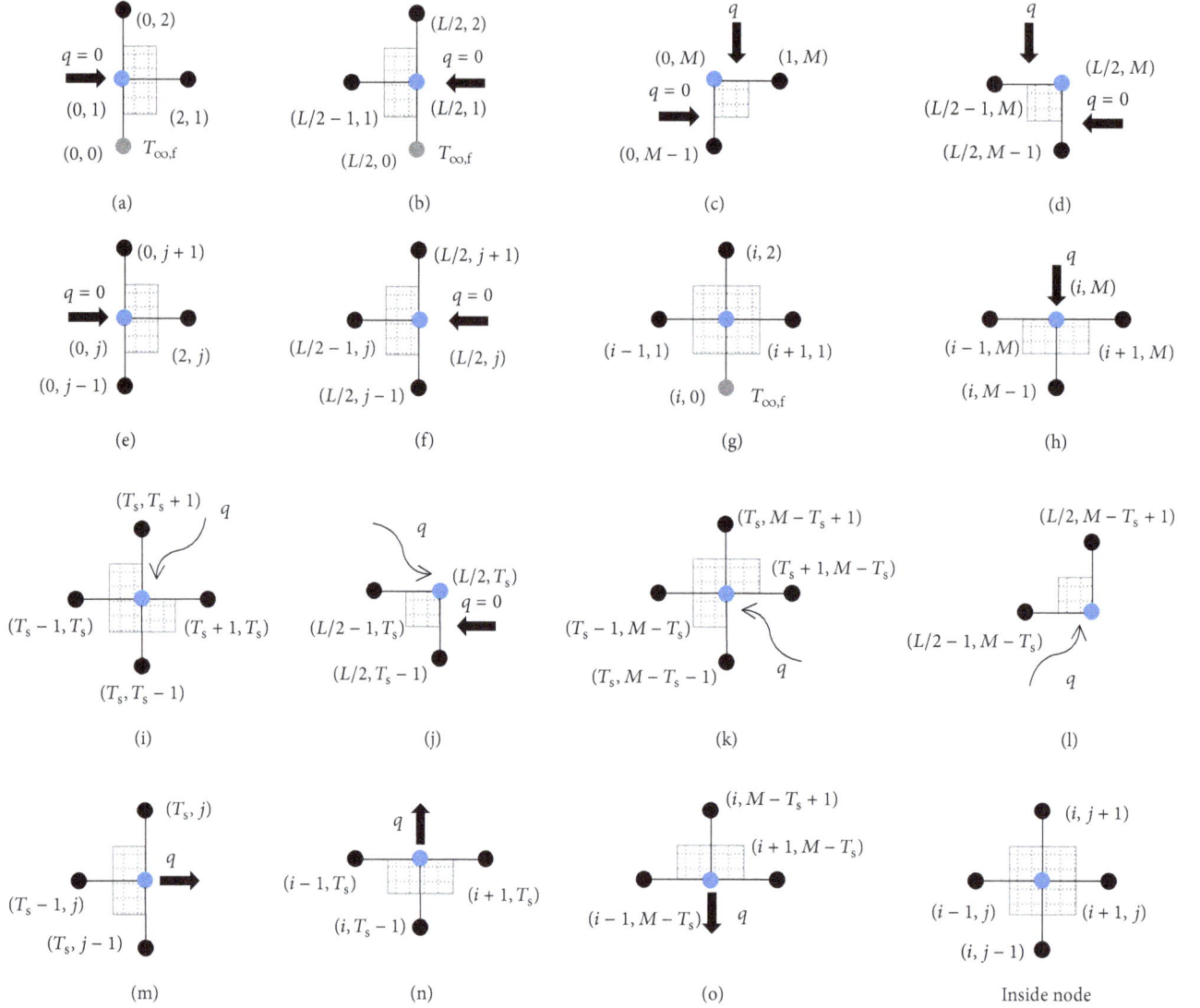

FIGURE 2: Schematic diagram of thermal boundary conditions.

The difference schemes of temperature-dependent Young's modulus $E_{i,j}$ at time step n can be expressed as

$$E_{i,j}^n = E_g \left(1 - \alpha_{g,i,j}^n\right) + E_r \cdot \alpha_{g,i,j}^n \left(1 - \alpha_{d,i,j}^n\right). \tag{20}$$

The schemes of the total bending stiffness EI at time step n can be expressed as

$$EI^n = \sum_{i=1}^{n_w} E_{w,i}^n I_{w,i}^n + \sum_{i=1}^{n_f} E_{tf,i}^n I_{tf,i}^n + E_{bf,i}^n I_{bf,i}^n, \tag{21}$$

where n_w and n_f were the number of layers of the webs and flanges, respectively. Therefore the midspan at each time step can be calculated:

$$\delta^n = \frac{PL^3}{24EI^n} \left(\frac{3a}{L} - \frac{4a^3}{L^3}\right). \tag{22}$$

4. Validation and Discussion

4.1. Basic Model. As described in Section 1, Bai et al. [7] conducted the fire structural experiment of pultruded E-glass/polyester composite square beam subjected to ISO-834 fire, and a 37 min fire resistance was reached. In this paper, this experiment is used to validate the 2D numerical model developed in Section 3.

For boundary conditions, the hot face of bottom flange was set as prescribed ISO-834 curve because the temperature progressions of hot face were very close to ISO-834 fire curve. The outer faces of webs were set as adiabatic condition. As shown in Figure 1, convection and radiation play an important role in the rest boundaries, and the heat convective coefficients at the cold face of bottom flange were estimated as $2 \, \text{W/m}^2 \, \text{K}$ by fitting the temperature profiles at the cold face of the bottom flange [34]. The convective coefficients at inner faces of web and top flange were set as 0 by minimizing the

TABLE 1: Parameters and variables used in numerical model.

Parameters and variables		Refs.
Thermophysical properties		
Density (kg/m^3)	1870 (ρ_b), 1141 (ρ_a)	[34]
Thermal conductivity (W/m K)	0.35 ($k_{\perp b}$), 0.1 ($k_{\perp a}$)	[34]
Specific heat capacity (J/kg K)	1170 ($C_{p,b}$), 840 ($C_{p,a}$)	[34]
Decomposition heat, C_d (kJ/kg)	234	[34]
Final mass fraction, $M_{a,e}/M_{a,v}$	0.61	[29]
Material mechanical properties		
Young's modulus (GPa)	31 (E_g), 5.8 (E_r)	[31, 34]
Chemical kinetic parameters		
Preexponential factor (min^{-1})	2.49×10^{15} (A_g), 2.72×10^{10} (A_d)	[34]
Activation energy (J/mol)	118,591 ($E_{A,g}$), 124,953 ($E_{A,d}$)	[34]
Reaction order	1.89 (n_g), 2.75 (n_d)	[34]
Gas constant, R (J/mol K)	8.314	[34]
Convective and radiant parameters		
Convective coefficient (W/m^2 K)	2 (h_{ib}), 0 (h_{il}), 0 (h_{iu}), Eq. (24) (h_{ou})	[29, 34]
Stefan-Boltzmann constant, σ (Wm^{-2} K^{-4})	5.67×10^{-8}	[29]
Emissivity of solid surface, ε_{ra}	Equation (23)	[29]

difference between modeling temperatures and test results for webs and top flanges. This indicated that convection can be omitted. In fact, the temperature of air in the inner cavity was higher than ambient temperature (20°C), and the convection was weak due to the close temperature for both webs and hot gas in the cavity. The effects of heat radiation, which was described by Stephan-Boltzmann law, were considered for all inner faces and out face of top flanges. The Stephan-Boltzmann constant was given in Table 1 and the solid emissivity can be seen from (23), which was adopted from [27] for modeling thermal response of a very similar E-glass fiber-reinforced polyester composite. The convection coefficient of at the outer face of upper flange can be obtained based on hydrodynamics proposed by Tracy (see (24)). The detailed modeling parameters were summarized in Table 1.

$$\varepsilon_{ra} = 0.75 + \frac{T_{bd} - 20}{1000 - 20} \times (0.95 - 0.75), \qquad (23)$$

$$h_{ou} = 0.14 k_g \left(\Pr \frac{g\beta}{\nu} \left(T_{Ubd} - T_\infty \right) \right)^{1/3}, \qquad (24)$$

where k_g is the thermal conductivity of air (0.03 W/m K); g is the acceleration of gravity (9.81 m/s^2); Pr is the Prandtl number (0.722); β denotes the volumetric coefficient of thermal expansion of air (3.43×10^{-3} K^{-1}); and ν represents the kinetic viscosity of air (1.57×10^{-5} m^2/s) [27].

4.2. Validation and Comparison. Figure 3 shows the comparisons of temperature progressions at different depths from the hot face of bottom flange. During the first 5 min, the heating rate of oven was much higher than the ISO-834 curve, which was used as the temperature of hot face, and this led to an underestimation of temperatures from 5 min to

20 min. This may attribute to the temperature lag resulting from the heat transfer from the oven to GFRP beam. In addition, the forming charred layers at the bottom flange may act as an insulator which producing a high experimental temperature due to the accumulation of hot gas in the pyrolysis zone. This insulation effect can be implemented in the model by adding a convection term of pyrolysis gas in (8) and a momentum equation (using Darcy's law) to simulate the gas transport in the porous charred layers. A complex computational fluid dynamics (CFD) model should be developed and solved. Nevertheless, for the duration from 0 min to 5 min and 20 min to 40 min, the modeling results agreed well with the test values. The temperature gradients at 5 min, 10 min, 20 min, and 40 min were presented in Figure 4. A large temperature gradient through the thickness of bottom flange was found at 5 min and 10 min; however, the gradient decreased from 10 min to 40 min. This indicated that a more even temperature distribution was found through the thickness of bottom flange, and the temperatures of hot face and inner face were approximately 750°C and 900°C at 40 min (Figure 3). The elastic modulus profiles at 5 min, 10 min, 20 min, and 40 min were given in Figure 5. At 5 min, the modulus was 5.8 GPa and the glass transition was finished. Besides, the modulus of web also was also degraded. At 40 min, the modulus of bottom flange was almost lost and the bottom flange was fully decomposed with the lower part of web was in rubbery state.

The time-dependent elastic modulus profiles of web can be seen from Figure 6(a), and the decreasing modulus of elements resulted in an upper migration of neutral axis (Figure 6(b)). In the fire duration, the distance from the neutral axis at hot face was increasing from 50 mm to 75.5 mm while the one at outer face of top flange was decreasing from 50 mm to 36.5 mm. As a result, the time-dependent second

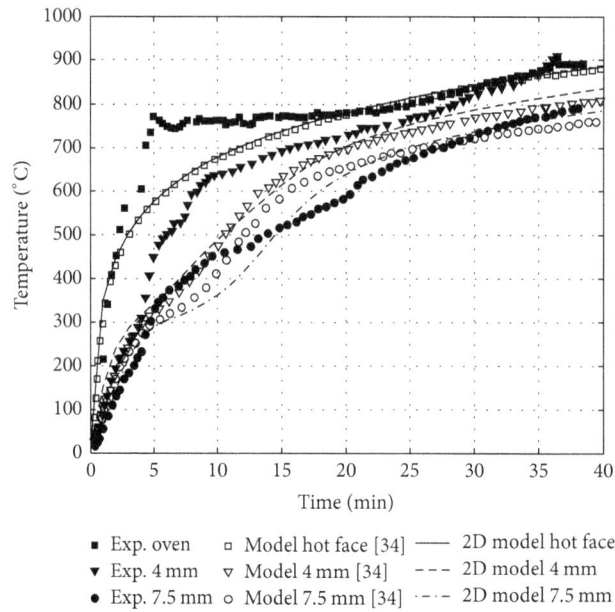

Figure 3: Comparisons of temperature progressions of bottom flange.

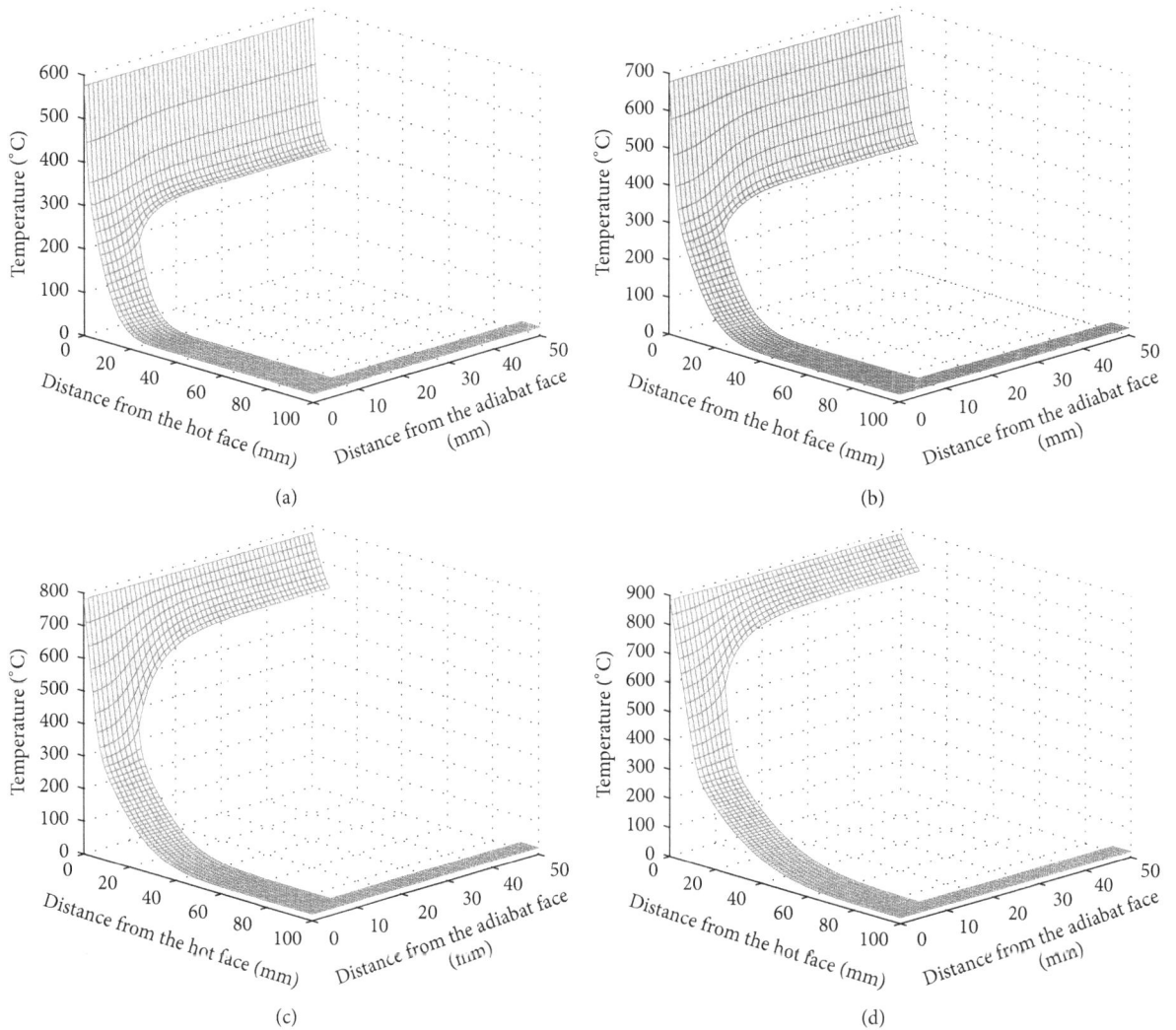

Figure 4: Temperature distribution of GFRP tube: (a) 5 min, (b) 10 min, (c) 20 min, and (d) 40 min.

(a)

(b)

(c)

(d)

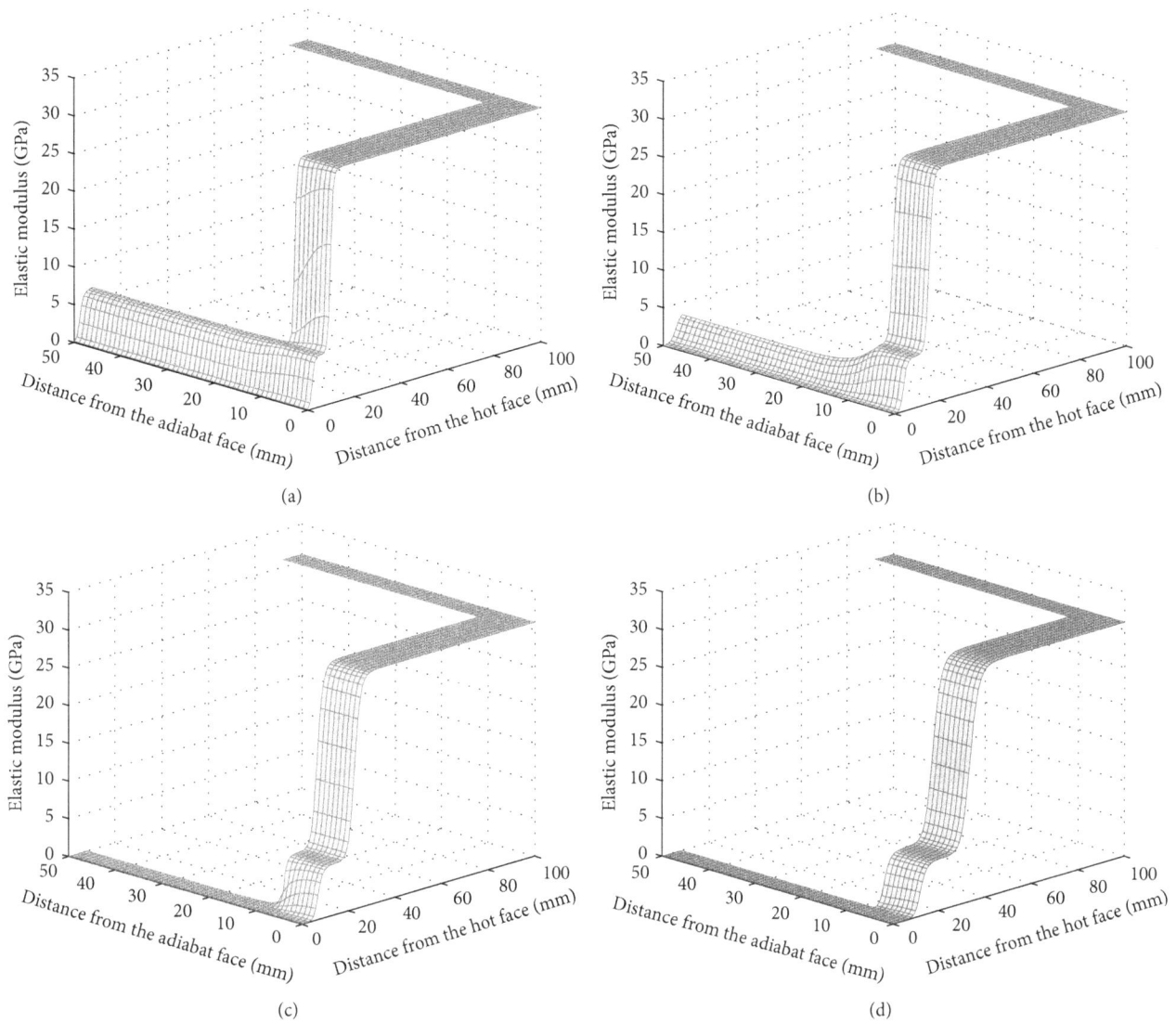

FIGURE 5: Elastic modulus distribution of GFRP tube: (a) 5 min, (b) 10 min, (c) 20 min, and (d) 40 min.

moment of area of each element and bending stiffness was shown in Figures 6(c) and 6(d). A 78% reduction of bending stiffness was found during the whole fire durations. Figure 7 shows comparison of the midspan deflection progressions between the 2D modeling results, test values and 1D modeling results from [34]. As can be seen from Figure 7, the 1D thermomechanical model developed by the Bai et al. can accurately predict the midspan deflections during the first 12 min. However, due to the limitation of 1D model, the thermal responses of webs cannot be predicted; therefore mechanical responses after the onset of web degradation were not presented. The 2D modeling results agreed well with the test values in the first 8 min; however, overestimation was found from 8 min to 37 min. This may be attributed to two main reasons. First, it was assumed that the total bending stiffness was calculated as the sum of independent individual element. However, in fact each element in the cross section was constrained by the adjacent elements. Therefore this led to underestimation of total bending stiffness. In addition, here it was assumed that the whole span was

subjected to the same loading and exhibited the same stiffness degradation, but only 0.95 m (64% of the span of 1.51 m) was directly exposed to the oven [34]. Therefore the overall bending stiffness was underestimated and the deflections were overestimated.

5. Conclusions

In this study, a 2D model to predict the thermomechanical responses of rectangular GFRP tubes were developed and solved by ADI finite difference scheme. The modeling results were compared with the temperature and mechanical responses from experiment on GFRP rectangular beam under the four-point bending and one-side fire exposure. The main conclusion remarks were drawn:

(1) A 2D thermomechanical model based on finite difference method was developed and solved by Gauss-Seidel iterative approach. The detailed ADI schemes considering complex boundary conditions were also

(a)

(b)

(c)

(d)

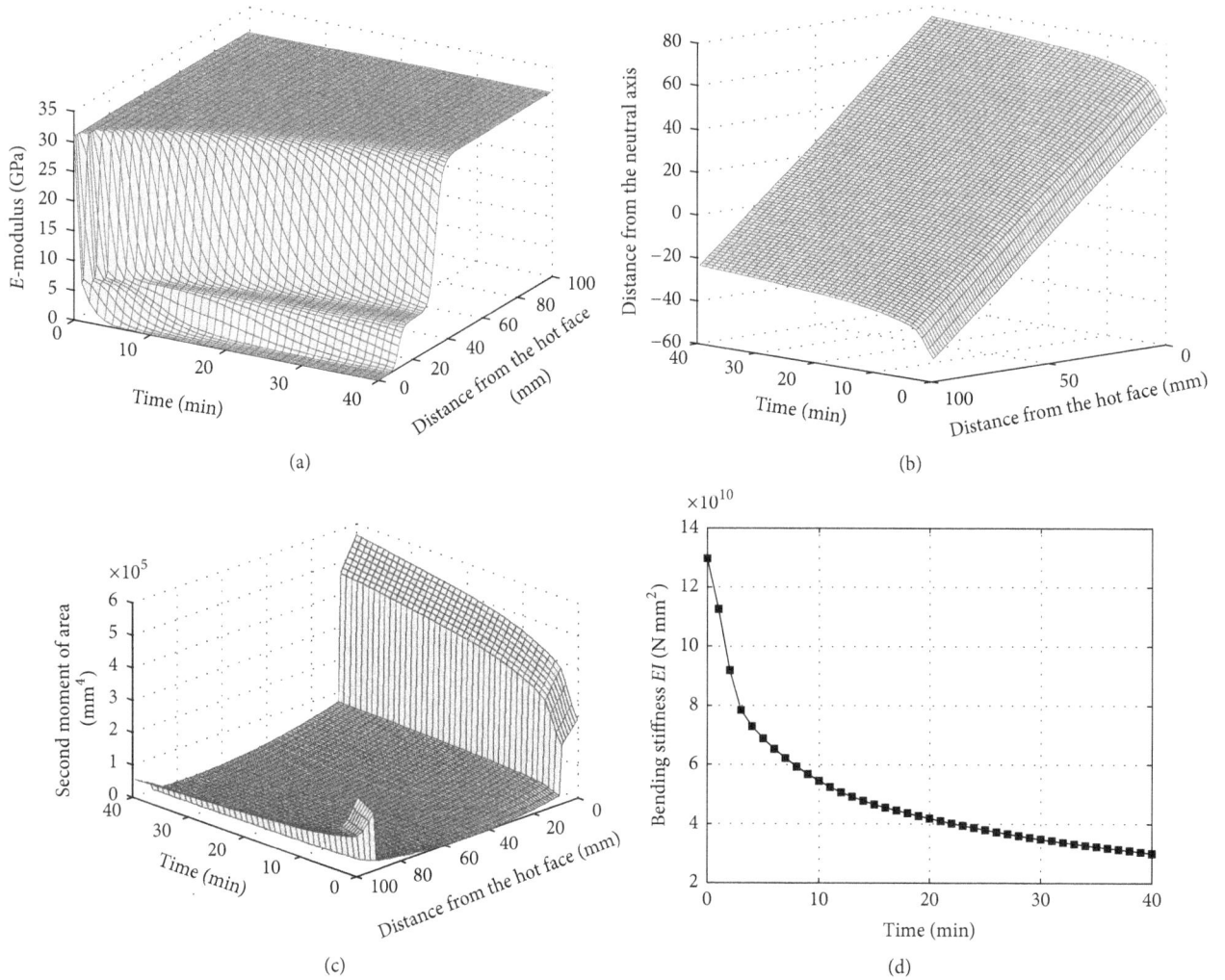

FIGURE 6: Mechanical responses of GFRP tube: (a) elastic modulus, (b) neutral axis, (c) second moment of area, and (d) bending stiffness.

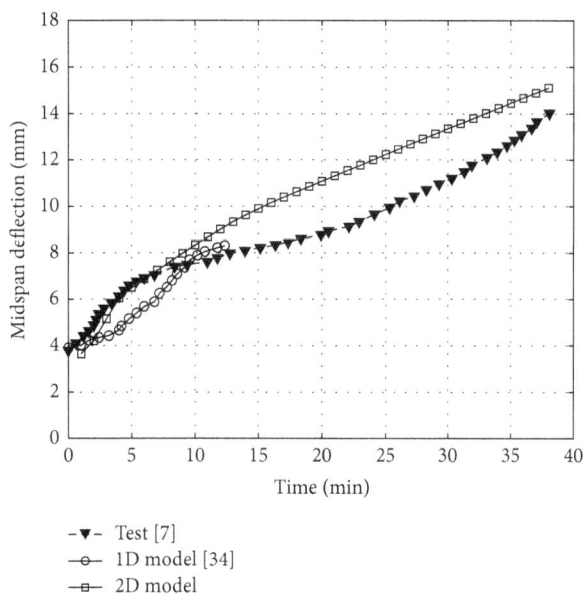

FIGURE 7: Comparison of mechanical responses of GFRP tube.

presented, and the numerical results were stable due to the stability of this difference format.

(2) The proposed 2D thermal model can reasonably predict the temperatures of GFRP rectangular tube. Based on the calculated temperature, the temperature-dependent conversion degree of glass transition and decomposition as well as thermophysical properties of the whole cross section can also be obtained.

(3) The evolution of elastic modulus, neutral axis, second moment of area, and bending stiffness can also be captured by the thermomechanical model, and midspan deflections of GFRP beam under four-point bending and fire exposure from one side can be reasonably predicted.

(4) The overestimation of the midspan deflections may be resulted from the underestimation of bending stiffness and this may be attributed to two reasons: (a) the individual element in the cross section, in fact, was enhanced by the constrain of the adjacent elements; (b) it was assumed that the whole span was

subjected to the same loading and exhibited the same stiffness degradation, but only part of the span was exposed to the oven directly [34], and this also make a contribution of underestimation of stiffness.

(5) The 2D thermomechanical model can also be extended to simulate the thermomechanical responses of beam subjected to four-point bending and three-side fire loading. The 2D thermal also could be used to model the thermal responses of GFRP tubular columns subjected to multiside fire exposure, and this makes it possible to provide a base for estimating the mechanical performance of the GFRP tubular column under mechanical and multiside fire loading.

Appendix

A. $i=0, \ j=1$

$$\text{Step I:} \ \frac{k_{x,i,j}^{n} + k_{x,i+1,j}^{n}}{2} \cdot \frac{T_{i+1,j}^{n+1/2} - T_{i,j}^{n+1/2}}{\Delta x} \Delta y$$

$$+ \frac{k_{y,i,j-1}^{n} + k_{y,i,j}^{n}}{2} \cdot \frac{T_{\infty,f}^{n} - T_{i,j}^{n}}{\Delta y} \frac{\Delta x}{2}$$

$$+ \frac{k_{y,i,j}^{n} + k_{y,i,j+1}^{n}}{2} \cdot \frac{T_{i,j+1}^{n} - T_{i,j}^{n}}{\Delta y} \frac{\Delta x}{2}$$

$$= C_{p,i,j}^{n} \rho_{i,j}^{n} \frac{T_{i,j}^{n+1/2} - T_{i,j}^{n}}{\Delta t/2} \frac{\Delta x \Delta y}{2}$$

$$\text{Step II:} \ \frac{k_{x,i,j}^{n+1/2} + k_{x,i+1,j}^{n+1/2}}{2} \cdot \frac{T_{i+1,j}^{n+1/2} - T_{i,j}^{n+1/2}}{\Delta x} \Delta y$$

$$+ \frac{k_{y,i,j}^{n+1/2} + k_{y,i,j+1}^{n+1/2}}{2} \cdot \frac{T_{i,j+1}^{n+1} - T_{i,j}^{n+1}}{\Delta y} \frac{\Delta x}{2}$$

$$+ \frac{k_{y,i,j-1}^{n+1/2} + k_{y,i,j}^{n+1/2}}{2} \cdot \frac{T_{\infty,f}^{n+1} - T_{i,j}^{n+1}}{\Delta y} \frac{\Delta x}{2}$$

$$= C_{p,i,j}^{n+1/2} \rho_{i,j}^{n+1/2} \frac{T_{i,j}^{n+1} - T_{i,j}^{n+1/2}}{\Delta t/2} \frac{\Delta x \Delta y}{2}.$$

(A.1)

B. $i=L/2, \ j=1$

$$\text{Step I:} \ \frac{k_{x,i-1,j}^{n} + k_{x,i,j}^{n}}{2} \cdot \frac{T_{i-1,j}^{n+1/2} - T_{i,j}^{n+1/2}}{\Delta x} \Delta y$$

$$+ \frac{k_{y,i,j-1}^{n} + k_{y,i,j}^{n}}{2} \cdot \frac{T_{\infty,f}^{n} - T_{i,j}^{n}}{\Delta y} \frac{\Delta x}{2} + \frac{k_{y,i,j}^{n} + k_{y,i,j+1}^{n}}{2}$$

$$\cdot \frac{T_{i,j+1}^{n} - T_{i,j}^{n}}{\Delta y} \frac{\Delta x}{2} = C_{p,i,j}^{n} \rho_{i,j}^{n} \frac{T_{i,j}^{n+1/2} - T_{i,j}^{n}}{\Delta t/2} \frac{\Delta x \Delta y}{2}$$

$$\text{Step II:} \ \frac{k_{x,i-1,j}^{n+1/2} + k_{x,i,j}^{n+1/2}}{2} \cdot \frac{T_{i-1,j}^{n+1/2} - T_{i,j}^{n+1/2}}{\Delta x} \Delta y$$

$$+ \frac{k_{y,i,j}^{n+1/2} + k_{y,i,j+1}^{n+1/2}}{2} \cdot \frac{T_{i,j+1}^{n+1} - T_{i,j}^{n+1}}{\Delta y} \frac{\Delta x}{2}$$

$$+ \frac{k_{y,i,j-1}^{n+1/2} + k_{y,i,j}^{n+1/2}}{2} \cdot \frac{T_{\infty,f}^{n+1} - T_{i,j}^{n+1}}{\Delta y} \frac{\Delta x}{2}$$

$$= C_{p,i,j}^{n+1/2} \rho_{i,j}^{n+1/2} \frac{T_{i,j}^{n+1} - T_{i,j}^{n+1/2}}{\Delta t/2} \frac{\Delta x \Delta y}{2}.$$

(B.1)

C. $i=0, \ j=M$

$$\text{Step I:} \ \frac{k_{x,i,j}^{n} + k_{x,i+1,j}^{n}}{2} \cdot \frac{T_{i+1,j}^{n+1/2} - T_{i,j}^{n+1/2}}{\Delta x} \frac{\Delta y}{2}$$

$$+ \frac{k_{y,i,j-1}^{n} + k_{y,i,j}^{n}}{2} \cdot \frac{T_{i,j-1}^{n} - T_{i,j}^{n}}{\Delta y} \frac{\Delta x}{2}$$

$$+ \left[h_{ao} \left(T_{\infty,ao} - T_{i,j}^{n} \right) + \varepsilon_r \sigma_r \left(T_{\infty,ao}^{4} - \left(T_{i,j}^{n} \right)^{4} \right) \right]$$

$$\cdot \frac{\Delta x}{2} = C_{p,i,j}^{n} \rho_{i,j}^{n} \frac{T_{i,j}^{n+1/2} - T_{i,j}^{n}}{\Delta t/2} \frac{\Delta x \Delta y}{4}$$

$$\text{Step II:} \ \frac{k_{x,i,j}^{n+1/2} + k_{x,i+1,j}^{n+1/2}}{2} \cdot \frac{T_{i+1,j}^{n+1/2} - T_{i,j}^{n+1/2}}{\Delta x} \frac{\Delta y}{2}$$

$$+ \frac{k_{y,i,j-1}^{n+1/2} + k_{y,i,j}^{n+1/2}}{2} \cdot \frac{T_{i,j-1}^{n+1} - T_{i,j}^{n+1}}{\Delta y} \frac{\Delta x}{2}$$

$$+ \left[h_{ao} \left(T_{\infty,ao} - T_{i,j}^{n+1} \right) + \varepsilon_r \sigma_r \left(T_{\infty,ao}^{4} - \left(T_{i,j}^{n+1} \right)^{4} \right) \right]$$

$$\cdot \frac{\Delta x}{2} = C_{p,i,j}^{n+1/2} \rho_{i,j}^{n+1/2} \frac{T_{i,j}^{n+1} - T_{i,j}^{n+1/2}}{\Delta t/2} \frac{\Delta x \Delta y}{4}.$$

(C.1)

D. $i = L/2, \ j = M$

$$\text{Step I:} \ \frac{k_{x,i-1,j}^{n} + k_{x,i,j}^{n}}{2} \cdot \frac{T_{i-1,j}^{n+1/2} - T_{i,j}^{n+1/2}}{\Delta x} \frac{\Delta y}{2}$$

$$+ \frac{k_{y,i,j-1}^{n} + k_{y,i,j}^{n}}{2} \cdot \frac{T_{i,j-1}^{n} - T_{i,j}^{n}}{\Delta y} \frac{\Delta x}{2}$$

$$+ \left[h_{ao} \left(T_{\infty,ao} - T_{i,j}^{n} \right) + \varepsilon_r \sigma_r \left(T_{\infty,ao}^{4} - \left(T_{i,j}^{n} \right)^{4} \right) \right]$$

$$\cdot \frac{\Delta x}{2} = C_{p,i,j}^{n} \rho_{i,j}^{n} \frac{T_{i,j}^{n+1/2} - T_{i,j}^{n}}{\Delta t/2} \frac{\Delta x \Delta y}{4}$$

Step II:
$$\frac{k_{x,i-1,j}^{n+1/2} + k_{x,i,j}^{n+1/2}}{2} \cdot \frac{T_{i-1,j}^{n+1/2} - T_{i,j}^{n+1/2}}{\Delta x}\frac{\Delta y}{2}$$
$$+ \frac{k_{y,i,j-1}^{n+1/2} + k_{y,i,j}^{n+1/2}}{2} \cdot \frac{T_{i,j-1}^{n+1} - T_{i,j}^{n+1}}{\Delta y}\frac{\Delta x}{2}$$
$$+ \left[h_{ao}\left(T_{\infty,ao} - T_{i,j}^{n+1}\right)\right.$$
$$\left. + \varepsilon_r\sigma_r\left(T_{\infty,ao}{}^4 - \left(T_{i,j}^{n+1}\right)^4\right)\right]\frac{\Delta x}{2} = C_{p,i,j}^{n+1/2}\rho_{i,j}^{n+1/2}$$
$$\cdot \frac{T_{i,j}^{n+1} - T_{i,j}^{n+1/2}}{\Delta t/2}\frac{\Delta x\Delta y}{4}. \tag{D.1}$$

E. $i=0, \; 1<j<M$

Step I:
$$\frac{k_{x,i,j}^{n} + k_{x,i+1,j}^{n}}{2} \cdot \frac{T_{i+1,j}^{n+1/2} - T_{i,j}^{n+1/2}}{\Delta x}\Delta y$$
$$+ \frac{k_{y,i,j-1}^{n} + k_{y,i,j}^{n}}{2} \cdot \frac{T_{i,j-1}^{n} - T_{i,j}^{n}}{\Delta y}\frac{\Delta x}{2}$$
$$+ \frac{k_{y,i,j}^{n} + k_{y,i,j+1}^{n}}{2} \cdot \frac{T_{i,j+1}^{n} - T_{i,j}^{n}}{\Delta y}\frac{\Delta x}{2}$$
$$= C_{p,i,j}^{n}\rho_{i,j}^{n}\frac{T_{i,j}^{n+1/2} - T_{i,j}^{n}}{\Delta t/2}\frac{\Delta x\Delta y}{2} \tag{E.1}$$

Step II:
$$\frac{k_{x,i,j}^{n+1/2} + k_{x,i+1,j}^{n+1/2}}{2} \cdot \frac{T_{i+1,j}^{n+1/2} - T_{i,j}^{n+1/2}}{\Delta x}\Delta y$$
$$+ \frac{k_{y,i,j}^{n+1/2} + k_{y,i,j+1}^{n+1/2}}{2} \cdot \frac{T_{i,j+1}^{n+1} - T_{i,j}^{n+1}}{\Delta y}\frac{\Delta x}{2}$$
$$+ \frac{k_{y,i,j-1}^{n+1/2} + k_{y,i,j}^{n+1/2}}{2} \cdot \frac{T_{i,j-1}^{n+1} - T_{i,j}^{n+1}}{\Delta y}\frac{\Delta x}{2}$$
$$= C_{p,i,j}^{n+1/2}\rho_{i,j}^{n+1/2}\frac{T_{i,j}^{n+1} - T_{i,j}^{n+1/2}}{\Delta t/2}\frac{\Delta x\Delta y}{2}.$$

F. $i = L/2, \; 1 < j < T_s \cup M - T_s < j < M$

Step I:
$$\frac{k_{x,i-1,j}^{n} + k_{x,i,j}^{n}}{2} \cdot \frac{T_{i-1,j}^{n+1/2} - T_{i,j}^{n+1/2}}{\Delta x}\Delta y$$
$$+ \frac{k_{y,i,j-1}^{n} + k_{y,i,j}^{n}}{2} \cdot \frac{T_{i,j-1}^{n} - T_{i,j}^{n}}{\Delta y}\frac{\Delta x}{2}$$
$$+ \frac{k_{y,i,j}^{n} + k_{y,i,j+1}^{n}}{2} \cdot \frac{T_{i,j+1}^{n} - T_{i,j}^{n}}{\Delta y}\frac{\Delta x}{2}$$
$$= C_{p,i,j}^{n}\rho_{i,j}^{n}\frac{T_{i,j}^{n+1/2} - T_{i,j}^{n}}{\Delta t/2}\frac{\Delta x\Delta y}{2}$$

Step II:
$$\frac{k_{x,i-1,j}^{n+1/2} + k_{x,i,j}^{n+1/2}}{2} \cdot \frac{T_{i-1,j}^{n+1/2} - T_{i,j}^{n+1/2}}{\Delta x}\Delta y$$
$$+ \frac{k_{y,i,j}^{n+1/2} + k_{y,i,j+1}^{n+1/2}}{2} \cdot \frac{T_{i,j+1}^{n+1} - T_{i,j}^{n+1}}{\Delta y}\frac{\Delta x}{2}$$
$$+ \frac{k_{y,i,j-1}^{n+1/2} + k_{y,i,j}^{n+1/2}}{2} \cdot \frac{T_{i,j-1}^{n+1} - T_{i,j}^{n+1}}{\Delta y}\frac{\Delta x}{2}$$
$$= C_{p,i,j}^{n+1/2}\rho_{i,j}^{n+1/2}\frac{T_{i,j}^{n+1} - T_{i,j}^{n+1/2}}{\Delta t/2}\frac{\Delta x\Delta y}{2}. \tag{F.1}$$

G. $0<i<L/2, \; j=1$

Step I:
$$\frac{k_{x,i-1,j}^{n} + k_{x,i,j}^{n}}{2} \cdot \frac{T_{i-1,j}^{n+1/2} - T_{i,j}^{n+1/2}}{\Delta x}\Delta y$$
$$+ \frac{k_{x,i,j}^{n} + k_{x,i+1,j}^{n}}{2} \cdot \frac{T_{i+1,j}^{n+1/2} - T_{i,j}^{n+1/2}}{\Delta x}\Delta y$$
$$+ \frac{k_{y,i,j-1}^{n} + k_{y,i,j}^{n}}{2} \cdot \frac{T_{\infty,f}^{n} - T_{i,j}^{n}}{\Delta y}\Delta x + \frac{k_{y,i,j}^{n} + k_{y,i,j+1}^{n}}{2}$$
$$\cdot \frac{T_{i,j+1}^{n} - T_{i,j}^{n}}{\Delta y}\Delta x = C_{p,i,j}^{n}\rho_{i,j}^{n}\frac{T_{i,j}^{n+1/2} - T_{i,j}^{n}}{\Delta t/2}\Delta x\Delta y$$

Step II:
$$\frac{k_{x,i-1,j}^{n+1/2} + k_{x,i,j}^{n+1/2}}{2} \cdot \frac{T_{i-1,j}^{n+1/2} - T_{i,j}^{n+1/2}}{\Delta x}\Delta y \tag{G.1}$$
$$+ \frac{k_{x,i,j}^{n+1/2} + k_{x,i+1,j}^{n+1/2}}{2} \cdot \frac{T_{i+1,j}^{n+1/2} - T_{i,j}^{n+1/2}}{\Delta x}\Delta y$$
$$+ \frac{k_{y,i,j-1}^{n+1/2} + k_{y,i,j}^{n+1/2}}{2} \cdot \frac{T_{\infty,f}^{n+1} - T_{i,j}^{n+1}}{\Delta y}\Delta x$$
$$+ \frac{k_{y,i,j}^{n+1/2} + k_{y,i,j+1}^{n+1/2}}{2} \cdot \frac{T_{i,j+1}^{n+1} - T_{i,j}^{n+1}}{\Delta y}\Delta x$$
$$= C_{p,i,j}^{n+1/2}\rho_{i,j}^{n+1/2}\frac{T_{i,j}^{n+1} - T_{i,j}^{n+1/2}}{\Delta t/2}\Delta x\Delta y.$$

H. $0 < i < L/2, \; j = M$

Step I:
$$\frac{k_{x,i-1,j}^{n} + k_{x,i,j}^{n}}{2} \cdot \frac{T_{i-1,j}^{n+1/2} - T_{i,j}^{n+1/2}}{\Delta x}\frac{\Delta y}{2}$$
$$+ \frac{k_{x,i,j}^{n} + k_{x,i+1,j}^{n}}{2} \cdot \frac{T_{i+1,j}^{n+1/2} - T_{i,j}^{n+1/2}}{\Delta x}\frac{\Delta y}{2}$$
$$+ \frac{k_{y,i,j-1}^{n} + k_{y,i,j}^{n}}{2} \cdot \frac{T_{i,j-1}^{n} - T_{i,j}^{n}}{\Delta y}\Delta x$$
$$+ \left[h_{ao}\left(T_{\infty,ao} - T_{i,j}^{n}\right) + \varepsilon_r\sigma_r\left(T_{\infty,ao}{}^4 - \left(T_{i,j}^{n}\right)^4\right)\right]$$
$$\cdot \Delta x = C_{p,i,j}^{n}\rho_{i,j}^{n}\frac{T_{i,j}^{n+1/2} - T_{i,j}^{n}}{\Delta t/2}\frac{\Delta x\Delta y}{2}$$

Step II: $\dfrac{k_{x,i-1,j}^{n+1/2} + k_{x,i,j}^{n+1/2}}{2} \cdot \dfrac{T_{i-1,j}^{n+1/2} - T_{i,j}^{n+1/2}}{\Delta x} \dfrac{\Delta y}{2}$

$+ \dfrac{k_{x,i,j}^{n+1/2} + k_{x,i+1,j}^{n+1/2}}{2} \cdot \dfrac{T_{i+1,j}^{n+1/2} - T_{i,j}^{n+1/2}}{\Delta x} \dfrac{\Delta y}{2}$

$+ \dfrac{k_{y,i,j-1}^{n+1/2} + k_{y,i,j}^{n+1/2}}{2} \cdot \dfrac{T_{i,j-1}^{n+1} - T_{i,j}^{n+1}}{\Delta y} \Delta x$

$+ \Big[h_{ao} \left(T_{i,j}^{n+1} - T_{\infty,ao} \right)$

$+ \varepsilon_r \sigma_r \left(T_{\infty,ao}{}^4 - \left(T_{i,j}^{n+1} \right)^4 \right) \Big] \Delta x = C_{p,i,j}^{n+1/2} \rho_{i,j}^{n+1/2}$

$\cdot \dfrac{T_{i,j}^{n+1} - T_{i,j}^{n+1/2}}{\Delta t/2} \dfrac{\Delta x \Delta y}{2}.$

(H.1)

I. $i = T_s, \quad j = T_s$

Step I: $\dfrac{k_{x,i-1,j}^{n} + k_{x,i,j}^{n}}{2} \cdot \dfrac{T_{i-1,j}^{n+1/2} - T_{i,j}^{n+1/2}}{\Delta x} \Delta y$

$+ \dfrac{k_{x,i,j}^{n} + k_{x,i+1,j}^{n}}{2} \cdot \dfrac{T_{i+1,j}^{n+1/2} - T_{i,j}^{n+1/2}}{\Delta x} \dfrac{\Delta y}{2}$

$+ \dfrac{k_{y,i,j-1}^{n} + k_{y,i,j}^{n}}{2} \cdot \dfrac{T_{i,j-1}^{n} - T_{i,j}^{n}}{\Delta y} \Delta x + \dfrac{k_{y,i,j}^{n} + k_{y,i,j+1}^{n}}{2}$

$\cdot \dfrac{T_{i,j+1}^{n} - T_{i,j}^{n}}{\Delta y} \dfrac{\Delta x}{2} + \Big[h_{al} \left(T_{\infty,al} - T_{i,j}^{n+1/2} \right)$

$+ \varepsilon_r \sigma_r \left(T_{\infty,al}{}^4 - \left(T_{i,j}^{n+1/2} \right)^4 \right) \Big] \dfrac{\Delta y}{2}$

$+ \Big[h_{ab} \left(T_{\infty,ab} - T_{i,j}^{n} \right) + \varepsilon_r \sigma_r \left(T_{\infty,ab}{}^4 - \left(T_{i,j}^{n} \right)^4 \right) \Big] \dfrac{\Delta x}{2}$

$= C_{p,i,j}^{n} \rho_{i,j}^{n} \dfrac{T_{i,j}^{n+1/2} - T_{i,j}^{n}}{\Delta t/2} \dfrac{3\Delta x \Delta y}{4}$

Step II: $\dfrac{k_{x,i-1,j}^{n+1/2} + k_{x,i,j}^{n+1/2}}{2} \cdot \dfrac{T_{i-1,j}^{n+1/2} - T_{i,j}^{n+1/2}}{\Delta x} \Delta y$

$+ \dfrac{k_{x,i,j}^{n+1/2} + k_{x,i+1,j}^{n+1/2}}{2} \cdot \dfrac{T_{i+1,j}^{n+1/2} - T_{i,j}^{n+1/2}}{\Delta x} \dfrac{\Delta y}{2}$

$+ \dfrac{k_{y,i,j-1}^{n+1/2} + k_{y,i,j}^{n+1/2}}{2} \cdot \dfrac{T_{i,j-1}^{n+1} - T_{i,j}^{n+1}}{\Delta y} \Delta x$

$+ \dfrac{k_{y,i,j}^{n+1/2} + k_{y,i,j+1}^{n+1/2}}{2} \cdot \dfrac{T_{i,j+1}^{n+1} - T_{i,j}^{n+1}}{\Delta y} \dfrac{\Delta x}{2}$

$+ \Big[h_{al} \left(T_{\infty,al} - T_{i,j}^{n+1/2} \right)$

$+ \varepsilon_r \sigma_r \left(T_{\infty,al}{}^4 - \left(T_{i,j}^{n+1/2} \right)^4 \right) \Big] \dfrac{\Delta y}{2}$

$+ \Big[h_{ab} \left(T_{\infty,ab} - T_{i,j}^{n+1} \right) + \varepsilon_r \sigma_r \left(T_{\infty,ab}{}^4 - \left(T_{i,j}^{n+1} \right)^4 \right) \Big]$

$\cdot \dfrac{\Delta x}{2} = C_{p,i,j}^{n+1/2} \rho_{i,j}^{n+1/2} \dfrac{T_{i,j}^{n+1} - T_{i,j}^{n+1/2}}{\Delta t/2} \dfrac{3\Delta x \Delta y}{4}.$

(I.1)

J. $i = L/2, \quad j = T_s$

Step I: $\dfrac{k_{x,i-1,j}^{n} + k_{x,i,j}^{n}}{2} \cdot \dfrac{T_{i-1,j}^{n+1/2} - T_{i,j}^{n+1/2}}{\Delta x} \dfrac{\Delta y}{2}$

$+ \dfrac{k_{y,i,j-1}^{n} + k_{y,i,j}^{n}}{2} \cdot \dfrac{T_{i,j-1}^{n} - T_{i,j}^{n}}{\Delta y} \dfrac{\Delta x}{2}$

$+ \Big[h_{ab} \left(T_{\infty,ab} - T_{i,j}^{n} \right) + \varepsilon_r \sigma_r \left(T_{\infty,ab}{}^4 - \left(T_{i,j}^{n} \right)^4 \right) \Big] \dfrac{\Delta x}{2}$

$= C_{p,i,j}^{n} \rho_{i,j}^{n} \dfrac{T_{i,j}^{n+1/2} - T_{i,j}^{n}}{\Delta t/2} \dfrac{\Delta x \Delta y}{4}$

Step II: $\dfrac{k_{x,i-1,j}^{n+1/2} + k_{x,i,j}^{n+1/2}}{2} \cdot \dfrac{T_{i-1,j}^{n+1/2} - T_{i,j}^{n+1/2}}{\Delta x} \dfrac{\Delta y}{2}$

$+ \dfrac{k_{y,i,j-1}^{n+1/2} + k_{y,i,j}^{n+1/2}}{2} \cdot \dfrac{T_{i,j-1}^{n+1} - T_{i,j}^{n+1}}{\Delta y} \dfrac{\Delta x}{2}$

$+ \Big[h_{ab} \left(T_{\infty,ab} - T_{i,j}^{n+1} \right) + \varepsilon_r \sigma_r \left(T_{\infty,ab}{}^4 - \left(T_{i,j}^{n+1} \right)^4 \right) \Big]$

$\cdot \dfrac{\Delta x}{2} = C_{p,i,j}^{n+1/2} \rho_{i,j}^{n+1/2} \dfrac{T_{i,j}^{n+1} - T_{i,j}^{n+1/2}}{\Delta t/2} \dfrac{\Delta x \Delta y}{4}.$

(J.1)

K. $i = T_s, \quad j = M - T_s$

Step I: $\dfrac{k_{x,i-1,j}^{n} + k_{x,i,j}^{n}}{2} \cdot \dfrac{T_{i-1,j}^{n+1/2} - T_{i,j}^{n+1/2}}{\Delta x} \Delta y$

$+ \dfrac{k_{x,i,j}^{n} + k_{x,i+1,j}^{n}}{2} \cdot \dfrac{T_{i+1,j}^{n+1/2} - T_{i,j}^{n+1/2}}{\Delta x} \dfrac{\Delta y}{2}$

$+ \dfrac{k_{y,i,j-1}^{n} + k_{y,i,j}^{n}}{2} \cdot \dfrac{T_{i,j-1}^{n} - T_{i,j}^{n}}{\Delta y} \dfrac{\Delta x}{2}$

$+ \dfrac{k_{y,i,j}^{n} + k_{y,i,j+1}^{n}}{2} \cdot \dfrac{T_{i,j+1}^{n} - T_{i,j}^{n}}{\Delta y} \Delta x$

$+ \Big[h_{al} \left(T_{\infty,al} - T_{i,j}^{n+1/2} \right)$

$+ \varepsilon_r \sigma_r \left(T_{\infty,al}{}^4 - \left(T_{i,j}^{n+1/2} \right)^4 \right) \Big] \dfrac{\Delta y}{2}$

$+ \Big[h_{at} \left(T_{\infty,at} - T_{i,j}^{n} \right) + \varepsilon_r \sigma_r \left(T_{\infty,at}{}^4 - \left(T_{i,j}^{n} \right)^4 \right) \Big] \dfrac{\Delta x}{2}$

$= C_{p,i,j}^{n} \rho_{i,j}^{n} \dfrac{T_{i,j}^{n+1/2} - T_{i,j}^{n}}{\Delta t/2} \dfrac{3\Delta x \Delta y}{4}$

Step II: $\dfrac{k_{x,i-1,j}^{n+1/2} + k_{x,i,j}^{n+1/2}}{2} \cdot \dfrac{T_{i-1,j}^{n+1/2} - T_{i,j}^{n+1/2}}{\Delta x} \Delta y$

$+ \dfrac{k_{x,i,j}^{n+1/2} + k_{x,i+1,j}^{n+1/2}}{2} \cdot \dfrac{T_{i+1,j}^{n+1/2} - T_{i,j}^{n+1/2}}{\Delta x} \dfrac{\Delta y}{2}$

$+ \dfrac{k_{y,i,j-1}^{n+1/2} + k_{y,i,j}^{n+1/2}}{2} \cdot \dfrac{T_{i,j-1}^{n+1} - T_{i,j}^{n+1}}{\Delta y} \dfrac{\Delta x}{2}$

$+ \dfrac{k_{y,i,j}^{n+1/2} + k_{y,i,j+1}^{n+1/2}}{2} \cdot \dfrac{T_{i,j+1}^{n+1} - T_{i,j}^{n+1}}{\Delta y} \Delta x$

$+ \left[h_{al}\left(T_{\infty,al} - T_{i,j}^{n+1/2} \right) \right.$

$+ \left. \varepsilon_r \sigma_r \left(T_{\infty,al}{}^4 - \left(T_{i,j}^{n+1/2} \right)^4 \right) \right] \dfrac{\Delta y}{2}$

$+ \left[h_{at}\left(T_{\infty,at} - T_{i,j}^{n+1} \right) + \varepsilon_r \sigma_r \left(T_{\infty,at}{}^4 - \left(T_{i,j}^{n+1} \right)^4 \right) \right]$

$\cdot \dfrac{\Delta x}{2} = C_{p,i,j}^{n+1/2} \rho_{i,j}^{n+1/2} \dfrac{T_{i,j}^{n+1} - T_{i,j}^{n+1/2}}{\Delta t/2} \dfrac{3\Delta x \Delta y}{4}.$

$\hfill \text{(K.1)}$

L. $i = L/2, \ j = M - T_s$

Step I: $\dfrac{k_{x,i-1,j}^{n} + k_{x,i,j}^{n}}{2} \cdot \dfrac{T_{i-1,j}^{n+1/2} - T_{i,j}^{n+1/2}}{\Delta x} \dfrac{\Delta y}{2}$

$+ \dfrac{k_{y,i,j}^{n} + k_{y,i,j+1}^{n}}{2} \cdot \dfrac{T_{i,j+1}^{n} - T_{i,j}^{n}}{\Delta y} \dfrac{\Delta x}{2}$

$+ \left[h_{at}\left(T_{\infty,at} - T_{i,j}^{n} \right) + \varepsilon_r \sigma_r \left(T_{\infty,at}{}^4 - \left(T_{i,j}^{n} \right)^4 \right) \right] \dfrac{\Delta x}{2}$

$= C_{p,i,j}^{n} \rho_{i,j}^{n} \dfrac{T_{i,j}^{n+1/2} - T_{i,j}^{n}}{\Delta t/2} \dfrac{\Delta x \Delta y}{4}$

$\hfill \text{(L.1)}$

Step II: $\dfrac{k_{x,i-1,j}^{n+1/2} + k_{x,i,j}^{n+1/2}}{2} \cdot \dfrac{T_{i-1,j}^{n+1/2} - T_{i,j}^{n+1/2}}{\Delta x} \dfrac{\Delta y}{2}$

$+ \dfrac{k_{y,i,j}^{n+1/2} + k_{y,i,j+1}^{n+1/2}}{2} \cdot \dfrac{T_{i,j+1}^{n+1} - T_{i,j}^{n+1}}{\Delta y} \dfrac{\Delta x}{2}$

$+ \left[h_{at}\left(T_{\infty,at} - T_{i,j}^{n+1} \right) + \varepsilon_r \sigma_r \left(T_{\infty,at}{}^4 - \left(T_{i,j}^{n+1} \right)^4 \right) \right]$

$\cdot \dfrac{\Delta x}{2} = C_{p,i,j}^{n+1/2} \rho_{i,j}^{n+1/2} \dfrac{T_{i,j}^{n+1} - T_{i,j}^{n+1/2}}{\Delta t/2} \dfrac{\Delta x \Delta y}{4}.$

M. $i = T_s, \ T_s < j < M - T_s$

Step I: $\dfrac{k_{x,i-1,j}^{n} + k_{x,i,j}^{n}}{2} \cdot \dfrac{T_{i-1,j}^{n+1/2} - T_{i,j}^{n+1/2}}{\Delta x} \Delta y$

$+ \dfrac{k_{y,i,j-1}^{n} + k_{y,i,j}^{n}}{2} \cdot \dfrac{T_{i,j-1}^{n} - T_{i,j}^{n}}{\Delta y} \dfrac{\Delta x}{2}$

$+ \dfrac{k_{y,i,j}^{n} + k_{y,i,j+1}^{n}}{2} \cdot \dfrac{T_{i,j+1}^{n} - T_{i,j}^{n}}{\Delta y} \dfrac{\Delta x}{2}$

$+ \left[h_{al}\left(T_{\infty,al} - T_{i,j}^{n+1/2} \right) \right.$

$+ \left. \varepsilon_r \sigma_r \left(T_{\infty,al}{}^4 - \left(T_{i,j}^{n+1/2} \right)^4 \right) \right] \Delta y = C_{p,i,j}^{n} \rho_{i,j}^{n}$

$\cdot \dfrac{T_{i,j}^{n+1/2} - T_{i,j}^{n}}{\Delta t/2} \dfrac{\Delta x \Delta y}{2}$

Step II: $\dfrac{k_{x,i-1,j}^{n+1/2} + k_{x,i,j}^{n+1/2}}{2} \cdot \dfrac{T_{i-1,j}^{n+1/2} - T_{i,j}^{n+1/2}}{\Delta x} \Delta y$

$+ \dfrac{k_{y,i,j-1}^{n+1/2} + k_{y,i,j}^{n+1/2}}{2} \cdot \dfrac{T_{i,j-1}^{n+1} - T_{i,j}^{n+1}}{\Delta y} \dfrac{\Delta x}{2}$

$+ \dfrac{k_{y,i,j}^{n+1/2} + k_{y,i,j+1}^{n+1/2}}{2} \cdot \dfrac{T_{i,j+1}^{n+1} - T_{i,j}^{n+1}}{\Delta y} \dfrac{\Delta x}{2}$

$+ \left[h_{al}\left(T_{\infty,al} - T_{i,j}^{n+1/2} \right) \right.$

$+ \left. \varepsilon_r \sigma_r \left(T_{\infty,al}{}^4 - \left(T_{i,j}^{n+1/2} \right)^4 \right) \right] \Delta y = C_{p,i,j}^{n+1/2} \rho_{i,j}^{n+1/2}$

$\cdot \dfrac{T_{i,j}^{n+1} - T_{i,j}^{n+1/2}}{\Delta t/2} \dfrac{\Delta x \Delta y}{2}.$

$\hfill \text{(M.1)}$

N. $T_s < i < L/2, \ j = T_s$

Step I: $\dfrac{k_{x,i-1,j}^{n} + k_{x,i,j}^{n}}{2} \cdot \dfrac{T_{i-1,j}^{n+1/2} - T_{i,j}^{n+1/2}}{\Delta x} \dfrac{\Delta y}{2}$

$+ \dfrac{k_{x,i,j}^{n} + k_{x,i+1,j}^{n}}{2} \cdot \dfrac{T_{i+1,j}^{n+1/2} - T_{i,j}^{n+1/2}}{\Delta x} \dfrac{\Delta y}{2}$

$+ \dfrac{k_{y,i,j-1}^{n} + k_{y,i,j}^{n}}{2} \cdot \dfrac{T_{i,j-1}^{n} - T_{i,j}^{n}}{\Delta y} \Delta x$

$+ \left[h_{ab}\left(T_{\infty,ab} - T_{i,j}^{n} \right) + \varepsilon_r \sigma_r \left(T_{\infty,ab}{}^4 - \left(T_{i,j}^{n} \right)^4 \right) \right]$

$\cdot \Delta x = C_{p,i,j}^{n} \rho_{i,j}^{n} \dfrac{T_{i,j}^{n+1/2} - T_{i,j}^{n}}{\Delta t/2} \dfrac{\Delta x \Delta y}{2}$

Step II: $\dfrac{k_{x,i-1,j}^{n+1/2} + k_{x,i,j}^{n+1/2}}{2} \cdot \dfrac{T_{i-1,j}^{n+1/2} - T_{i,j}^{n+1/2}}{\Delta x} \dfrac{\Delta y}{2}$ $\hfill \text{(N.1)}$

$+ \dfrac{k_{x,i,j}^{n+1/2} + k_{x,i+1,j}^{n+1/2}}{2} \cdot \dfrac{T_{i+1,j}^{n+1/2} - T_{i,j}^{n+1/2}}{\Delta x} \dfrac{\Delta y}{2}$

$+ \dfrac{k_{y,i,j-1}^{n+1/2} + k_{y,i,j}^{n+1/2}}{2} \cdot \dfrac{T_{i,j-1}^{n+1} - T_{i,j}^{n+1}}{\Delta y} \Delta x$

$+ \left[h_{ab}\left(T_{\infty,ab} - T_{i,j}^{n+1} \right) \right.$

$+ \left. \varepsilon_r \sigma_r \left(T_{\infty,ab}{}^4 - \left(T_{i,j}^{n+1} \right)^4 \right) \right] \Delta x = C_{p,i,j}^{n+1/2} \rho_{i,j}^{n+1/2}$

$\cdot \dfrac{T_{i,j}^{n+1} - T_{i,j}^{n+1/2}}{\Delta t/2} \dfrac{\Delta x \Delta y}{2}.$

O. $T_s < i < L/2, \ j = M - T_s$

Step I:
$$\frac{k_{x,i-1,j}^n + k_{x,i,j}^n}{2} \cdot \frac{T_{i-1,j}^{n+1/2} - T_{i,j}^{n+1/2}}{\Delta x} \frac{\Delta y}{2}$$

$$+ \frac{k_{x,i,j}^n + k_{x,i+1,j}^n}{2} \cdot \frac{T_{i+1,j}^{n+1/2} - T_{i,j}^{n+1/2}}{\Delta x} \frac{\Delta y}{2}$$

$$+ \frac{k_{y,i,j}^n + k_{y,i,j+1}^n}{2} \cdot \frac{T_{i,j+1}^n - T_{i,j}^n}{\Delta y} \Delta x$$

$$+ \left[h_{at} \left(T_{\infty,at} - T_{i,j}^n \right) + \varepsilon_r \sigma_r \left(T_{\infty,at}^{\ 4} - \left(T_{i,j}^n \right)^4 \right) \right] \Delta x$$

$$= C_{p,i,j}^n \rho_{i,j}^n \frac{T_{i,j}^{n+1/2} - T_{i,j}^n}{\Delta t/2} \frac{\Delta x \Delta y}{2}$$

(O.1)

Step II:
$$\frac{k_{x,i-1,j}^{n+1/2} + k_{x,i,j}^{n+1/2}}{2} \cdot \frac{T_{i-1,j}^{n+1/2} - T_{i,j}^{n+1/2}}{\Delta x} \frac{\Delta y}{2}$$

$$+ \frac{k_{x,i,j}^{n+1/2} + k_{x,i+1,j}^{n+1/2}}{2} \cdot \frac{T_{i+1,j}^{n+1/2} - T_{i,j}^{n+1/2}}{\Delta x} \frac{\Delta y}{2}$$

$$+ \frac{k_{y,i,j}^{n+1/2} + k_{y,i,j+1}^{n+1/2}}{2} \cdot \frac{T_{i,j+1}^{n+1} - T_{i,j}^{n+1}}{\Delta y} \Delta x$$

$$+ \left[h_{at} \left(T_{\infty,at} - T_{i,j}^{n+1} \right) + \varepsilon_r \sigma_r \left(T_{\infty,at}^{\ 4} - \left(T_{i,j}^{n+1} \right)^4 \right) \right]$$

$$\cdot \Delta x = C_{p,i,j}^{n+1/2} \rho_{i,j}^{n+1/2} \frac{T_{i,j}^{n+1} - T_{i,j}^{n+1/2}}{\Delta t/2} \frac{\Delta x \Delta y}{2},$$

where $T_{\infty,f}$ and $T_{\infty,a}$ represent the surrounding temperatures of fire and air, respectively. h denotes the convective coefficients, and the subscript ao denotes the outer face of top flange; the subscripts ab, al, and at represent the inner face of bottom flange, web, and top flange, respectively.

Nomenclature

a: Distance between the support and load
A: Preexponential factor
C_p: Apparent specific heat capacity
E: Young's (elastic) modulus
E_A: Activation energy
f: Mass fraction of material
g: Acceleration of gravity
h: Convective coefficient
I: Second moment of area of the section
k: Thermal conductivity
l: Width of tube
L: Number of elements in width direction
L_s: Span
m: Height of tube
M: Number of elements in height direction
Ma: Mass
N: Number of time steps

P: Half of total load
Pr: Prandtl number
q: Heat flux
R: Universal gas constant
t: Time
T: Temperature
T_s: Number of elements of flanges or webs
v: Kinetic viscosity of air
V: Volume fraction of material.

Greek

α: Conversion degree
β: Volumetric coefficient of thermal expansion of air
δ: Midspan deflection
∞:   Surrounding
ε: Emissivity of solid surface
ρ: Density
σ: Stefan-Boltzmann constant.

Subscripts

A: Material after decomposition (char)
ab: Inner surface of bottom flange
al: Inner surface of web
at: Inner surface of top flange
b: Material before decomposition (virgin)
bd: The boundaries of tube
bf: Bottom flange
d: Decomposition
e: Final mass
f: Fiber
g: Glass transition
ga: Gas
i: Space step in x direction
j: Space step in y direction
m: Matrix resin
ou: Outer surface of upper flange
r: Rubbery and leathery states
ra: Radiation
tf: Top flange
v: Initial mass
w: Web
$\perp$:   Through thickness direction.

Superscripts

n: Time step
nr: Reaction order.

Conflicts of Interest

The authors declare that they have no conflicts of interest.

Acknowledgments

The research described here was supported by the National Natural Science Foundation (Grants nos. 51678297, 51408305, and 51238003) and Natural Science Foundation of Jiangsu Province (Grant no. BK20140946).

References

[1] F. C. Campbell, "Structural composite materials," in *Proceedings of the ASM international*, 2010.

[2] J. M. Davies and D. W. Dewhurst, "The fire performance of GRP pipes in empty and dry, stagnant water filled and flowing water filled conditions," in *Proceedings of the International Conference on the Response of Composite Materials*, A. G. Gibson, Ed.

[3] T. Keller, A. Zhou, C. Tracy, E. Hugi, and P. Schnewlin, "Experimental study on the concept of liquid cooling for improving fire resistance of FRP structures for construction," *Composites Part A: Applied Science and Manufacturing*, vol. 36, no. 11, pp. 1569–1580, 2005.

[4] T. Keller, C. Tracy, and E. Hugi, "Fire endurance of loaded and liquid-cooled GFRP slabs for construction," *Composites Part A: Applied Science and Manufacturing*, vol. 37, no. 7, pp. 1055–1067, 2006.

[5] C. D. Tracy, *Fire endurance of multicellular panels in an FRP building system [Ph.D. thesis]*, April 2005.

[6] J. R. Correia, F. A. Branco, J. G. Ferreira, Y. Bai, and T. Keller, "Fire protection systems for building floors made of pultruded GFRP profiles: Part 1: Experimental investigations," *Composites Part B: Engineering*, vol. 41, no. 8, pp. 617–629, 2010.

[7] Y. Bai, E. Hugi, C. Ludwig, and T. Keller, "Fire performance of water-cooled GFRP columns. I: Fire endurance investigation," *Journal of Composites for Construction*, vol. 15, no. 3, pp. 404–412, 2011.

[8] T. Morgado, J. R. Correia, A. Moreira, F. A. Branco, and C. Tiago, "Experimental study on the fire resistance of GFRP pultruded tubular columns," *Composites Part B: Engineering*, vol. 69, pp. 201–211, 2014.

[9] C. A. Griffis, R. A. Masumura, and C. I. Chang, "Thermal Response of Graphite Epoxy Composite Subjected to Rapid Heating," *Journal of Composite Materials*, vol. 15, no. 5, pp. 427–442, 1981.

[10] G. S. Springer, "Model for predicting the mechanical properties of composites at elevated temperatures," *Journal of Reinforced Plastics and Composites*, vol. 3, no. 1, pp. 85–95, 1984.

[11] J. B. Henderson, J. A. Wiebelt, and M. R. Tant, "A model for the thermal response of polymer composite materials with experimental verification," *Journal of Composite Materials*, vol. 19, no. 6, pp. 579–595, 1985.

[12] A. G. Gibson, Y. S. Wu, H. W. Chandler, J. A. Wilcox, and P. Bettess, "A model for the thermal performance of thick composite laminates in hydrocarbon fires," *Revue de l'Institut Français du Pétrole*, vol. 50, no. 1, pp. 69–74, 1995.

[13] M. R. E. Looyeh, P. Bettess, and A. G. Gibson, "A one-dimensional finite element simulation for the fire-performance of GRP panels for offshore structures," *International Journal of Numerical Methods for Heat and Fluid Flow*, vol. 7, no. 5, pp. 609–625, 1997.

[14] N. Dodds, A. G. Gibson, D. Dewhurst, and J. M. Davies, "Fire behaviour of composite laminates," *Composites Part A: Applied Science and Manufacturing*, vol. 31, no. 7, pp. 689–702, 2000.

[15] B. Y. Lattimer and J. Ouellette, "Properties of composite materials for thermal analysis involving fires," *Composites Part A: Applied Science and Manufacturing*, vol. 37, no. 7, pp. 1068–1081, 2006.

[16] A. G. Gibson, P. N. H. Wright, Y.-S. Wu, A. P. Mouritz, Z. Mathys, and C. P. Gardiner, "Modelling residual mechanical properties of polymer composites after fire," *Plastics, Rubber and Composites*, vol. 32, no. 2, pp. 81–90, 2003.

[17] A. G. Gibson, P. N. H. Wright, Y.-S. Wu, A. P. Mouritz, Z. Mathys, and C. P. Gardiner, "The integrity of polymer composites during and after fire," *Journal of Composite Materials*, vol. 38, no. 15, pp. 1283–1307, 2004.

[18] A. P. Mouritz and Z. Mathys, "Post-fire mechanical properties of marine polymer composites," *Composite Structures*, vol. 47, no. 1-4, pp. 643–653, 1999.

[19] A. G. Gibson, T. Browne, S. Feih, and A. P. Mouritz, "Modeling composite high temperature behavior and fire response under load," *Journal of Composite Materials*, vol. 46, no. 16, pp. 2005–2022, 2012.

[20] V. U. Miano and A. G. Gibson, "FIRE model for fibre reinforced plastic composites using Apparent thermal diffusivity (ATD)," *Plastics, Rubber and Composites*, vol. 38, no. 2-4, pp. 87–92, 2009.

[21] V. U. Miano, *Modelling composite fire behaviour using apparent thermal diffusivity [Ph.D. thesis]*, Newcastle University, September 2011.

[22] S. Feih, Z. Mathys, A. G. Gibson, and A. P. Mouritz, "Modelling the compression strength of polymer laminates in fire," *Composites Part A: Applied Science and Manufacturing*, vol. 38, no. 11, pp. 2354–2365, 2007.

[23] S. Feih, Z. Mathys, A. G. Gibson, and A. P. Mouritz, "Modelling the tension and compression strengths of polymer laminates in fire," *Composites Science and Technology*, vol. 67, no. 3-4, pp. 551–564, 2007.

[24] A. P. Mouritz, S. Feih, Z. Mathys, and A. G. Gibson, "Modeling compressive skin failure of sandwich composites in fire," *Journal of Sandwich Structures and Materials*, vol. 10, no. 3, pp. 217–245, 2008.

[25] A. Anjang, V. S. Chevali, E. Kandare, A. P. Mouritz, and S. Feih, "Tension modelling and testing of sandwich composites in fire," *Composite Structures*, vol. 113, no. 1, pp. 437–445, 2014.

[26] A. Anjang, V. S. Chevali, B. Y. Lattimer, S. W. Case, S. Feih, and A. P. Mouritz, "Post-fire mechanical properties of sandwich composite structures," *Composite Structures*, vol. 132, pp. 1019–1028, 2015.

[27] T. Keller, C. Tracy, and A. Zhou, "Structural response of liquid-cooled GFRP slabs subjected to fire—Part II: thermo-chemical and thermo-mechanical modeling," *Composites Part A: Applied Science and Manufacturing*, vol. 37, no. 9, pp. 1296–1308, 2006.

[28] Y. Bai, T. Vallée, and T. Keller, "Modeling of thermo-physical properties for FRP composites under elevated and high temperature," *Composites Science and Technology*, vol. 67, no. 15-16, pp. 3098–3109, 2007.

[29] Y. Bai, T. Vallée, and T. Keller, "Modeling of thermal responses for FRP composites under elevated and high temperatures," *Composites Science and Technology*, vol. 68, no. 1, pp. 47–56, 2008.

[30] Y. Bai, T. Keller, and T. Vallée, "Modeling of stiffness of FRP composites under elevated and high temperatures," *Composites Science and Technology*, vol. 68, no. 15-16, pp. 3099–3106, 2008.

[31] Y. Bai and T. Keller, "Modeling of mechanical response of FRP composites in fire," *Composites Part A: Applied Science and Manufacturing*, vol. 40, no. 6-7, pp. 731–738, 2009.

[32] Y. Bai and T. Keller, "Modeling of strength degradation for Fiber-reinforced polymer composites in fire," *Journal of Composite Materials*, vol. 43, no. 21, pp. 2371–2385, 2009.

[33] S. Shi, L. Li, G. Fang, J. Liang, F. Yi, and G. Lin, "Three-dimensional modeling and experimental validation of thermo-mechanical response of FRP composites exposed to one-sided heat flux," *Materials and Design*, vol. 99, pp. 565–573, 2016.

[34] Y. Bai, T. Keller, J. R. Correia, F. A. Branco, and J. G. Ferreira, "Fire protection systems for building floors made of pultruded GFRP profiles - Part 2: Modeling of thermomechanical responses," *Composites Part B: Engineering*, vol. 41, no. 8, pp. 630–636, 2010.

[35] A. P. Mouritz, S. Feih, E. Kandare et al., "Review of fire structural modelling of polymer composites," *Composites Part A: Applied Science and Manufacturing*, vol. 40, no. 12, pp. 1800–1814, 2009.

[36] B. Mutnuri, *Thermal conductivity characterization of composite materials*, West Virginia University Libraries, 2006.

[37] Alternating direction implicit method. Wikipedia, 2016, https://en.wikipedia.org/wiki/Alternating_direction_implicit_method&gt.

[38] D. W. Peaceman and J. Rachford, "The numerical solution of parabolic and elliptic differential equations," *Society for Industrial and Applied Mathematics*, vol. 3, no. 1, pp. 28–41, 1955.

[39] J. Douglas and J. Rachford, "On the numerical solution of heat conduction problems in two and three space variables," *Transactions of the American Mathematical Society*, vol. 82, pp. 421–439, 1956.

4

Synthesis and Characterization of Porous Fly Ash-Based Geopolymers Using Si as Foaming Agent

D. Kioupis,[1] **Ch. Kavakakis,**[2] **S. Tsivilis** ⓘ**,**[1] **and G. Kakali**[1]

[1]*National Technical University of Athens, School of Chemical Engineering, 9 Heroon Polytechniou St., 15773 Athens, Greece*
[2]*Hellenic Open University, School of Science and Technology, 18 Parodos Aristotelous St., 26335 Patras, Greece*

Correspondence should be addressed to S. Tsivilis; stsiv@central.ntua.gr

Academic Editor: Fernando Lusquiños

This paper concerns the synthesis of foamed geopolymers using fly ash and metallic Si as the binder and the porogent agent, respectively. The Taguchi approach was applied in order to study the effect of some significant synthesis parameters such as the Si foaming agent content (% w/w fly ash based), the alkali type (R : Na or K), and the alkalinity (R/Al molar ratio) of the activation solution on the compressive strength and apparent density of the foamed geopolymers. The final products were characterized by means of XRD, FTIR, and SEM, while optical microscopy was applied for the evaluation of the porosity. Lightweight geopolymers with a compressive strength of 2.08–14.88 MPa and an apparent density of 0.84–1.55 g/cm^3 were prepared by introducing a Si content up to 0.2% w/w on fly ash basis.

1. Introduction

Geopolymers are a class of inorganic materials which comprises a three-dimensional polymeric-type structure [1]. They are formed by the reaction of an alkaline solution with an aluminosilicate material at ambient or slightly elevated temperature [2–4]. The synthesis and chemical composition of geopolymers are similar to those of zeolites, but their microstructure is amorphous to semicrystalline. The aluminosilicate source is usually an industrial mineral, waste, or by-product where silicon and aluminum ions are preferably located in amorphous phases [5–10]. Fly ash produced at coal-fired thermal plants has, recently, attracted a lot of research attention on its alkali activation and transformation into cementitious materials (geopolymer) [11–15]. The fly ash exploitation in geopolymer technology leads to the reuse of this material and the drastic reduction of soil, water, and air pollution.

Geopolymers have a very low embodied energy and CO_2 footprint compared to conventional building materials [16, 17], but their greatest advantage is that based on the choice of the raw materials and the design of the processing, geopolymers can meet a variety of requirements. This flexibility of geopolymer synthesis is of great importance when products with specific properties are required.

The application of geopolymers as insulating materials for construction sector is a new research field. The main challenge on inorganic insulation materials' development is related to the production of a high porous material with sufficient mechanical properties. Geopolymer foams can be produced by the introduction of foaming agents such as hydrogen peroxide [18–21], metal powders [22–24], or silica fumes [25–27]. Their foaming action is due to the oxygen or hydrogen formation in alkaline conditions according to the following reactions:

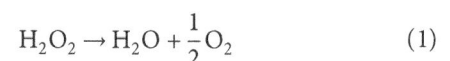

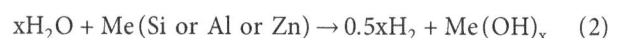

$$H_2O_2 \rightarrow H_2O + \frac{1}{2}O_2 \qquad (1)$$

$$xH_2O + Me\,(Si \text{ or } Al \text{ or } Zn) \rightarrow 0.5xH_2 + Me(OH)_x \quad (2)$$

The major advantage of the inorganic foaming is its suitability to produce ceramic foams without utilizing high temperature treatments (such as burn out of organics and sintering) in contrast to the conventional production of macroporous ceramics [28].

On the other hand, the lightweight geopolymers developed through the chemical reaction of foaming agents usually exhibit limited mechanical behavior because of the large pore size distribution (in the range of 0.5–3.0 mm).

If the pore size distribution is efficiently controlled, the mechanical properties of the foamed geopolymers can be improved [24].

Factors such as the mineral and chemical composition of raw materials, the curing conditions and the ratios of the starting materials, strongly affect the structure and properties of geopolymers. Therefore, the optimization of the synthesis is mandatory prior to any development of geopolymeric material.

This paper presents a study on the synthesis of foamed geopolymers. The synthesis optimization was performed via the Taguchi experimental design method that provides a simple and efficient tool for optimizing parameters in complicated systems [29–33]. Fly ash was used as the aluminosilicate precursor and Si was added as the foaming agent. The aim of this work is to record the combined effect of some significant synthesis parameters such as the Si foaming agent content (% w/w fly ash based), the alkali type (R : Na or K), and the alkalinity (R/Al molar ratio) of the activation solution on the compressive strength, the apparent density, and the porosity development of the foamed fly ash-based geopolymers.

2. Experimental

2.1. Materials and Methods. The fly ash used in this study comes from the power station of Megalopolis in Greece. This material can be classified as calcareous fly ash (Type W) according to EN 197-1, as it contains CaO between 10 and 15% w/w and SiO_2 higher than 25% w/w. Fly ash was previously ground and its mean particle size (d_{50}) was approximately 20 μm. This is a typical fineness of fly ash when used as the main constituent in blended cements. The chemical composition of fly ash is presented in Table 1. Silicon metal powder (99+%, 325 mesh, CAS No. 7440-21-3) was used as the foaming agent.

The synthesis of the foamed geopolymers includes three steps: (a) the preparation of the activation solution, (b) the mixing of the raw materials (fly ash and foaming agent) with the activation solution, and (c) the molding and the curing of the specimens. The activation solution contains soluble Si in the form of alkali silicates (Na_2SiO_3 with $SiO_2 = 27.56$–28.39% w/w and $Na_2O = 8.53$–8.79% w/w, K_2SiO_3 with $SiO_2 = 27.67$–28.47% w/w and $Na_2O = 14.03$–14.83% w/w, Multiplass SA, Greece) and NaOH ($\geq 98.0\%$, CAS No. 1310-73-2) or KOH ($\geq 85.0\%$, CAS No. 1310-58-3) in order to provide the alkaline environment that is necessary for the activation of the geopolymerization reactions. The activation solutions were stored for a minimum of 24 h prior to use to allow equilibrium. The foaming agent was effectively mixed with the fly ash in order to obtain a homogenous blend. Then, this blend and the activation solution were mechanically mixed (standard mortar mixer: Controls 65-L0005) to form an homogenous slurry that was transferred into 50 × 50 × 50 mm cubic molds and mildly vibrated. In all mixtures, the silicon to aluminum (Si/Al) molar ratio in the starting mixture and the solids to liquids (s/l) mass ratio were kept constant to 2.4 and 2.8, respectively [32]. The specimens were left at room

TABLE 1: Chemical composition of fly ash (% w/w).

SiO_2	Al_2O_3	Fe_2O_3	CaO	MgO	K_2O	Na_2O	SO_3	L.O.I.
45.52	22.13	8.99	14.75	2.84	1.55	0.72	1.69	1.49

temperature for 2 h and then were transferred to an oven at 80°C for 24 h. Compression tests were carried out on a Tonitechnik uniaxial testing press, 7 days after the specimens' preparation (load rate 1.5 kN/s, according to the EN196-1 requirements). The apparent density of the geopolymers was measured by the specimens' dimensions and mass. The presented value for each synthesis is the average density value of three specimens.

The final products were characterized by means of XRD, FTIR, SEM, and optical microscopy. The phase distribution of the powders was investigated by X-ray diffraction, on a Bruker D8 ADVANCE X-ray diffractometer. The data were collected in a 2θ range 2–70° with 0.02° step size and 0.2 sec per step and were evaluated using Diffrac.Eva v3.1 software.

FTIR measurements were carried out using a Fourier Transform IR Spectrophotometer (Jasco 4200 Type A). The FTIR spectra were obtained by the KBr pellet technique in the wavenumber range from 400 to 4000 cm^{-1} and resolution 4 cm^{-1}. The pellets were prepared by pressing a mixture of the sample and dried KBr (sample: KBr equals to 1 : 200 approximately) at 7.5 t/cm^2.

The morphology as well as the phase stoichiometry of the products was examined using a JEOL JSM-5600 scanning electron microscope equipped with an OXFORD LINK ISIS 300 energy dispersive X-ray spectrometer (EDX). The samples were examined as powders or cross sections and in all cases were gold or graphite coated prior to the measurement. In order to evaluate the morphology and porosity of the products by image analysis, a Zeiss Stemi 2000 C high resolution optical microscope equipped with an Axio Cam ErcS5 digital camera was used. The samples were carefully cut down into 3 cross sections of specified dimensions (50 × 50 × 10 mm) by the use of a Buhler IsoMet™ low-speed precision cutter equipped with a diamond saw disk. The cross section of the samples was mildly polished by the use of appropriate emery papers (Grade: P220, P80).

2.2. Image Analysis. Image analysis was applied in order to evaluate the porosity of the samples. In this study, image analysis was carried out using the ImageJ® software.

The specimens (3 cross sections for each sample) were scanned using an optical microscope, and images were generated. The images were converted into grayscale 8-bit files, and the threshold adjustment took place. In any image, the dark-colored areas account for the pores of the sample while the bright ones indicate the geopolymer matrix. Through threshold adjustment, the area of dark objects was computed in pixels, and the porosity was calculated based on the area of the pores and the total area of the image. As representative pore size of the samples, the pore areas with diameter greater than 0.1 mm were chosen. This choice was adopted as image processing records a large number of features with diameter less than 0.1 mm, and their consideration in the analysis results in unrealistic values of porosity. The circularity of the pores was kept above 0.2.

2.3. Experimental Design.

Geopolymerization is a complicated and dynamic process, and in the case of a full factorial design, the experiments are numerous and practically, as well as economically, not possible to be carried out. In the Taguchi method [30], the implementation of a multifactorial experimental designing model allows the investigation of the combining effect of selected parameters by conducting the minimum number of experiments. This method involves the use of a special set of arrays called orthogonal arrays to arrange the factors affecting the product/process as well as the levels at which the selected factors are varied. The selection of the array is related with the number and the variation level of the tested factors.

After conducting the experimental designing model, the collected data are analyzed by means of ANOVA (ANalysis Of VAriance) in order to determine the percent contribution of each factor on the chosen quality performance characteristics of the product or process. In this work, the studied characteristics are the compressive strength and the apparent density of the produced geopolymers.

The factors selected to be investigated are (i) the foaming agent content (Si% w/w on fly ash basis), (ii) the alkali to aluminum molar ratio (R/Al, R : Na or K) that associates the alkali quantity in the activation solution with the aluminum content of the raw material, and (iii) the (Na/(Na + K)) ratio which is related to the kind of alkali in the activation solution. The experimental design involved the variation of the above factors in four levels selected according to physical and chemical restrictions as well as to the previously published literature [32, 34]. The variation levels of the selected factors are presented in Table 2. A full factorial design requires the conduction of 64 experiments (3 factors, 4 levels per factor, experiments = (levels)$^{\text{(factors)}}$ = 4^3 = 64). The application of a L16 orthogonal array, based on Taguchi method, reduces the number of the experiments to 16.

3. Results and Discussion

3.1. Synthesis Optimization.

The results of the Taguchi model are presented in Table 3 which shows the synthesis parameters along with the compressive strength and apparent density of the corresponding geopolymers. In each experiment, three specimens were prepared and tested under compression. In case a value varies by more than ±10% from the mean, the result is discarded and new specimens are prepared. Table 3 also presents the maximum % variation from the mean and the standard deviation of the compressive strength measurements. The maximum % variation from the mean and the standard deviation of the apparent density are near zero and thus are not presented in Table 3. As it is seen, the produced geopolymers have a wide variation of strength and apparent density. More specifically, the foams exhibit compressive strength from 2.08 to 14.88 MPa and apparent density from 0.84 to 1.55 g/cm^3. The wide variation of the strength and the apparent density indicates

TABLE 2: Parameters and their variation levels.

Parameters	Level 1	Level 2	Level 3	Level 4
Silicon content (Si% w/w)	0.050	0.085	0.120	0.150
Alkali to aluminum ratio (R/Al)	0.8	1.2	1.6	2.0
Alkali species (Na/(Na + K))	0.00	0.35	0.65	1.00

that the chosen parameters strongly affect these properties. The reference geopolymer prepared with no Si addition (Si/Al = 2.4, R/Al = 0.8, (Na/(Na + K))= 1.00, and s/l = 2.8) possesses enhanced strength and apparent density values (53.70 MPa and 1.77 g/cm^3).

The effect of each parameter on the measured properties is presented in Figure 1 which has been based on the processing of the data presented in Table 3. For example the compressive strength corresponding to Factor 1, Level 1 (0.050% w/w Si) is the average strength of all samples containing 0.050% w/w Si. The impact of each factor on the development of compressive strength and apparent density was estimated through ANOVA and is presented in Table 4.

The factor with the greatest impact on both properties is the alkali to aluminum molar ratio. The increase of the R/Al ratio from 0.8 to 1.2 leads to the drastic reduction of both strength and density. Further addition of alkali (R/Al > 1.2) does not have a major effect on compressive strength while in the case of the apparent density, the produced foams are more compact. It seems that the alkali environment of geopolymer paste promotes the foaming action of Si powder, with R/Al = 1.2 being the optimal level. Higher alkalinity (R/Al > 1.2) creates mixing problems between the fly ash and the activation solutions, leading to the drastic coagulation of the geopolymeric pastes and therefore the limited action of the foaming agent.

The second more influential parameter is the foaming agent content (Si% w/w). An increase in the Si metal content (from 0.050 to 0.120% w/w) causes a decrease of the compressive strength and apparent density values by 50 and 20%, respectively. This is expected since higher foaming agent content leads to higher H$_2$ formation and therefore to higher porosity and lower compressive strength. The kind of the alkali ion (Na/(Na + K)) has only a marginal effect on the geopolymers' properties.

The optimal synthesis conditions for achieving the highest strength are Si% w/w = 0.050, R/Al = 0.8, and (Na/(Na + K)) = 0.35, while those for the lowest apparent density are Si% w/w = 0.120, R/Al = 1.2, and (Na/(Na + K)) = 0.00. The optimized value of the compressive strength and apparent density for a 95% confidence interval was predicted to be 12.42 ± 1.44 MPa and 0.7342 ± 0.0699 g/cm^3, respectively. In order to confirm the model prediction geopolymer specimens with Si% w/w = 0.050, R/Al = 0.8, (Na/(Na + K)) = 0.35, and Si% w/w = 0.120, R/Al = 1.2, (Na/(Na + K)) = 0.00 were prepared appropriately cured for 7 days and subjected to compression and density measurements. The compressive strength (mean of 3 specimens) of the first synthesis was found to be 11.04 MPa while the apparent density of the

TABLE 3: Synthesis parameters, compressive strength, and apparent density of Si foamed geopolymers.

Experiment	Si% w/w	R/Al	(Na/(Na + K))	Compressive strength (MPa)						Apparent density (g/cm³)
				(1)	(2)	(3)	Mean	Maximum variation from mean (%)	Standard deviation	
1	0.050	0.8	0.00	13.49	16.32	14.82	14.88	9.32	1.42	1.30
2	0.050	1.2	0.35	6.40	5.88	6.45	6.24	5.78	0.31	1.10
3	0.050	1.6	0.65	6.26	6.11	6.90	6.43	7.43	0.42	1.25
4	0.050	2.0	1.00	3.82	4.11	3.48	3.80	8.56	0.32	1.55
5	0.085	0.8	0.35	9.73	10.66	9.73	10.04	6.17	0.54	1.18
6	0.085	1.2	0.00	2.36	2.41	2.72	2.50	8.87	0.19	0.84
7	0.085	1.6	1.00	5.47	5.63	5.20	5.43	4.29	0.22	1.44
8	0.085	2.0	0.65	4.23	4.24	4.19	4.22	0.71	0.03	1.35
9	0.120	0.8	0.65	6.79	6.84	6.83	6.82	0.43	0.03	1.05
10	0.120	1.2	1.00	3.80	4.06	4.42	4.09	7.94	0.31	0.86
11	0.120	1.6	0.00	2.19	2.03	2.02	2.08	5.38	0.10	0.98
12	0.120	2.0	0.35	3.32	3.42	3.45	3.40	2.32	0.07	1.26
13	0.150	0.8	1.00	6.20	6.46	7.21	6.62	8.88	0.53	0.98
14	0.150	1.2	0.65	4.18	4.14	4.08	4.13	0.16	0.05	0.94
15	0.150	1.6	0.35	3.73	3.98	4.40	4.04	9.12	0.34	1.21
16	0.150	2.0	0.00	2.62	2.90	3.13	2.88	9.13	0.26	1.14

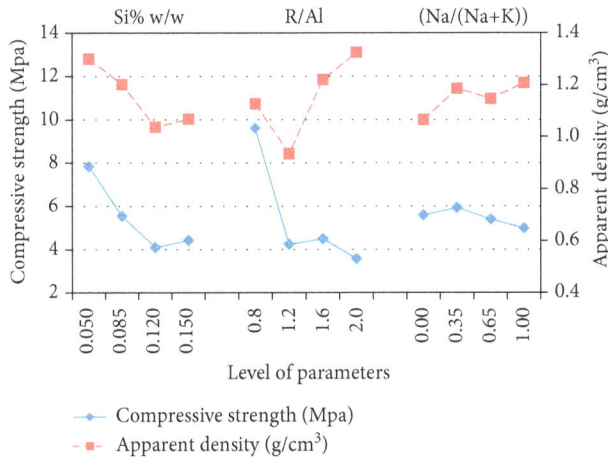

FIGURE 1: Effect of the studied parameters on the development of the compressive strength and apparent density.

TABLE 4: Contribution (%) of the studied parameters on the development of the compressive strength and apparent density.

Parameters	Si content	R/Al	(Na/(Na + K))
Compressive strength	26.9	71.7	1.4
Apparent density	32.0	59.5	8.5

second synthesis was found to be 0.72 g/cm³. Both values fall within the predicting range confirming in this way the validity of the used experimental design model.

The above results show, as it was expected, that the compressive strength of foamed geopolymers is in close relationship with the apparent density: the higher the density, the higher the compressive strength. The Taguchi approach can be very useful in practice since it allows the optimization of the synthesis on the basis of specific requirements that must be met by the final products (e.g., strength and density). When a lightweight geopolymer must be designed for a specific application, it is recommended to define the optimum synthesis parameters using multicriteria analysis and introducing appropriate weighting factors for each criterion (strength, density).

3.2. Characterization of Geopolymers. The XRD patterns of the fly ash and some representative foamed geopolymers are presented in Figure 2. The selection of the XRD patterns was made on the basis of the R/Al ratio and the Si% w/w, which were defined by the Taguchi method as the most influential parameters. The main crystallographic phases in the fly ash pattern are quartz ($d = 3.34$ Å), cristobalite ($d = 4.04$ Å), feldspars ($d = 3.19$ Å), gehlenite ($d = 2.85$ Å), maghemite ($d = 2.52$ Å), and anhydrite ($d = 3.49$ Å). Moreover the broad peak between 20 and 40° indicates the presence of an amorphous phase. In the XRD patterns of the geopolymeric products, anhydrite and cristobalite phases are consumed while the other phases of fly ash are still present. The 20–40° broad peak becomes more intense, indicating the formation of the amorphous geopolymer phase. As R/Al ratio is increased, the formation of zeolitic phases is favored. Particularly, exp. 3 (R/Al = 1.6) contains small amounts of hydroxysodalite while in exp. 4 (R/Al = 2.0) high amounts of hydroxysodalite and hydroxycancrinite are identified. Besides, it is observed that the increase of R/Al rate favors the formation of alkali carbonates. The gradual introduction of the foaming agent does not affect the crystallographic phases as it can be seen by the pattern of exp. 13.

Figure 3 shows the FTIR spectra of fly ash and selected geopolymer samples. The grouping of the samples is the

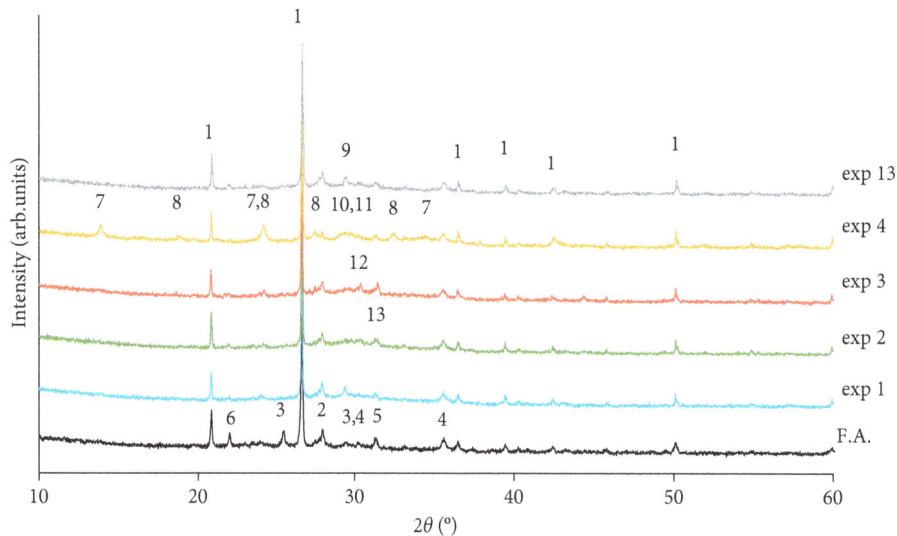

Figure 2: XRD patterns of fly ash and selected Si foamed geopolymers. 1: quartz (PDF: 46-1045), 2: albite (PDF: 20-0554), 3: anhydrate (PDF: 03-0377), 4: maghemite (PDF: 04-0755), 5: ghelenite (PDF: 04-0690), 6: cristobalite (PDF: 03-0276), 7: hydroxysodalite (PDF: 89-9098), 8: hydroxycancrinite (PDF: 31-1272), 9: natrite (PDF: 37-0451), 10: calcite (PDF: 47-1743), 11: potassium carbonate (PDF: 70-0292), 12: sodium bicarbonate (PDF: 15-0475), and 13: potassium bicarbonate (PDF: 70-0995).

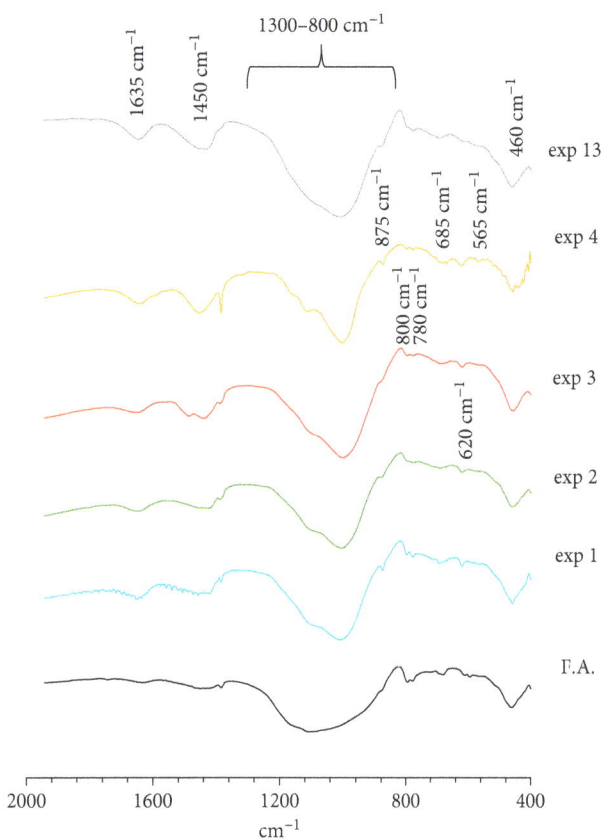

Figure 3: FTIR spectra of the fly ash and Si foamed geopolymers.

same as in the case of XRD analysis. The fly ash spectrum exhibits, relatively, shapeless peaks due to the semicrystalline state of this material. The region of interest in both fly ash and geopolymer products lies in the range between 800 and

$1300 \, cm^{-1}$. This broad hump exhibits overlapped peaks which are associated with the asymmetric stretching vibrations of Si-O-T (T: tetrahedral Si or Al) bonds. In the case of the fly ash, the maximum of this hump is at around $1110 \, cm^{-1}$ while in the case of the geopolymers this hump becomes narrower and shifts to lower wavenumbers ($\sim 1000 \, cm^{-1}$). This alteration is directly linked with the substitution of Si by tetrahedral Al and therefore the formation of an aluminosilicate network [35]. The Si-O-T asymmetric bending vibrations at $\sim 460 \, cm^{-1}$ are not as sensitive as the stretching vibrations to changes of the aluminosilicate network. The band at $\sim 1635 \, cm^{-1}$ is related to O-H bending vibrations of molecular water. All samples contain carbonate species pointed out by the presence of the band at around $1450 \, cm^{-1}$, related to antisymmetric vibrations of CO_3^{2-} ions. Samples with high R/Al ratios (exp. 3 and exp. 4) tend to exhibit more pronounced carbonate ion vibrations. This observation is in well accordance with the XRD results which showed that samples of high alkalinity contain carbonate phases such as natrite, calcite, and potassium carbonate. The absorbance at $875 \, cm^{-1}$ is related to symmetric stretching vibrations of Al–O and it constitutes sign of geopolymerization. As for 800, 780, and $685 \, cm^{-1}$ peaks, these are connected with symmetric stretching vibrations of Si-O-T bonds. In case of geopolymeric products, the latter band is located at higher wavenumbers in relation to the fly ash showing the enrichment of the aluminosilicate network in silicon [35]. In all geopolymeric products, a vibration at $620 \, cm^{-1}$ is observed. This absorbance peak is assigned to hexasilicate ring vibrations. This network formation indicates the participation of silicon in a more organized structure than the fly ash and refers to nanocrystalline zeolitic phase formation into geopolymeric matrix [36]. In the case of the geopolymer with high alkali content (exp. 4), there is a peak at $\sim 565 \, cm^{-1}$ which indicates

(a)

(b)

(c)

(d)

(e)

(f)

(g)

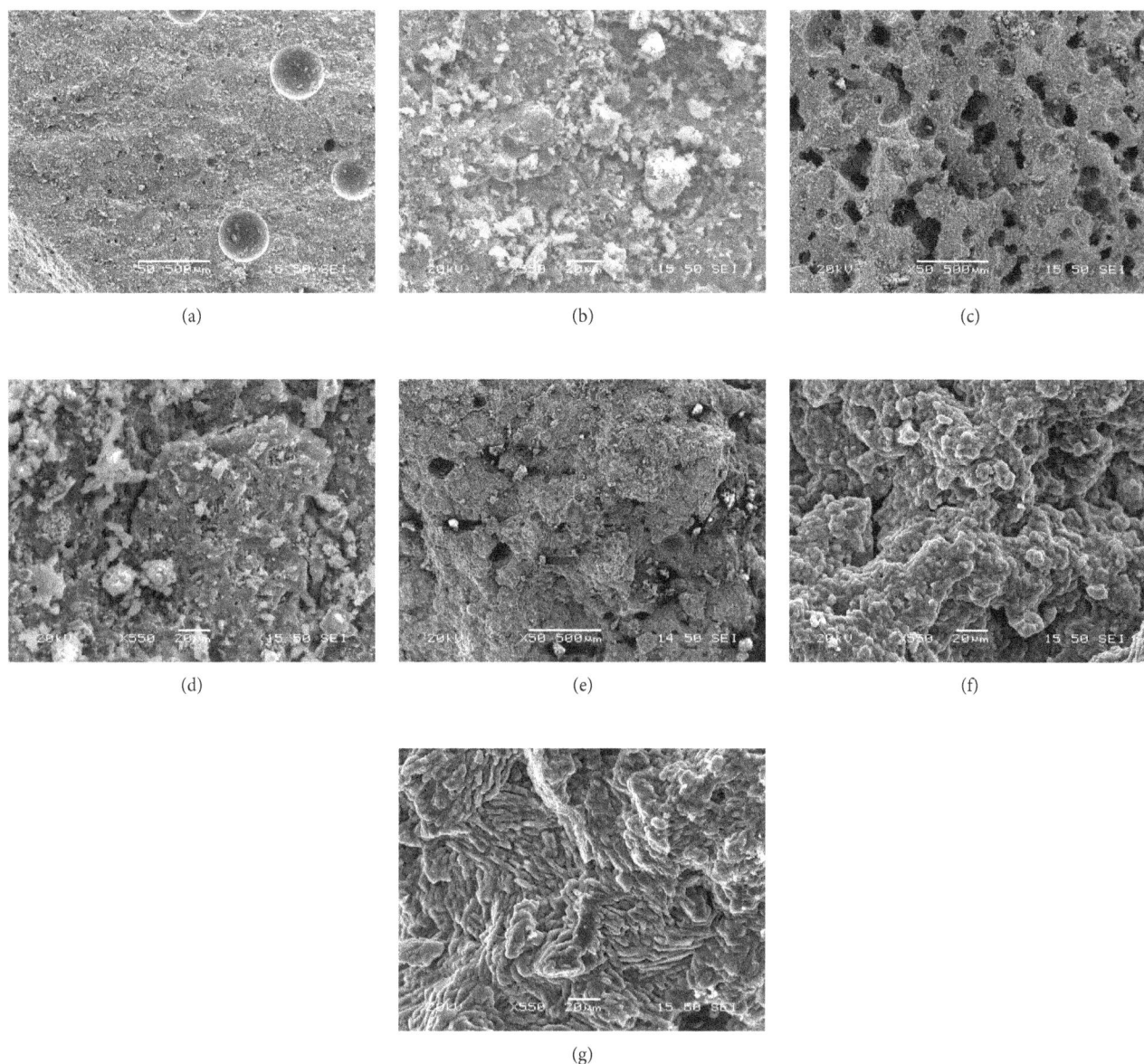

FIGURE 4: SEM images of selected geopolymer samples. (a, b) Reference geopolymer; (c, d) exp. 1; (e, f, g) exp. 4.

the double ring vibrations and characteristics of zeolite formation.

Figure 4 shows SEM images of selected samples. In particular, Figures 4(a)–4(b) present the microstructure of the reference geopolymer ($Si/Al = 2.4$, $R/Al = 0.8$, $(Na/(Na + K)) = 0.00$, and $s/l = 2.8$) which was prepared with the same synthetic parameters as exp. 1 but without any foaming agent.

As it is seen, there are unreacted particles of fly ash well embodied in the geopolymer matrix. The reference geopolymer exhibits microstructure homogeneity and constitutes a very compact material. Figures 4(c)–4(d) correspond to exp. 1-geopolymer which contains 0.050% w/w Si. In contrast to the reference geopolymer, this material (exp. 1) possesses a very porous structure with the size of pores being approximately $350\,\mu m$. A more intense study of the sample shows its good homogenous microstructure. The pore shape

is regular, while the dispersion of the pores seems to be uniform. However, the lightweight products differ in the coherence of the binding material and the quantity of the unreacted fly ash particles. In every case the compressive strength and the density are close related to the microstructure of the final products. In Figures 4(e)–4(g), SEM images of the exp. 4 are presented. This geopolymer differs from the exp. 1 in the extent of the alkali ions ($R/Al = 2.0$). A comparison with the exp. 1 shows that this material is more compact and its microstructure is rather heterogeneous. This can be clarified by Figures 4(f)–4(g) that were taken from different points of sample's surface. In one point, the sample exhibit polygonal shape of particles while in the other the particles are prismatic.

The reduced porosity of the exp. 4-geopolymer is in well accordance with its apparent density value which is the highest among the Si geopolymer series. As it is shown by

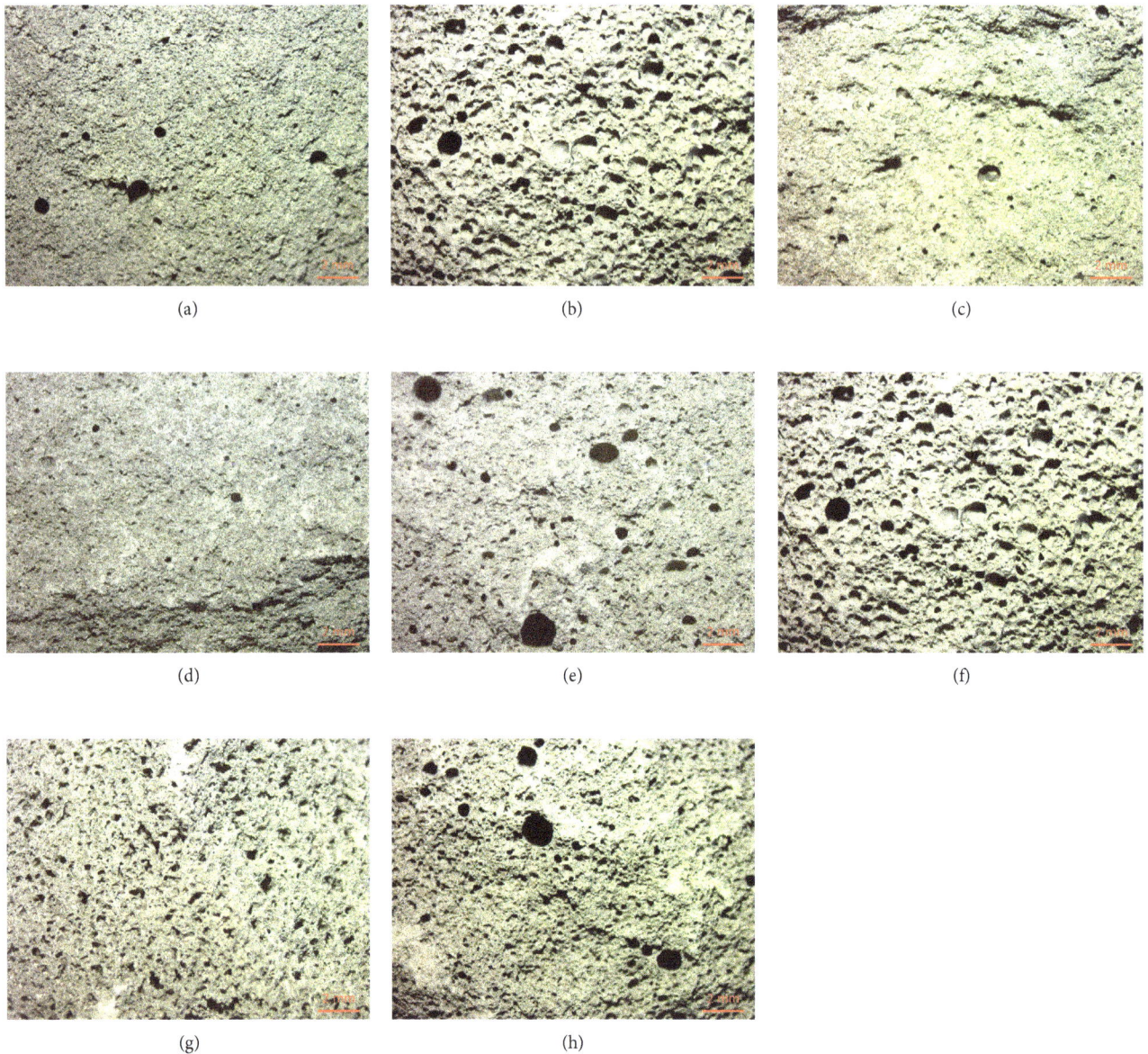

FIGURE 5: Optical microscope images of foamed geopolymers. (a) exp. 5 (R/Al = 0.8, Si = 0.085%); (b) exp. 6 (R/Al = 1.2, Si = 0.085%); (c) exp. 7 (R/Al = 1.6, Si = 0.085%); (d) exp. 8 (R/Al = 2.0, Si = 0.085%); (e) exp. 2 (R/Al = 1.2, Si = 0.050% w/w); (f) exp. 6 (R/Al = 1.2, Si 0.085% w/w); (g) exp. 10 (R/Al = 1.2, Si = 0.120% w/w); (h) exp. 14 (R/Al = 1.2, Si = 0.150% w/w).

XRD and FTIR measurements, the high alkalinity of the starting mixture favors the formation of crystalline phases (zeolites and alkali carbonates) which inhibit the geopolymerization reaction. Furthermore, high R/Al ratios tend to produce more compact pastes, limiting in this way the foaming action of Si.

The foamed geopolymers were examined by means of a high resolution optical microscope in order to evaluate the porosity of the samples. Figure 5 presents images of selected Si foamed geopolymers. Particularly, Figures 5(a)–5(d) show geopolymers with varying R/Al ratios and constant foaming agent content (0.085% w/w Si), while those from e to h correspond to geopolymers with varying foaming agent content and constant R/Al ratio (R/Al = 1.2). Once more the grouping of materials was chosen keeping in

mind the most influential parameters of Taguchi model (R/Al and Si %w/w).

An increase of the alkalinity of the samples leads to a drastic expansion of pores till the R/Al = 1.2 (Figure 5(b)). Further increase of the alkali content has a negative effect on the porosity of the samples (Figures 5(c)–5(d)). As it was stated before, the presence of alkali ions promotes the foaming reaction, producing geopolymers with increased porosity. However, high alkali contents (R/Al > 1.2) create drastic coagulation of the pastes and therefore restrain the foaming agent action. These observations are in well accordance with the Taguchi results where the optimum R/Al value for the preparation of lightweight products was found to be 1.2. In general, the high alkali content tends to produce compact materials. The increase of the foaming agent

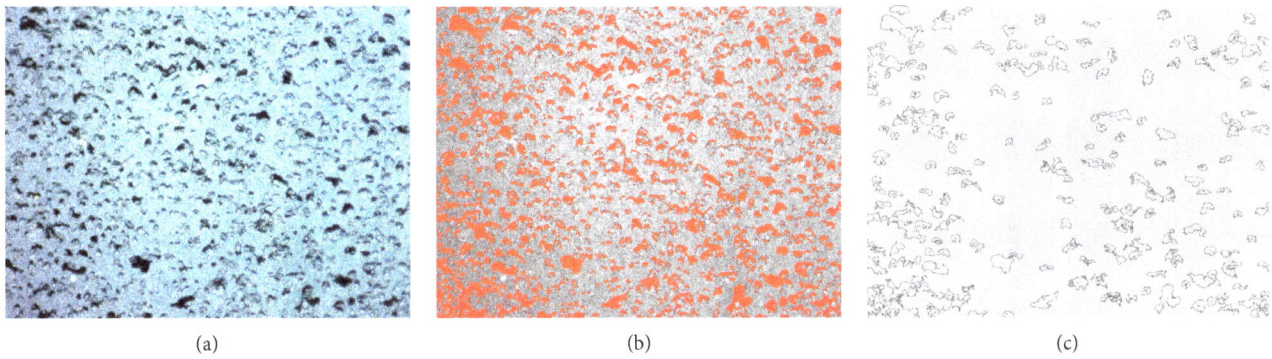

FIGURE 6: Steps of image analysis. (a) Cross-sectional image; (b) grayscale 8-bit image adjusted by the threshold option; (c) drawing of the pores.

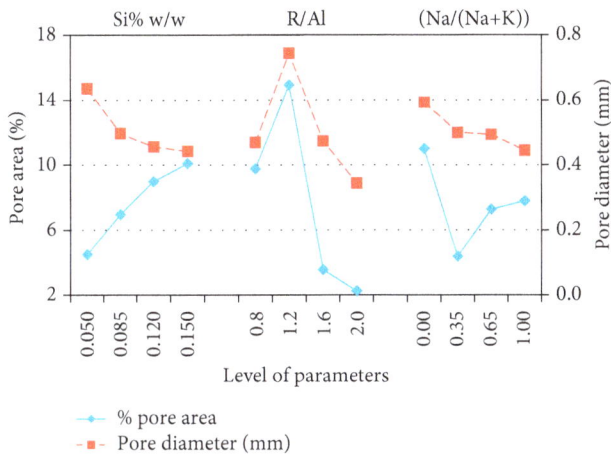

FIGURE 7: Effect of the studied parameters on the porosity of the Si foams.

TABLE 5: Contribution (%) of the studied parameters to the pore area and pore diameter of the samples.

Parameters	Si% w/w	R/Al	(Na/(Na + K))
Pore area	12.7	71.9	15.4
Pore diameter	19.5	70.9	9.6

content in these samples does not have any significant effect on the porosity.

The gradual incorporation of metallic Si into the pastes leads to the increase of the number of pores and therefore the overall pore volume (Figures 5(e)–5(h)). At the same time the size of the pores shows a tendency to decrease. This copes well with the fact that Si rich samples exhibit the lowest values of apparent density and compressive strength. For the whole Si foam series, the shape of the pores seems to be spherical. However, an increase in silicon content changes the shape of pores from spherical to irregular, possibly, due to the more intense conditions of the gaseous hydrogen release.

Image analysis was applied in order to have a semi-quantitative evaluation of the porosity. The image processing correlates the surface of the pores and 2-d porosity of the samples and converts this correlation to a percentage of the total surface of the image. As an example Figure 6 presents images of the exp. 9 before and after the image processing. In particular, Figure 6(a) stands for the cross section of the exp. 9 before analysis, Figure 6(b) presents the grayscale image generated after threshold adjustment, and Figure 6(c) shows a drawing of the pores which took place in the porosity calculation (pores with diameter higher than 0.1 mm

and circularity above 0.2). Corresponding images were generated for all the samples.

The image processing showed that Si foamed geopolymers exhibit pore area and pore diameter in the range of 1–21% and 0.34–0.90 mm, respectively. The collected data were analyzed by means of ANOVA analysis as in the case of the compressive strength and apparent density values. The effect of each parameter on the pore area and pore diameter is presented in Figure 7, while their contribution to the measured properties is shown in Table 5.

The alkali content is the most significant factor for both the pore area (71.9%) and pore diameter (70.9%). Particularly, the increase of R/Al ratio up to 1.2 leads to a boost of the pore area by almost 50% (Figure 7). The enhancement of the pore diameter is even higher and close to 60%. However, further increase of the alkali content leads to a considerable reduction of the pore area and size. The foaming agent content has a contradictory effect on the pore area and the pore size. The increase of Si content increases the porosity of the sample and at the same time decreases the mean pore size. The increase of the Si content from 0.050 to 0.150% w/w results in a 125% increase of the pore surface while the pore diameter is reduced by 32% (Figure 7). The alkali selection has a limited effect on porosity. In any case, the potassium anions tend to produce more porous structures in relation to sodium anions. This observation is related to the effect of each ion on pastes' workability. It is well known that potassium ions exhibit greater levels of reactivity due to their higher basicity, and they produce more workable pastes in relation to the sodium ions [37].

The pore size distribution was also evaluated through the image analysis of the samples. Figure 8 shows the pore size distribution of Si foams in relation to the R/Al (a, b, and c) and in relation to the Si content (d, e, and f). Histograms (a) to (c) of Figure 8 reveal the effect of R/Al molar ratio on the

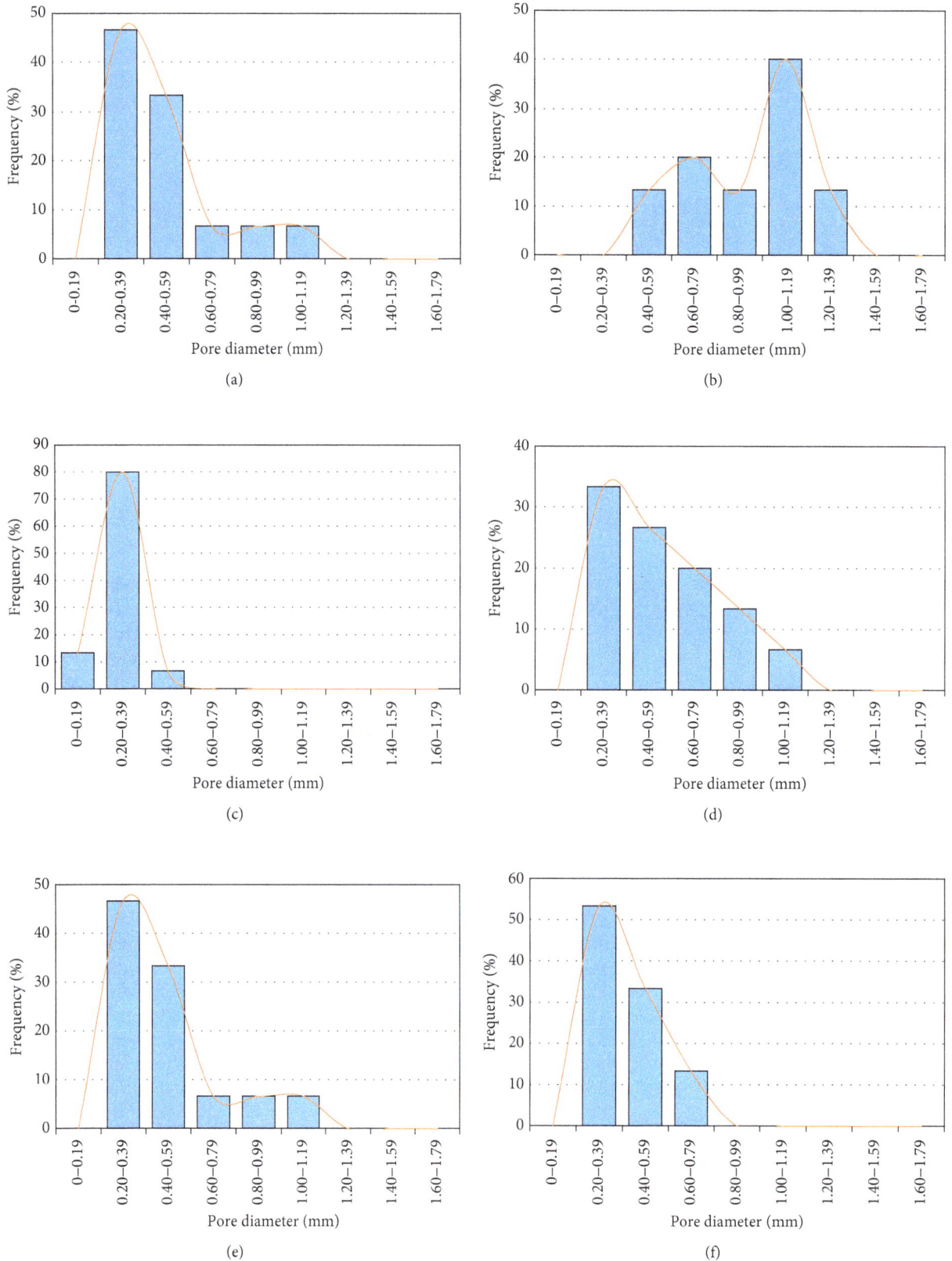

FIGURE 8: Pore size distribution of Si samples in relation to the R/Al molar ratio (a, b, and c) and the foaming agent content (d, e, and f). (a) exp. 5 (R/Al = 0.8, Si = 0.085%); (b) exp. 6 (R/Al = 1.2, Si = 0.085%); (c) exp. 8 (R/Al = 2.0, Si = 0.085%); (d) exp. 1 (R/Al = 0.8, Si = 0.050%); (e) exp. 5 (R/Al = 0.8, Si = 0.085% w/w); (f) exp. 13 (R/Al = 0.8, Si = 0.150% w/w).

pore size distribution. Samples of modest alkalinity (Figure 8(a)) exhibit wide distributions with the peak size being located in low values (47% of pores below 0.39 mm for the exp. 5). When R/Al ratio equals to 1.2 (Figure 8(b)), the size distribution remains wide but at the same time shifts to higher diameters. Identically, for the exp. 6, the peak of pore diameter (40%) lies beneath 1.00 and 1.19 mm. Higher alkali contents (R/Al > 1.2) increase the viscosity of the paste, inhibit the foaming agent action, and generate narrower distributions shifted to smaller diameters (Figure 8(c)).

The pore size distribution of the samples with varying Si content (Figures 8(d)–(f)) show that increasing the Si incorporation, the distributions become gradually narrower and shift to lower pore diameters. It seems that the introduction of more Si to geopolymer pastes produces foams of enhanced porosity based on small and more homogenous pores.

4. Conclusions

This work came up with the following results:

(i) Fly ash (Type W, EN 197-1) was successfully recycled into the production of lightweight geopolymers. Metallic silicon, added in the starting mixture even in small quantities, acts effectively as a foaming agent. Lightweight geopolymers with a compressive strength of 2.08–14.88 MPa and an apparent density of 0.84–1.55 g/cm^3 were prepared by introducing a Si content up to 0.2 % w/w on fly ash basis. This wide range of properties values reveals the great applicability potential of geopolymer foams in the building sector.

(ii) The compressive strength of the foamed geopolymers is in close relation with the porosity of the final products, but even in the case of highly porous structures, compressive strength higher than 2 MPa can be easily achieved.

(iii) Apart from the foaming agent's content in the raw mixture, the development of the porosity and strength is also affected by the synthesis parameters of the geopolymer matrix. The alkali to aluminum molar ratio (R/Al) in the starting mixture is the most influential factor, while the kind of the alkali (Na or K) has only a marginal effect on foams' properties.

(iv) The Taguchi design methodology was applied in order to evaluate the impact of the synthesis parameters on the development of the porosity and strength. The Taguchi approach can also be very useful in practice since it allows the optimization of the synthesis on the basis of specific requirements that must be met by the final products.

(v) The applied analytic techniques (XRD, FTIR, and SEM) led to the complete characterization of the foamed geopolymers. As it is shown, the introduction of the foaming agent does not affect the mineral composition of the geopolymer matrix, while the increase of the alkali content in the starting mixture favors the formation of zeolites and alkali carbonates.

(vi) Optical microscope image analysis provided an estimation of the pore size distribution. The R/Al molar ratio has an effect on the pore size. An increase in the alkali content till 1.2 value leads to the expansion of the pores. Higher alkalinity (R/Al > 1.2) tends to reduce the workability of the pastes having as a consequence the restrain of pores formation. Metallic Si seems to affect the total number of the pores. High introduction levels produce foams with increased number of pores.

Conflicts of Interest

The authors declare that there are no conflicts of interest regarding the publication of this paper.

Acknowledgments

This research work has been financed by the European Union Seventh Framework Program (FP7-2013-NMPENV-EeB, "Multifunctional facades of reduced thickness for fast and cost-effective retrofitting, MF-Retrofit," Grant no. 609345).

References

[1] J. Davidovits, "Geopolymers: inorganic polymeric new materials," *Journal of Thermal Analysis*, vol. 37, no. 8, pp. 1633–1656, 1991.

[2] J. L. Provis and J. S. J. van Deventer, *Alkali-Activated Materials: State-of-the-Art Report*, Springer/RILEM, Berlin, Germany, 2013.

[3] J. Davidovits, *Geopolymer Chemistry and Applications*, Institut Geopolymere, Saint-Quentin, France, 4th edition, 2008.

[4] P. Duxson and J. L. Provis, "Designing precursors for geopolymer cements," *Journal of American Ceramic Society*, vol. 91, no. 12, pp. 3864–3869, 2008.

[5] Z. Zhang, H. Wang, Y. Zhu, A. Reid, J. L. Provis, and F. Bullen, "Using fly ash to partially substitute metakaolin in geopolymer synthesis," *Applied Clay Science*, vol. 88–89, pp. 194–201, 2014.

[6] J. L. Provis and S. A. Bernal, "Geopolymers and related alkali-activated materials," *Annual Review of Materials Research*, vol. 44, no. 1, pp. 299–327, 2014.

[7] Z. Yunsheng, S. Wei, C. Qianli, and C. Lin, "Synthesis and heavy metal immobilization behaviors of slag based geopolymer," *Journal of Hazardous Materials*, vol. 143, no. 1-2, pp. 206–213, 2007.

[8] P. Duxson, A. Fernández-Jiménez, J. L. Provis, G. C. Lukey, A. Palomo, and J. S. J. Van Deventer, "Geopolymer technology: the current state of the art," *Journal of Materials Science*, vol. 42, no. 9, pp. 2917–2933, 2007.

[9] N. K. Lee, G. H. An, K. T. Koh, and G. S. Ryu, "Improved reactivity of fly ash-slag geopolymer by the addition of silica fume," *Advances in Materials Science and Engineering*, vol. 2016, Article ID 2192053, 11 pages, 2016.

[10] M. Y. J. Liu, C. P. Chua, U. Johnson Alengaram, and M. Z. Jumaat, "Utilization of palm oil fuel ash as binder in lightweight oil palm shell geopolymer concrete," *Advances*

in Materials Science and Engineering, vol. 2014, Article ID 610274, 6 pages, 2014.

[11] J. C. Swanepoel and C. A. Strydom, "Utilisation of fly ash in a geopolymeric material," *Applied Geochemistry*, vol. 17, no. 8, pp. 1143–1148, 2002.

[12] J. L. Provis, R. M. Harrex, S. A. Bernal, P. Duxson, and J. S. J. van Deventer, "Dilatometry of geopolymers as a means of selecting desirable fly ash sources," *Journal of Non-Crystalline Solids*, vol. 358, no. 16, pp. 1930–1937, 2012.

[13] K. T. Nguyen, N. Ahn, T. A. Le, and K. Lee, "Theoretical and experimental study on mechanical properties and flexural strength of fly ash-geopolymer concrete," *Construction and Building Materials*, vol. 106, pp. 65–77, 2016.

[14] J. G. Jang and H. K. Lee, "Effect of fly ash characteristics on delayed high-strength development of geopolymers," *Construction and Building Materials*, vol. 102, pp. 260–269, 2016.

[15] G. Lavanya and J. Jegan, "Durability study on high calcium fly ash based geopolymer concrete," *Advances in Materials Science and Engineering*, vol. 2015, Article ID 731056, 7 pages, 2015.

[16] J. S. J. Van Deventer, J. L. Provis, and P. Duxson, "Technical and commercial progress in the adoption of geopolymer cement," *Minerals Engineering*, vol. 29, pp. 89–104, 2012.

[17] E. U. Weizsäcker, K. C. Hargroves, M. H. Smith, C. Desha, and P. Stasinopoulos, "Factor five: transforming the global economy through 80% improvements in resource productivity," in in *Ernst Ulrich von Weizsäcker: A Pioneer on Environmental, Climate and Energy Policies*, E. U. von Weizsäcker, Ed., pp. 192–213, Springer International Publishing, Cham, Switzerland, 2014.

[18] Z. Abdollahnejad, F. Pacheco-Torgal, T. Félix, W. Tahri, and J. Barroso Aguiar, "Mix design, properties and cost analysis of fly ash-based geopolymer foam," *Construction and Building Materials*, vol. 80, pp. 18–30, 2015.

[19] P. Palmero, A. Formia, P. Antonaci, S. Brini, and J.-M. Tulliani, "Geopolymer technology for application-oriented dense and lightened materials. Elaboration and characterization," *Ceramics International*, vol. 41, no. 10, pp. 12967–12979, 2015.

[20] V. Ducman and L. Korat, "Characterization of geopolymer fly-ash based foams obtained with the addition of Al powder or H_2O_2 as foaming agents," *Materials Characterization*, vol. 113, pp. 207–213, 2016.

[21] V. Vaou and D. Panias, "Thermal insulating foamy geopolymers from perlite," *Minerals Engineering*, vol. 23, no. 14, pp. 1146–1151, 2010.

[22] V. Medri, E. Papa, J. Dedecek et al., "Effect of metallic Si addition on polymerization degree of in situ foamed alkali-aluminosilicates," *Ceramics International*, vol. 39, no. 7, pp. 7657–7668, 2013.

[23] E. Kamseu, Z. N. M. Ngouloure, B. N. Ali et al., "Cumulative pore volume, pore size distribution and phases percolation in porous inorganic polymer composites: relation microstructure and effective thermal conductivity," *Energy and Buildings*, vol. 88, pp. 45–56, 2015.

[24] T. Y. Yang and C. C. Chien, "A study on chemical foaming geopolymer building materials," *Advanced Materials Research*, vol. 853, pp. 202–206, 2014.

[25] J. Henon, A. Alzina, J. Absi, D. S. Smith, and S. Rossignol, "Potassium geopolymer foams made with silica fume pore forming agent for thermal insulation," *Journal of Porous Materials*, vol. 20, no. 1, pp. 37–46, 2012.

[26] E. Prud'homme, P. Michaud, E. Joussein, C. Peyratout, A. Smith, and S. Rossignol, "In situ inorganic foams prepared from various clays at low temperature," *Applied Clay Science*, vol. 51, no. 1-2, pp. 15–22, 2011.

[27] S. Delair, É. Prud'homme, C. Peyratout et al., "Durability of inorganic foam in solution: the role of alkali elements in the geopolymer network," *Corrosion Science*, vol. 59, pp. 213–221, 2012.

[28] A. R. Studart, U. T. Gonzenbach, E. Tervoort, and L. J. Gauckler, "Processing routes to macroporous ceramics: a review," *Journal of the American Ceramic Society*, vol. 89, no. 6, pp. 1771–1789, 2006.

[29] M. Olivia and H. Nikraz, "Properties of fly ash geopolymer concrete designed by Taguchi method," *Materials & Design*, vol. 36, pp. 191–198, 2012.

[30] K. R. Ranjit, *A Primer on the Taguchi Method*, Van Nostrand Reinhold, New York, NY, USA, 1990.

[31] E. Ayan, Ö. Saatçioğlu, and L. Turanli, "Parameter optimization on compressive strength of steel fiber reinforced high strength concrete," *Construction and Building Materials*, vol. 25, no. 6, pp. 2837–2844, 2011.

[32] C. Panagiotopoulou, S. Tsivilis, and G. Kakali, "Application of the Taguchi approach for the composition optimization of alkali activated fly ash binders," *Construction and Building Materials*, vol. 91, pp. 17–22, 2015.

[33] M. Ghanbari, A. M. Hadian, A. A. Nourbakhsh, and K. J. D. MacKenzie, "Modeling and optimization of compressive strength and bulk density of metakaolin-based geopolymer using central composite design: a numerical and experimental study," *Ceramics International*, vol. 43, no. 1, pp. 324–335, 2017.

[34] C. Panagiotopoulou, A. Asprogerakas, G. Kakali, and S. Tsivilis, "Synthesis and thermal properties of fly-ash based geopolymer pastes and mortars," in *Developments in Strategic Materials and Computational Design II*, pp. 17–28, John Wiley & Sons, Hoboken, NJ, USA, 2011.

[35] T. Bakharev, "Geopolymeric materials prepared using Class F fly ash and elevated temperature curing," *Cement and Concrete Research*, vol. 35, no. 6, pp. 1224–1232, 2005.

[36] I. Lecomte, C. Henrist, M. Liégeois, F. Maseri, A. Rulmont, and R. Cloots, "(Micro)-structural comparison between geopolymers, alkali-activated slag cement and Portland cement," *Journal of the European Ceramic Society*, vol. 26, no. 16, pp. 3789–3797, 2006.

[37] M. Criado, A. Fernández-Jiménez, A. G. de la Torre, M. A. G. Aranda, and A. Palomo, "An XRD study of the effect of the SiO_2/Na_2O ratio on the alkali activation of fly ash," *Cement and Concrete Research*, vol. 37, no. 5, pp. 671–679, 2007.

5

Role of Dielectric Constant on Ion Transport: Reformulated Arrhenius Equation

Shujahadeen B. Aziz

Advanced Polymeric Materials Research Laboratory, School of Science-Department of Physics,
Faculty of Science and Science Education, University of Sulaimani, Sulaimani, Kurdistan Region, Iraq

Correspondence should be addressed to Shujahadeen B. Aziz; shujaadeen78@yahoo.com

Academic Editor: Somchai Thongtem

Solid and nanocomposite polymer electrolytes based on chitosan have been prepared by solution cast technique. The XRD results reveal the occurrence of complexation between chitosan (CS) and the LiTf salt. The deconvolution of the diffractogram of nanocomposite solid polymer electrolytes demonstrates the increase of amorphous domain with increasing alumina content up to 4 wt.%. Further incorporation of alumina nanoparticles (6 to 10 wt.% Al_2O_3) results in crystallinity increase (large crystallite size). The morphological (SEM and EDX) analysis well supported the XRD results. Similar trends of DC conductivity and dielectric constant with Al_2O_3 concentration were explained. The TEM images were used to explain the phenomena of space charge and blocking effects. The reformulated Arrhenius equation ($\sigma_{(\varepsilon',T)} = \sigma_o \exp^{(-E_a/K_B\varepsilon'T)}$) was proposed from the smooth exponential behavior of DC conductivity versus dielectric constant at different temperatures. The more linear behavior of DC conductivity versus $1000/(\varepsilon' \times T)$ reveals the crucial role of dielectric constant in Arrhenius equation. The drawbacks of Arrhenius equation can be understood from the less linear behavior of DC conductivity versus $1000/T$. The relaxation processes have been interpreted in terms of Argand plots.

1. Introduction

Chitin, poly(β-(1→4)-N-acetyl-D-glucosamine), is a natural polysaccharide of major importance, first identified in 1884. This biopolymer is synthesized by an enormous number of living organisms; and considering the amount of chitin produced annually in the world, it is the most abundant polymer after cellulose [1]. On the other hand chitosan is a cationic polysaccharide, which contains β-1-4-linked 2-amino-2 deoxy-D-glucopyranose repeat units and is readily obtained by alkaline N-acetylation of chitin [2]. Chitosan has great potential as a biomaterial because of [3] good biocompatibility, biodegradability, low toxicity, low cost [4], antimicrobial activity [5], hydrophilicity which certainly be of benefit to the fuel cell operation, and chemical and thermal stability at higher temperatures [6]. Thus chitosan has become of great interest not only as an underutilized resource but also as a new functional biomaterial of high potential in various fields [7]. The unique properties which separate chitosan from other biopolymers are the presence of amino groups [8] as depicted in Scheme 1(b).

Thus, the polycationic nature of chitosan due to the abundance of free amino (NH_2) groups (Scheme 1(b)) on the backbone of chitosan makes it possible to produce ion-conducting polymer electrolyte [9]. This is related to the fact that the amine groups in chitosan structure can act as electron donors and interact with inorganic salts [10], and therefore chitosan meets an important requirement for acting as a polymer host for the solvation of salts [11]. One of the major reasons for poor ionic conductivity and concentration polarization has been attributed to the possibility of an ion-association (ion-pairing) effect. This is due to weak dielectric permittivity of host polymers [12]. A number of approaches have been proposed to solve the state-of-the-art problems of ion-conducting polymers. Nanocomposite formation is the latest suggestion to overcome the disadvantages and improve the properties of SPEs; the method applied mostly is to disperse inorganic inert fillers in the SPE [13]. Electrical

(a)

(b)

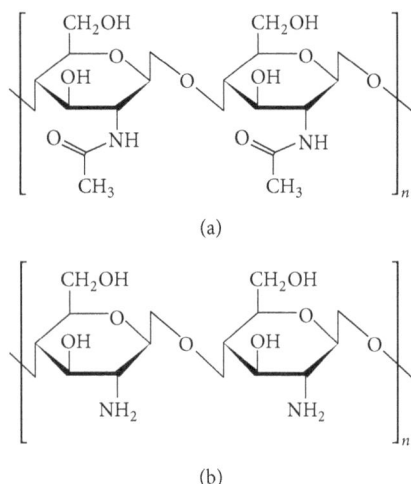

SCHEME 1: Structure of (a) chitin and (b) chitosan [16].

TABLE 1: Composition of $(1-x)(0.9CS:0.1LiTf)-xAl_2O_3$ $(0.02 \leq x \leq 0.1)$ NCPEs.

Designation	CS : LiTf = 90 : 10 (g)	Al_2O_3 (g)	Al_2O_3 (wt.%)
CSNC1	1.1111	0.0227	2.0
CSNC2	1.1111	0.0463	4.0
CSNC3	1.1111	0.0709	6.0
CSNC4	1.1111	0.0966	8.0
CSNC5	1.1111	0.1235	10

CSNC1 to CSNC5 represent the sample code of $(1-x)(0.9CS:0.1LiTf)-xAl_2O_3$ $(0.02 \leq x \leq 0.1)$ NCPE compositions.

properties such as dielectric permittivity can be suitably adjusted, simply by controlling the type and the amount of ceramic inclusions. Ceramic materials are typically brittle, possess low dielectric strength, and in many cases are difficult to be processed requiring high temperatures [14]. On the other hand, polymers possess relatively low dielectric permittivity; they can withstand high fields and are flexible and easy to process. By combining the advantages of both, one can fabricate new hybrid materials with high dielectric permittivity [15]. When these inorganic fillers are of nanodimensions, the composite polymer electrolyte formed is called nanocomposite polymer electrolyte (NCPE). But till now there is no fully satisfactory explanation available regarding the role played by these nanoparticles on DC conductivity [13]. Ion transport mechanism is a complicated subject in solid and nanocomposite polymer electrolytes. Thus the main objective of the present work is to show experimentally the crucial role of dielectric constant on cation transport mechanism and the necessary reformulation of Arrhenius equation.

2. Experimental Details

2.1. Raw Materials. Chitosan (CS) from crab shells ($\geq 75\%$ deacetylated, Sigma Aldrich), lithium triflate (LiCF$_3$SO$_3$), and aluminum oxide (Al$_2$O$_3$, size <50 nm) were supplied by sigma Aldrich. Acetic acid (1%) was prepared using glacial acetic acid solution and used as the solvent in casting solid and nanocomposite polymer electrolytes.

2.2. Preparation of Solid and Nanocomposite Polymer Electrolytes. The SPE films were prepared by the solution cast technique. For this purpose 1 gm of chitosan (CS) was dissolved in 100 mL of 1% acetic acid solution. The mixture was stirred continuously with a magnetic stirrer for several hours at room temperature until the chitosan powder has completely dissolved in the acetic acid solution. To this solution 10 wt.% lithium triflate (LiTf) was added and the mixture was stirred continuously until homogeneous solution was obtained. The solution of CS : LiTf (90 : 10) was then cast into

clean and dry Petri dish and allowed to evaporate at room temperature to obtain a solvent-free film. To prepare NCPEs initially the chitosan and LiTf (90 : 10) were dissolved in acetic acid to obtain homogeneous solutions as mentioned above. Al$_2$O$_3$ were first dispersed in 20 mL acetic acid solution and stirred. The Al$_2$O$_3$ concentrations were varied from 2 wt.% to 10 wt.%. The Al$_2$O$_3$ dispersoids were mixed with CS : LiTf (90 : 10) solutions and then continuously stirred. The solutions were then cast into different clean and dry Petri dish and allowed to evaporate at room temperature until solvent-free NCSPE films were obtained. The films were kept in desiccators with blue silica gel desiccant for further drying. Table 1 shows the codes and the concentration of the prepared nanocomposite solid polymer electrolytes (NCPEs).

2.3. XRD Investigation. In the present work, XRD used to reveal the structure of the prepared films; that is, they transferred to amorphous or crystalline after doping. XRD was carried out using a Siemens D5000 X-ray diffractometer with operating voltage and current of 40 kV and 40 mA, respectively. The wavelength of the monochromatic, X-ray beam is 1.5406 A$^\circ$ and the glancing angles were in the range of $5^\circ \leq 2\theta \leq 80^\circ$ with a step size of 0.1°.

2.4. Morphology Characterization. The combined usage of SEM-EDAX is an attempt to understand the structural and compositional complexity of the samples. A scanning electron micrograph (SEM) was taken using Leica 440 scanning electron microscope to study morphological appearance. The microscope was fitted with energy dispersive analysis of X-rays (EDX), Oxford Instrument (LINK ISIS), to detect the overall chemical composition of NCSPE. Before observation, the NCSPE films were attached to aluminum holder using a conductive tape and then coated with a thin layer of gold.

2.5. Transmission Electron Microscopy (TEM). Transmission electron microscopy (TEM) images of the polymer electrolyte containing Al$_2$O$_3$ nanoparticle were recorded using a LEO LIBRA (accelerating voltage 120 kV) instrument. A drop of the solution of chitosan-silver triflate electrolyte was placed on a carbon coated copper grid and dried at room temperature after removal of excess solution using a filter paper.

2.6. Complex Impedance Spectroscopy Technique. Complex impedance spectroscopy gives information on electrical properties of materials and their interface with electronically conducting electrodes. The nanocomposite solid polymer electrolyte (NCSPE) films were cut into small discs (2 cm diameter) and sandwiched between two stainless steel electrodes under spring pressure. The impedance of the films was measured in the frequency range from 50 Hz to 1000 kHz using the HIOKI 3531 Z Hi-tester which was interfaced to a computer. Measurements were also made at temperatures ranging between 303 K and 363 K. The software controls the measurements and calculates the real (Z') and imaginary (Z'') parts of impedance. Z' and Z'' data were presented as a Nyquist plot and the bulk resistance was obtained from the intercept of the plot with the real impedance axis. The conductivity can be calculated from the following equation [17]:

$$\sigma_{DC} = \left(\frac{1}{R_b}\right) \times \left(\frac{t}{A}\right). \tag{1}$$

In (1), t is the film's thickness and A is its area.

The real (Z') and imaginary (Z'') parts of complex impedance (Z^*) were also used for the evaluation of dielectric constant, the real (M') part, and imaginary (M'') part of complex electric modulus (M^*) using the following equations [18, 19]:

$$\varepsilon' = \frac{Z_i}{\omega C_o \left(Z_r^2 + Z_i^2\right)},$$
$$M' = \omega C_o Z_i, \tag{2}$$
$$M'' = \omega C_o Z_r.$$

Here C_o is the vacuum capacitance and given by $\varepsilon_o A/t$, where ε_o is a permittivity of free space and is equal to 8.85×10^{-12} F/m. The angular frequency ω is equal to $\omega = 2\pi f$, where f is the frequency of applied field.

3. Results and Discussion

3.1. XRD Analysis. In this study, XRD was performed to show the complex formation occurrence between $LiCF_3SO_3$ and chitosan. The diffractograms of the pure chitosan film and chitosan : LiTf (90 : 10) complexes are illustrated in Figure 1. It can be seen that, with addition of lithium triflate (LiTf) into chitosan matrix, intensity of the crystalline peaks of pure chitosan decreased considerably (see Figure 1). This indicates the reduction in crystalline region of chitosan matrix. In other words with increasing salt concentration the amorphous domain of the chitosan electrolyte membrane increases. It is obvious that the peak of pure chitosan at 15.5° shifted to 12.6° in CS : LiTf systems. The effect of LiTf salt on crystal structure of chitosan can be seen elsewhere [20] in detail.

Figure 2 shows several crystalline peaks at around 32°, 34.4°, 36°, 39°, 41°, 45.7°, 61°, and 67.1° for pure Al_2O_3. The X-ray diffraction pattern of the Al_2O_3 nanoparticles depicted in Figure 2 is very similar to the Al_2O_3 pattern reported in the literature [21]. It is obvious that the Al_2O_3 nanoparticles exhibit two sharp crystalline peaks at 45.7° and 67.1°.

X-ray diffraction analysis was performed to observe the effect of Al_2O_3 filler on the crystalline structure of CS : LiTf

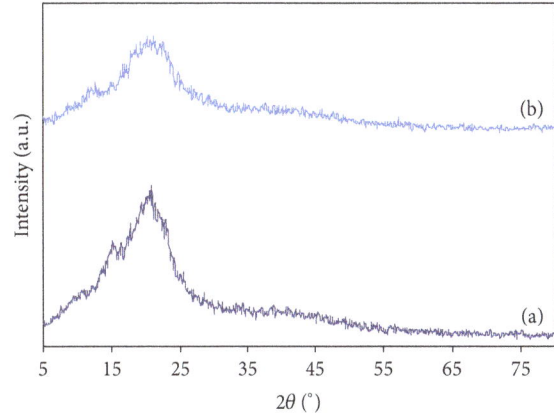

FIGURE 1: X-ray diffractogram of (a) pure chitosan (CS) and (b) chitosan : $LiCF_3SO_3$ (90 : 10) system.

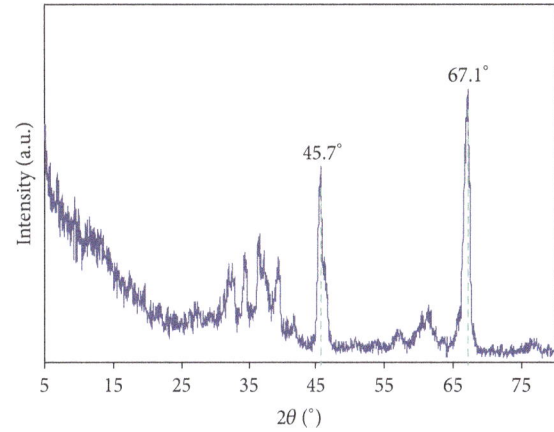

FIGURE 2: XRD patterns of pure Al_2O_3 nanoparticles.

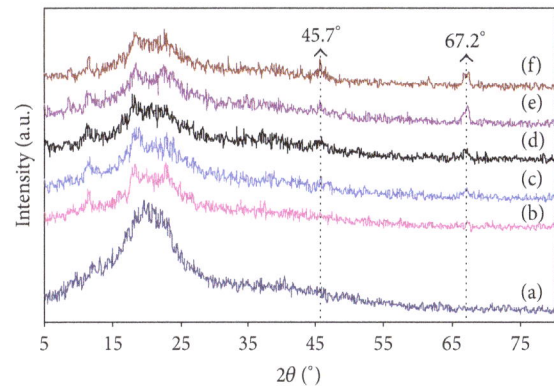

FIGURE 3: X-ray diffractogram of (a) CS : LiTf (90 : 10), (b) CSNC1, (c) CSNC2, (d) CSNC3, (e) CSNC4, and (f) CSNC5.

(90 : 10) system. It is interesting to note that the crystalline nature of NCPEs (Figure 3) is largely reduced compared to CS : LiTf (90 : 10) system (Figure 1(b)). In the diffractogram of Figure 3, the crystallinity for 2 and 4 wt.% of Al_2O_3 is largely reduced; however from 6 to 10 wt.% of Al_2O_3 the crystallinity is again increased. The absence of crystalline

(a)

(b)

(c)

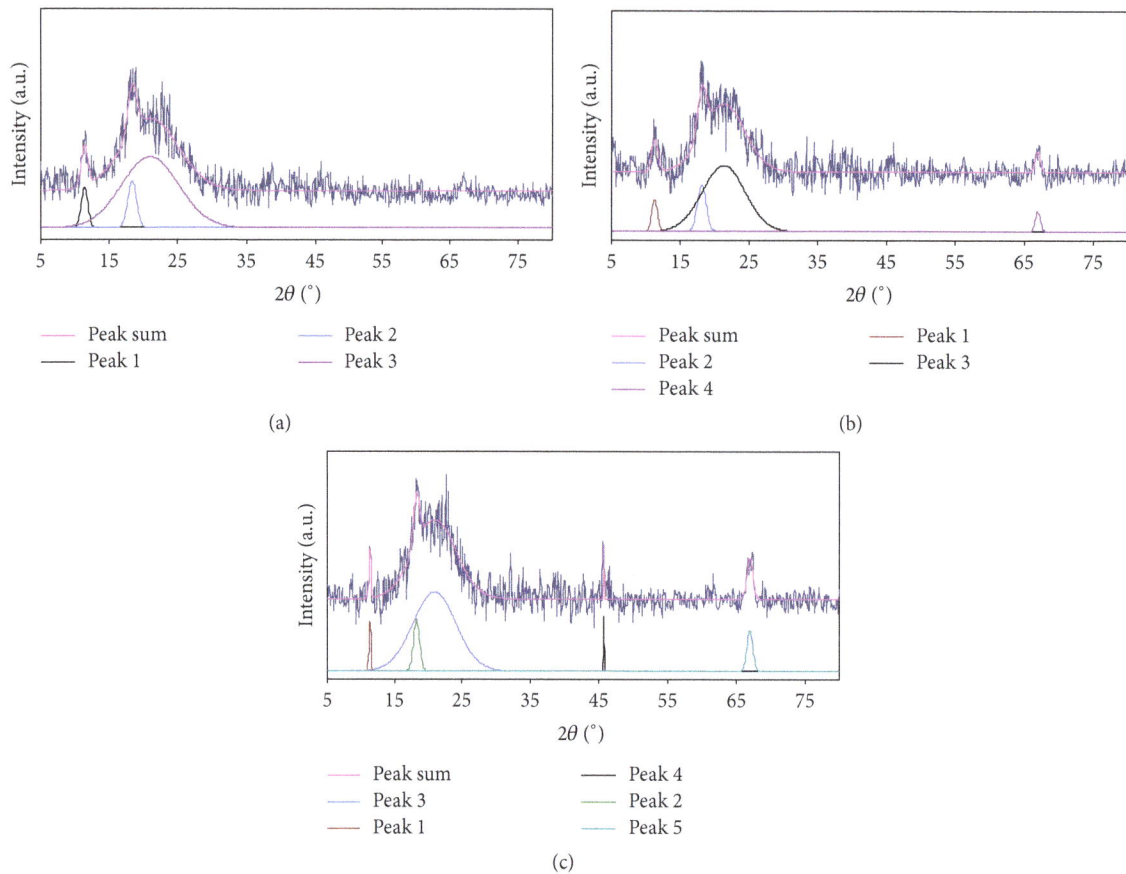

FIGURE 4: Gaussian fitting of XRD for (a) CSNC2, (b) CSNC3, and (c) CSNC5 samples.

peaks of pure Al_2O_3 nanoparticles at 2 and 4 wt.% indicates good dispersion and mixing between the polymer electrolyte and the nano-Al_2O_3 filler. The crystalline peaks at 45.7° and 67.2° degree for NCPEs above 4 wt.% are due to the crystalline peaks of Al_2O_3 nanoparticles.

To further study the crystalline structure of the samples Origin 8 software has been employed to obtain the full width at half maximum (FWHM). From the FWHM the crystallite size of the nanocomposite samples can be estimated by using the Debye-Scherrer formula [22]:

$$L = \frac{0.9\lambda}{\Delta 2\theta_b \cos\theta_b}, \tag{3}$$

where λ is X-ray wavelength, $\Delta 2\theta_b$ is full width at half maximum (FWHM), and θ_b is the peak location. Figure 4 depicts the Gaussian fitting on CSNC2, CSNC3, and CSNC5 samples. The obtained FWHM from Gaussian fitting and the calculated crystallite size from the Debye-Scherrer formula are presented in Table 2. From FWHM, crystallite size (L), and degree of crystallinity (χ) values (Table 2), it can be understood that CSNC2 is more amorphous. These results indicate the fact that high incorporation of Al_2O_3 nanoparticle concentration will increase the crystalline regions as a result of Al_2O_3 nanoparticle aggregation. The additional crystalline peaks at 45.7 and 67.2° for CSNC3 and CSNC5

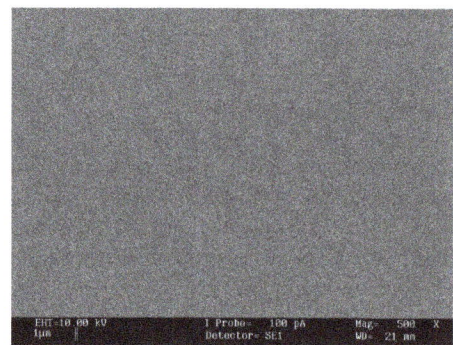

FIGURE 5: Scanning electron microscopy (SEM) image for CS : LiTf (90 : 10) sample.

samples can be ascribed to Al_2O_3 nanoparticle crystalline peaks. The other peaks (the peaks at 32°, 34.4°, 36°, and 39°) of Al_2O_3 nanoparticles also appeared in CSNC3 and CSNC5 samples but with smaller intensities.

3.2. Morphological Study. Figure 5 shows the surface morphology of CS : LiTf (90 : 10) sample. SEM is able to investigate surface structure and one of the advantages is that the range of magnification is wide allowing the investigator to

TABLE 2: $2\theta°$, FWHM, crystallite size (L), and degree of crystallinity (χ) for CSNC2, CSNC3, and CSNC5 NCPEs.

Sample designation	2θ (°)	FWHM (rad)	L (A°)	Degree of crystallinity (%)
CSNC2	21	0.143	10	8.4
CSNC3	21.2	0.108	12	9.7
CSNC5	20.9	0.109	13	12.3

easily focus on an area of interest of the sample [23]. Before observation, the SPE and NCPE films were attached to aluminum holder using a conductive tape and then coated with a thin layer of gold. The SEM characterizations were carried out in order to understand the morphological changes during LiTf addition to chitosan. Figure 5 shows a smooth and almost homogeneous surface for CS : LiTf (90 : 10) sample.

SEM and EDAX are carried out to detect the surface morphology of nanocomposite polymer electrolytes based on $(1 - x)(0.9CS : 0.1LiTf)$-xAl$_2$O$_3$ ($0.02 \leq x \leq 0.1$). The combined usage of SEM-EDX is an attempt to understand the structural and compositional complexity of the samples. Fortunately, the surface morphology is almost smooth and homogeneous for 2 and 4 wt.% of Al$_2$O$_3$ nanoparticle, as depicted in Figures 6(a) and 6(b), which strongly supports the XRD results. However, a large number of white clusters appeared on the surface of NCPEs from 6 to 10 wt.% of Al$_2$O$_3$ nanoparticle (Figures 6(c)–6(e)) indicating that Al$_2$O$_3$ nanoparticles at higher concentrations are not well distributed. Thus the morphology of NCPEs became rough when the concentration of Al$_2$O$_3$ nanoparticles increased beyond 4 wt.% and these results are in good agreement with the XRD results. The sharp peak of Al with a high intensity observed in EDAX results as presented in Figure 6(f) confirms that these white clusters are aggregates of Al$_2$O$_3$ nanoparticles. This again supports the XRD results.

3.3. DC Conductivity and Dielectric Constant Analysis. Figure 7 shows the variation of DC conductivity for CS : LiTf system with various concentrations of alumina nanoparticle. It is obvious that at low filler concentrations (2 to 4 wt.%) the DC conductivity increased. The increase of DC conductivity by four orders can be more understood from the dielectric constant study. However, at high filler concentrations (6 to 10 wt.%) the conductivity decreased. The high value of DC conductivity for $(1 - x)(0.9CS : 0.1LiTf)$-xAl$_2$O$_3$ ($0.02 \leq x \leq 0.1$) nanocomposites with respect to CS : LiTf (90 : 10) solid polymer electrolyte again can be attributable to the effect of alumina nanoparticles. According to Stephan et al. [24], the Lewis acid groups of the added inorganic filler may compete with the Lewis acid-lithium cations for the formation of complexes with the polar groups (NH$_2$ and OH) of chitosan chains as well as the anions of the added lithium salt. The occurrence of reaction between the Lewis acid-base interaction centers of the added filler and the electrolytic species results in lowering the ionic coupling and thus promotes the salt dissociation via a sort of "ion-filler complex" formation. The increase and decrease of DC conductivity can be interpreted in terms of space charge and blocking effect. It was reported that the addition of ceramic filler to the polymer matrix may create a space charge layer at the filler-polymer

interface and thus assists in ion transport [25]. At higher filler concentrations the blocking effect will occur. Consequently, the more abundant alumina grains could make the long polymer chains more "immobilized" leading to a decrease conductivity [26]. This can be more understood from the TEM images for 4 wt.% and 10 wt.% of Al$_2$O$_3$ nanoparticles as depicted in Figures 8 and 9, respectively. Transmission electron microscopy (TEM) is considered the best available imaging technique for analysis of the internal structure and views of the defect structure through direct visualization especially for thin films [27]. At lower filler concentration, the interaction of the filler with the polymer as a result of hydrogen bonding would alter inductive interchain interactions lowering the crystallinity of the macromolecule, enhancing the chain movement as a result of creating a space charge layer at the filler-polymer interface as depicted in Figure 8, and thus increasing the conductivity of the system [28]. However at higher concentration the fillers aggregate as shown in Figure 9, thus leaving the polymer chains to intermolecular and intramolecular hydrogen bonding, and consequently the increase in crystalline portion is expected. This explanation can be more understood from the XRD results. Figure 4(c) clearly showed that at 10 wt.% of of Al$_2$O$_3$ nanoparticles the crystalline peaks of chitosan are enhanced as a result of nanoparticle aggregations.

Figure 10 shows the frequency dependence of dielectric constant for different alumina concentrations. The high value of dielectric constant at low frequency can be ascribed to electrode polarization effect. These high dielectric constants at low frequencies are responsible for the suppression of high frequency dielectric constant for all the compositions to a single bundle. Thus the high frequency region must be plotted separately in another figure.

Figure 11 shows that the high frequency dielectric constant is also greatly influenced by the filler concentration. The high frequency dielectric constant is important because it represents the intrinsic property of the material and is directly related to DC conductivity. It can be seen that the dielectric constant is about 19 and is four times the dielectric constant of CS : LiTf (90 : 10) system which is about 4.8 as shown in Figure 12. Thus the increase of DC conductivity is associated with an increase of dielectric constant.

Figure 13 depicts the concentration dependence of bulk dielectric constant and DC conductivity. It is interesting to note that the behavior of bulk dielectric constant and DC conductivity (Figure 13) with filler concentration is almost the same. Similar trends of these two important parameters with filler concentration certify the existence of a strong relationship between DC conductivity and dielectric constant. Study of these two parameters at different temperature may give further information.

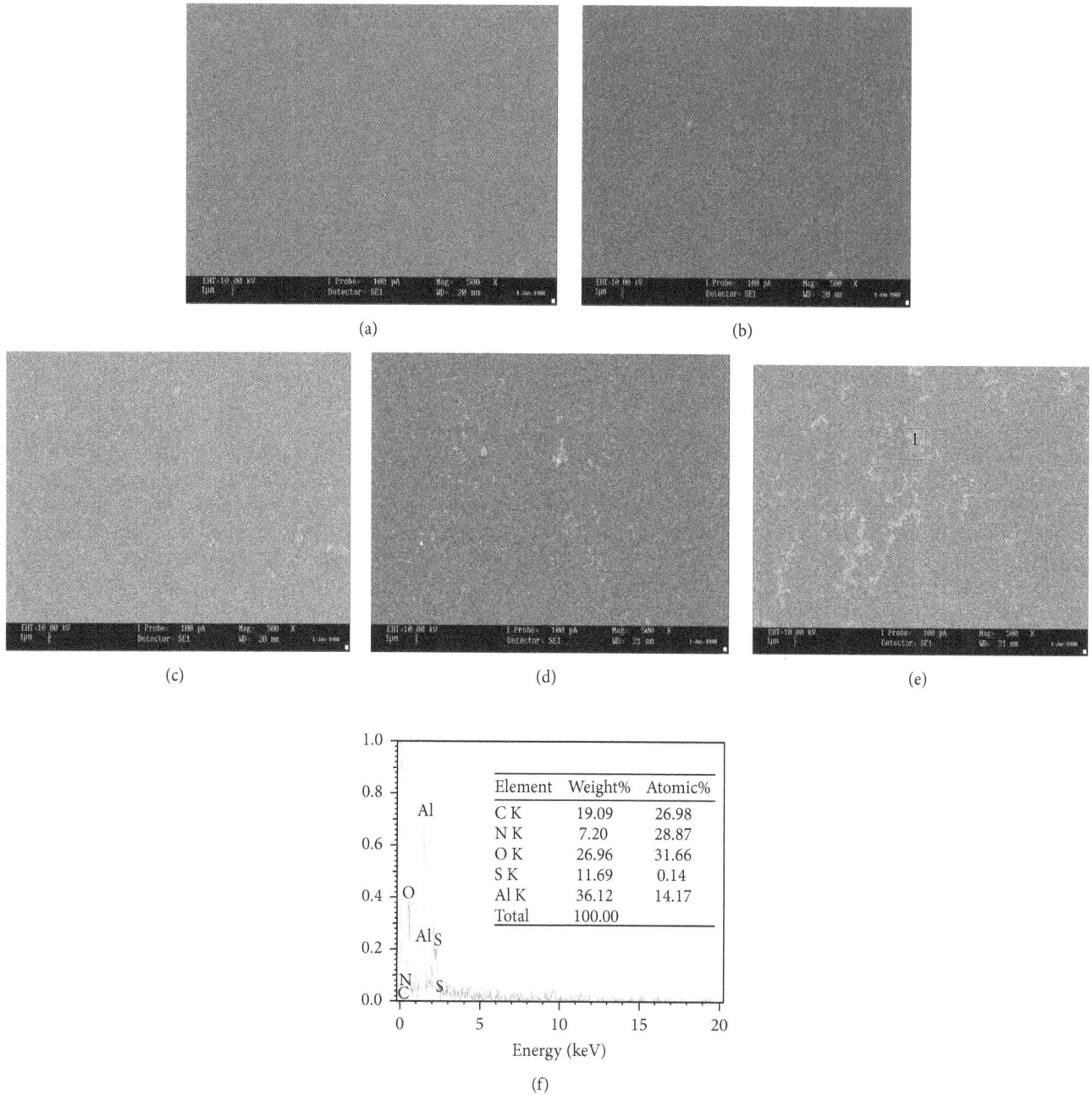

FIGURE 6: Scanning electron microscopy (SEM) image for (a) CSNC1, (b) CSNC2, (c) CSNC3, (d) CSNC4, (e) CSNC5, and (f) EDX for spot in box 1.

Figure 14 shows the smooth curve between DC conductivity and bulk dielectric constant at selected temperatures. It is obvious from the plot that the bulk DC conductivity is smoothly increased with increasing bulk dielectric constant at different temperatures. The concentration dependence of bulk DC conductivity and bulk dielectric constant together with the smooth curve between DC conductivity and dielectric constants confirms the dependence of DC conductivity on dielectric constant in NCPEs based on chitosan. The increase in dielectric constant means an increase in carrier density and hence an increase in DC conductivity. Extra research work is required in this field to confirm the

dependence of DC conductivity on dielectric constant, thus reformulating the Arrhenius equation. Recently, Petrowsky and Frech [29, 30] hypothesized that the DC conductivity is not only a function of temperature but also dependent on the dielectric constant for low salt concentration in organic liquid electrolytes. They have also interpreted the non-Arrhenius behavior of DC conductivity as a result of dependence of preexponential factor, σ_o, on the dielectric constant, $\sigma_{(\varepsilon',T)} = \sigma_{o(\varepsilon'(T))}\exp^{(-E_a/K_BT)}$. The difference of this hypothesis for solid electrolyte is that the preexponential factor, σ_o, is constant. From the results of this work it is understood that ion transport is not an easy task and can be considered

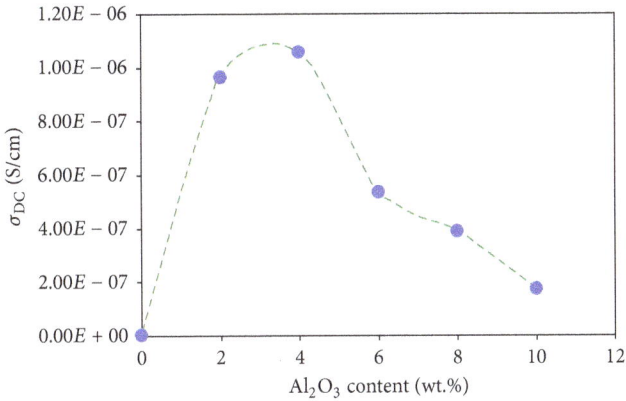

FIGURE 7: The ionic conductivity of chitosan : LiTf (CSC6) with various concentrations of Al_2O_3.

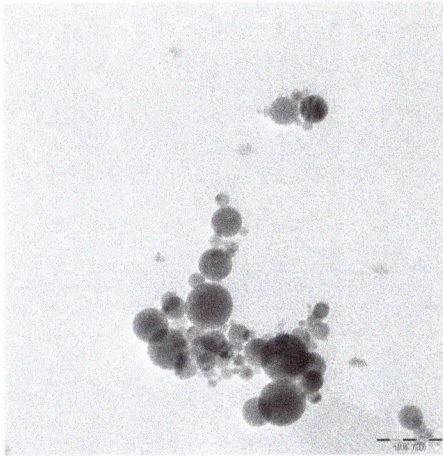

FIGURE 9: TEM image for CSNC5 (10 wt.% of Al_2O_3).

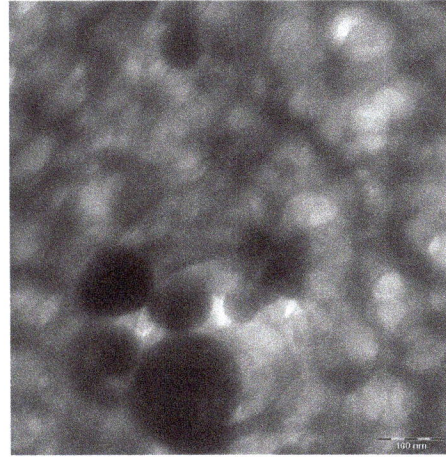

FIGURE 8: TEM image for CSNC2 (4 wt.% of Al_2O_3).

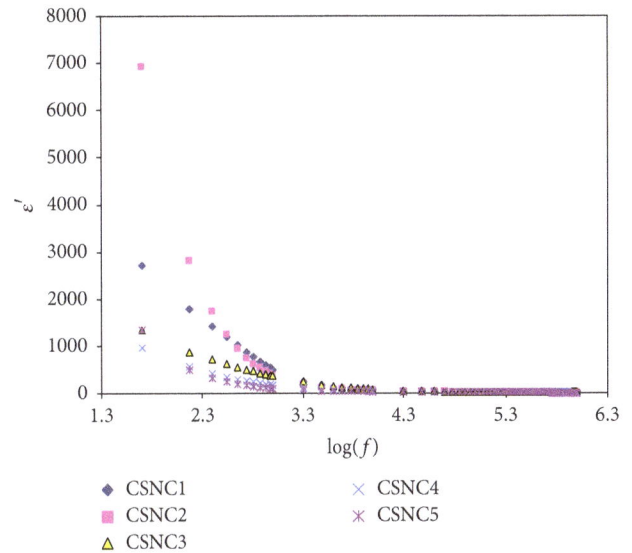

FIGURE 10: Composition dependence of dielectric constant for $(1 - x)(0.9CS : 0.1LiTf)$-xAl_2O_3 ($0.02 \leq x \leq 0.1$) NCPEs at room temperature.

as a complicated subject in solid polymer electrolytes as a branch of condensed matter physics. The temperature-dependent ionic conductivity for a solid or nanocomposite polymer electrolyte below the glass transition temperature is usually interpreted in terms of Arrhenius equation without any insight into the fundamental mechanisms governing ion transport. In our previous work [10, 19], we concluded the fact that the Arrhenius equation should be reformulated to another form which must contain both the temperature and dielectric constant is required. The distinguishable behavior between DC conductivity and dielectric constant at different temperatures as exhibited in Figure 14 motivates me to take a novel approach to explain a DC ionic conductivity. Thus my hypothesis is that in addition to temperature the dielectric constant must exist in Arrhenius equation as follows:

$$\sigma\left(T, \varepsilon'\right) = \sigma_o \exp^{(-E_a/K_B T \varepsilon')}. \tag{4}$$

This new empirical equation which is a modified or reformulated Arrhenius equation must be considered to calculate the activation energy. From the fundamental point of

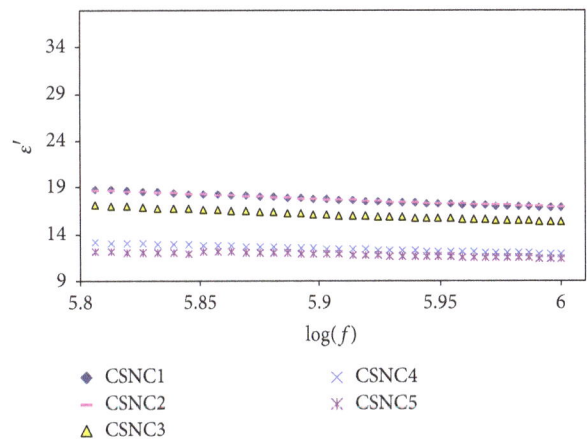

FIGURE 11: Composition dependence of bulk dielectric constant for CS : LiTf (90 : 10)-xAl_2O_3 ($2 \leq x \leq 0.1$) NCPEs.

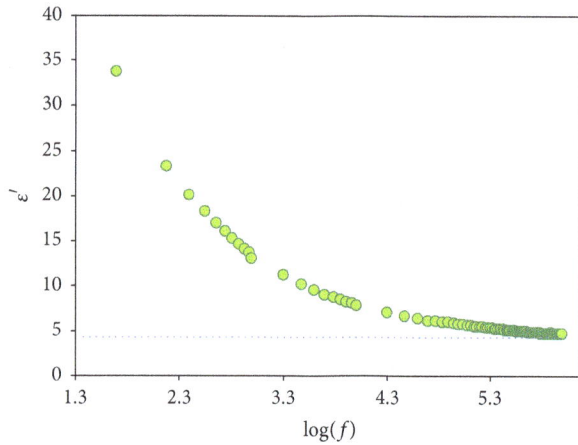

FIGURE 12: Dielectric constant as a function of frequency for CS : LiTf system at room temperature.

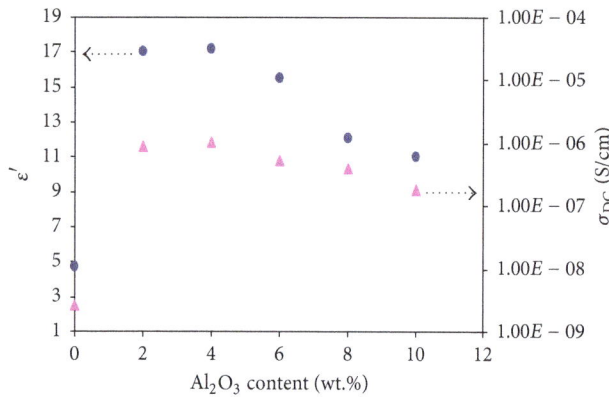

FIGURE 13: Dependence of bulk dielectric constant (ε' at 1 MHz) and DC conductivity of chitosan : LiTf (90 : 10) on Al_2O_3 concentration.

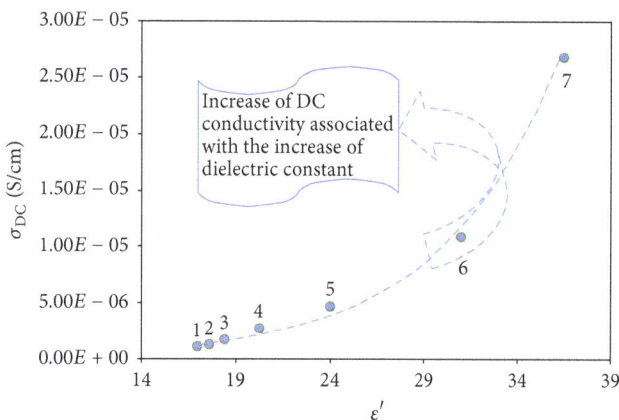

FIGURE 14: DC conductivity dependence on dielectric constant (ε' at 1 MHz) at (1) 303, (2) 308, (3) 313, (4) 318, (5) 323, (6) 328, and (7) 333 K for CSNC2 system.

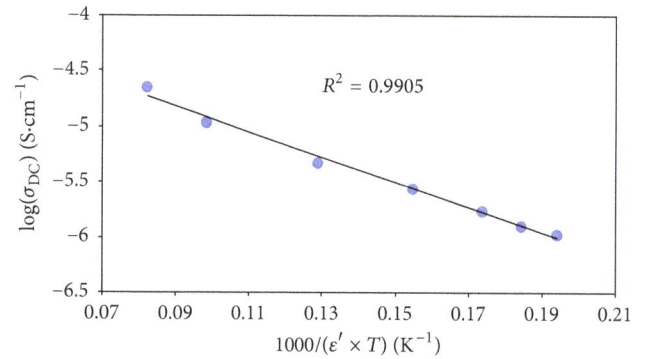

FIGURE 15: Temperature dependence of DC conductivity (reformulated Arrhenius equation) for CSNC2 system.

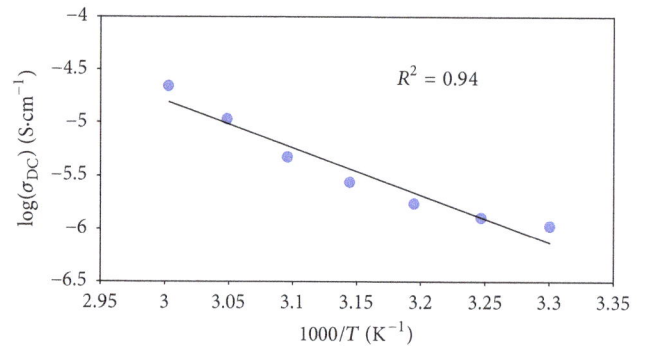

FIGURE 16: Temperature dependence of DC conductivity (Arrhenius equation) for CSNC2 system.

view, dielectric relaxation and ion conduction mechanism in solids are the most intensively researched topics in condensed matter physics and especially ionic transport in polymer electrolytes. Ion transport is complex and depends on factors such as salt concentration and dielectric constant of host polymer [10]. The increase of dielectric constant at higher temperatures can be ascribed to the fact that the movement of polymer chain segments and side groups becomes easier at elevated temperatures and thus the dissociation of ion pairs would increase and result in increase in DC ionic conductivity [19]. Figure 15 shows the DC conductivity versus $1000/(\varepsilon' \times T)$. The linear behavior of DC conductivity (Figure 15) with a regression value of 0.99 reveals the necessary existence of dielectric constant in Arrhenius equation. The DC conductivity versus $1000/T$ as exhibited in Figure 16 is less linear and the regression value is 0.94. The difference between Arrhenius and reformulated Arrhenius equation is distinguishable from Figures 15 and 16. The more linear behavior of Figure 15 reveals the validity of reformulated Arrhenius equation. The main conclusion in this work is that the dielectric constant can play a major role in ion transport and thus studies must be directed to develop or synthesize a high dielectric constant polymer in order to achieve a high DC conductivity at room temperature.

3.4. Relaxation Processes: Electric Modulus Study. Figure 17 depicts the frequency dependence of real part of electric

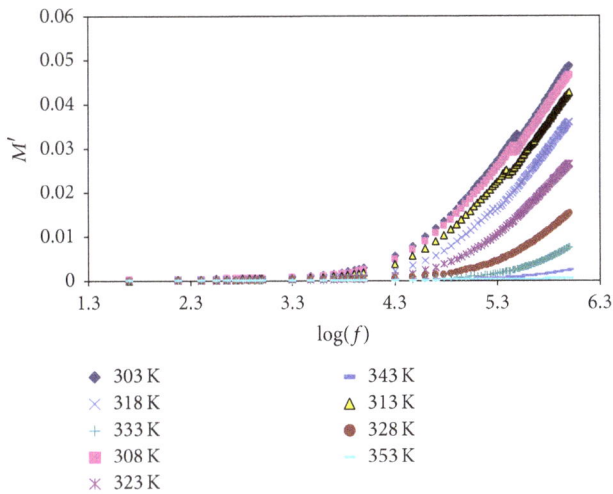

FIGURE 17: Frequency dependence of M' at different temperature for CSNC2 sample.

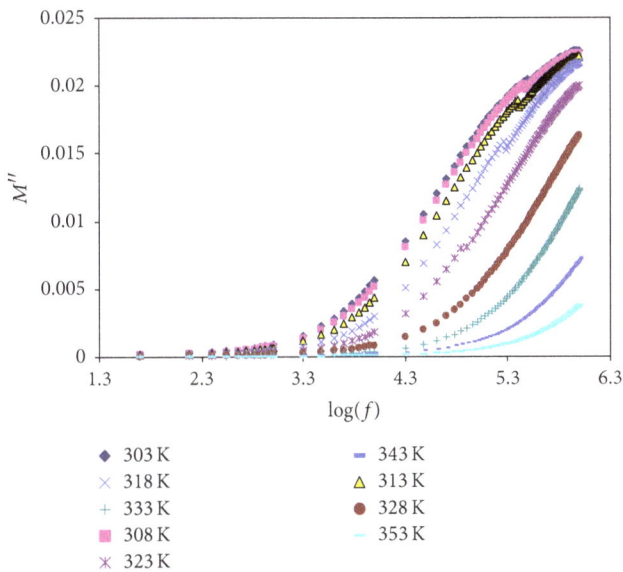

FIGURE 18: Frequency dependence of M'' at different temperature for CSNC2 sample.

the Argand plots exhibit incomplete semicircular arc. This indicates the non-Debye type relaxation process. The Debye model is developed to noninteracting identical dipoles [31]. Thus the non-Debye behavior is due to the fact that in real material there are more than one type of polarization mechanism and a lot of interactions between ions and dipoles. These result in a distribution of relaxation time. Moreover it can be seen that with increasing temperature M''-M' curve deviates more from the semicircular arc. This can be ascribed to the increase in conductivity. With increase in temperature Z_i and Z_r values decrease and thus M'' and M' values ($M'' = \omega C_o Z_r$, $M' = \omega C_o Z_i$) deviate more towards the origin.

4. Conclusions

In this research work solid and nanocomposite polymer electrolytes based on chitosan have been prepared by the well-known solution cast technique. The broadening of XRD peaks for CS : LiTf system reveals the disruption of crystalline nature of chitosan (CS). The deconvolution of the diffractogram of nanocomposite solid polymer electrolytes confirms the increase of amorphous domain for the samples which contains Al_2O_3 nanoparticle up to 4 wt.%. Further incorporation of alumina nanoparticles (6 to 10 wt.% Al_2O_3) results in crystallinity increase (large crystallite size). The SEM analysis reveals smooth surface for the samples containing lower concentration of alumina (2 to 4 wt.% of Al_2O_3). The morphological (SEM and EDX) analysis well supported the XRD results. Similar trends of DC conductivity with Al_2O_3 concentration are related to the behavior of dielectric constant with alumina content. Form the TEM results the phenomena of space charge and blocking effects were well understood. The smooth exponential curve between DC conductivity and dielectric constant at different temperatures indicates the necessary reformulation of Arrhenius equation. The modified or reformulated Arrhenius equation ($\sigma_{(\varepsilon',T)} = \sigma_o \exp^{(-E_a/K_B\varepsilon'T)}$) is achieved from the plot of DC conductivity versus dielectric constant. The more linear behavior of DC conductivity versus $1000/(\varepsilon' \times T)$ reveals the validity of reformulated Arrhenius equation. The incomplete semicircle of Argand plots indicates that the relaxation processes are of non-Debye type.

Competing Interests

The author declares that there are no competing interests regarding the publication of this paper.

Acknowledgments

The author gratefully acknowledges the financial support from University of Malaya in the form of grant (grant no. PS214/2009A) for this research project. The author also wishes to thank the Ministry of Higher Education and Scientific Research/Kurdistan Regional Government for the scholarship awarded.

modulus at different temperatures. The observed long tail at low frequencies is attributable to the large value of electrode polarization capacitance. The decrease in M' value at high temperature is due to the decrease of resistance of the sample and increase in electrode polarization.

Figure 18 shows the variation of imaginary part of electric modulus with frequency at selected temperatures. At low temperatures peaks were observed. The peaks are asymmetric and broad which indicate the distribution of relaxation times. However at high temperatures the peaks disappeared possibly due to the frequency limitation.

The Argand plots at different temperatures for the NCPE (CSNC2) are shown in Figures 19(a)–19(c). It can be seen that

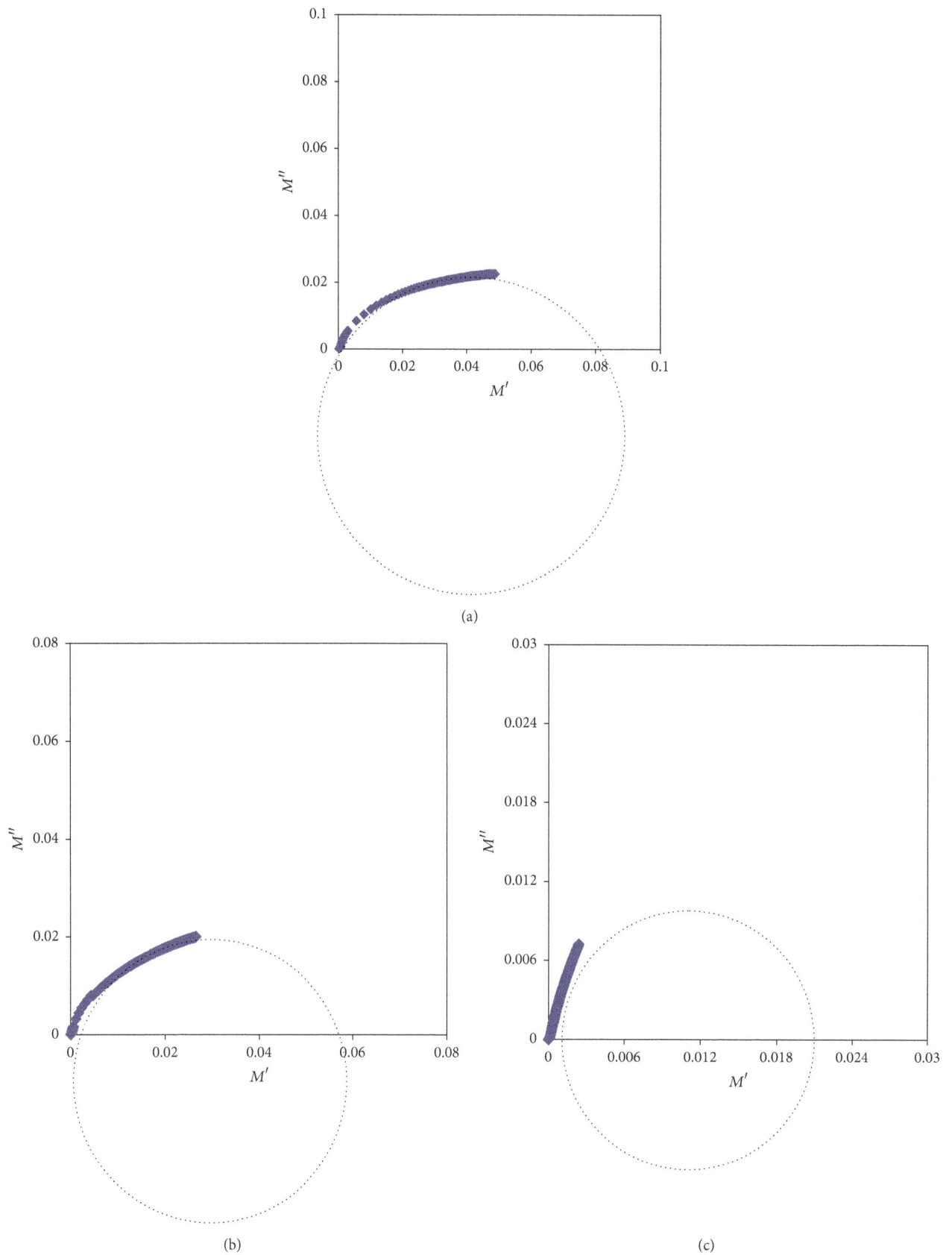

FIGURE 19: Argand plots for NCPE (CSNC2) at (a) 303 K, (b) 323 K, and (c) 343 K.

References

[1] M. Rinaudo, "Chitin and chitosan: properties and applications," *Progress in Polymer Science*, vol. 31, no. 7, pp. 603–632, 2006.

[2] C. G. A. Lima, R. S. de Oliveira, S. D. Figueiró, C. F. Wehmann, J. C. Góes, and A. S. B. Sombra, "DC conductivity and dielectric permittivity of collagen-chitosan films," *Materials Chemistry and Physics*, vol. 99, no. 2-3, pp. 284–288, 2006.

[3] P. Agrawal, G. J. Strijkers, and K. Nicolay, "Chitosan-based systems for molecular imaging," *Advanced Drug Delivery Reviews*, vol. 62, no. 1, pp. 42–58, 2010.

[4] M. Cheng, J. Deng, F. Yang, Y. Gong, N. Zhao, and X. Zhang, "Study on physical properties and nerve cell affinity of composite films from chitosan and gelatin solutions," *Biomaterials*, vol. 24, no. 17, pp. 2871–2880, 2003.

[5] H. Nagahama, H. Maeda, T. Kashiki, R. Jayakumar, T. Furuike, and H. Tamura, "Preparation and characterization of novel chitosan/gelatin membranes using chitosan hydrogel," *Carbohydrate Polymers*, vol. 76, no. 2, pp. 255–260, 2009.

[6] Y. Wan, K. A. M. Creber, B. Peppley, and V. T. Bui, "Chitosan-based solid electrolyte composite membranes: I. Preparation and characterization," *Journal of Membrane Science*, vol. 280, no. 1-2, pp. 666–674, 2006.

[7] C. K. S. Pillai, W. Paul, and C. P. Sharma, "Chitin and chitosan polymers: chemistry, solubility and fiber formation," *Progress in Polymer Science*, vol. 34, no. 7, pp. 641–678, 2009.

[8] N. M. El-Sawy, H. A. Abd El-Rehim, A. M. Elbarbary, and E.-S. A. Hegazy, "Radiation-induced degradation of chitosan for possible use as a growth promoter in agricultural purposes," *Carbohydrate Polymers*, vol. 79, no. 3, pp. 555–562, 2010.

[9] Y. Wan, K. A. M. Creber, B. Peppley, and V. T. Bui, "Ionic conductivity of chitosan membranes," *Polymer*, vol. 44, no. 4, pp. 1057–1065, 2003.

[10] S. B. Aziz and Z. H. Z. Abidin, "Electrical conduction mechanism in solid polymer electrolytes: new concepts to arrhenius equation," *Journal of Soft Matter*, vol. 2013, Article ID 323868, 8 pages, 2013.

[11] M. M. Costa, A. J. Terezo, A. L. Matos, W. A. Moura, J. A. Giacometti, and A. S. B. Sombra, "Impedance spectroscopy study of dehydrated chitosan and chitosan containing LiClO₄," *Physica B: Condensed Matter*, vol. 405, no. 21, pp. 4439–4444, 2010.

[12] S. R. Mohapatra, A. K. Thakur, and R. N. P. Choudhary, "Effect of nanoscopic confinement on improvement in ion conduction and stability properties of an intercalated polymer nanocomposite electrolyte for energy storage applications," *Journal of Power Sources*, vol. 191, no. 2, pp. 601–613, 2009.

[13] S. Mulmi, C. H. Park, H. K. Kim, C. H. Lee, H. B. Park, and Y. M. Lee, "Surfactant-assisted polymer electrolyte nanocomposite membranes for fuel cells," *Journal of Membrane Science*, vol. 344, no. 1-2, pp. 288–296, 2009.

[14] G. A. Kontos, A. L. Soulintzis, P. K. Karahaliou et al., "Electrical relaxation dynamics in TiO₂—polymer matrix composites," *eXPRESS Polymer Letters*, vol. 1, no. 12, pp. 781–789, 2007.

[15] P. Thomas, S. Satapathy, K. Dwarakanath, and K. B. R. Varma, "Dielectric properties of poly (vinylidene fluoride)/CaCu₃Ti₄O₁₂ nanocrystal composite thick films," *eXPRESS Polymer Letters*, vol. 4, no. 10, pp. 632–643, 2010.

[16] E. López-Chávez, J. M. Martínez-Magadán, R. Oviedo-Roa, J. Guzmán, J. Ramírez-Salgado, and J. Marín-Cruz, "Molecular modeling and simulation of ion-conductivity in chitosan membranes," *Polymer*, vol. 46, no. 18, pp. 7519–7527, 2005.

[17] S. B. Aziz, "Li⁺ ion conduction mechanism in poly (ε-caprolactone)-based polymer electrolyte," *Iranian Polymer Journal*, vol. 22, no. 12, pp. 877–883, 2013.

[18] S. B. Aziz, "Study of electrical percolation phenomenon from the dielectric and electric modulus analysis," *Bulletin of Materials Science*, vol. 38, no. 6, pp. 1597–1602, 2015.

[19] S. B. Aziz and Z. H. Z. Abidin, "Ion-transport study in nanocomposite solid polymer electrolytes based on chitosan: electrical and dielectric analysis," *Journal of Applied Polymer Science*, vol. 132, Article ID 41774, 2015.

[20] S. B. Aziz and Z. H. Z. Abidin, "Role of hard-acid/hard-base interaction on structural and dielectric behavior of solid polymer electrolytes based on chitosan-XCF₃SO₃ (X = Li⁺, Na⁺, Ag⁺)," *Journal of Polymers*, vol. 2014, Article ID 906780, 9 pages, 2014.

[21] S. Ahmad and S. A. Agnihotry, "Effect of nano γ-Al₂O₃ addition on ion dynamics in polymer electrolytes," *Current Applied Physics*, vol. 9, no. 1, pp. 108–114, 2009.

[22] S. B. Aziz, Z. H. Z. Abidin, and M. F. Z. Kadir, "Innovative method to avoid the reduction of silver ions to silver nanoparticles (Ag⁺ → Ag°) in silver ion conducting based polymer electrolytes," *Physica Scripta*, vol. 90, no. 3, Article ID 035808, 2015.

[23] C. T. K.-H. Stadtländer, "Scanning electron microscopy and transmission electron microscopy of mollicutes: challenges and opportunities," in *Modern Research and Educational Topics in Microscopy*, pp. 122–131, Formatex, 2007.

[24] A. M. Stephan, K. S. Nahm, M. A. Kulandainathan, G. Ravi, and J. Wilson, "Poly(vinylidene fluoride-hexafluoropropylene) (PVdF-HFP) based composite electrolytes for lithium batteries," *European Polymer Journal*, vol. 42, no. 8, pp. 1728–1734, 2006.

[25] R. Kumar, A. Subramania, N. T. K. Sundaram, G. V. Kumar, and I. Baskaran, "Effect of MgO nanoparticles on ionic conductivity and electrochemical properties of nanocomposite polymer electrolyte," *Journal of Membrane Science*, vol. 300, no. 1-2, pp. 104–110, 2007.

[26] M. A. K. L. Dissanayake, P. A. R. D. Jayathilaka, R. S. P. Bokalawala, I. Albinsson, and B.-E. Mellander, "Effect of concentration and grain size of alumina filler on the ionic conductivity enhancement of the (PEO)₉LiCF₃SO₃:Al₂O₃ composite polymer electrolyte," *Journal of Power Sources*, vol. 119–121, pp. 409–414, 2003.

[27] S. Sinha Ray and M. Okamoto, "Polymer/layered silicate nanocomposites: a review from preparation to processing," *Progress in Polymer Science*, vol. 28, no. 11, pp. 1539–1641, 2003.

[28] M. Moreno, M. A. S. Ana, G. Gonzalez, and E. Benavente, "Poly(acrylonitrile)-montmorillonite nanocomposites. Effects of the intercalation of the filler on the conductivity of composite polymer electrolytes," *Electrochimica Acta*, vol. 55, no. 4, pp. 1323–1327, 2010.

[29] M. Petrowsky and R. Frech, "Temperature dependence of ion transport: the compensated arrhenius equation," *The Journal of Physical Chemistry B*, vol. 113, no. 17, pp. 5996–6000, 2009.

[30] M. Petrowsky and R. Frech, "Salt concentration dependence of the compensated Arrhenius equation for alcohol-based electrolytes," *Electrochimica Acta*, vol. 55, no. 4, pp. 1285–1288, 2010.

[31] R. M. Hill and L. A. Dissado, "Debye and non-Debye relaxation," *Journal of Physics C: Solid State Physics*, vol. 18, no. 19, pp. 3829–3836, 1985.

6

Feasibility of Reprocessing Natural Fiber Filled Poly(lactic acid) Composites: An In-Depth Investigation

Sujal Bhattacharjee and Dilpreet S. Bajwa

Department of Mechanical Engineering, North Dakota State University, Fargo, ND 58108-6050, USA

Correspondence should be addressed to Dilpreet S. Bajwa; dilpreet.bajwa@ndsu.edu

Academic Editor: Luigi Nicolais

Poly(lactic acid) (PLA) based composites are biodegradable; their disposal after single use may be needless and uneconomical. Prodigal disposal of these composites could also create an environmental concern and additional demand for biobased feedstock. Under these circumstances, recycling could be an effective solution, since it will widen the composite service life and prevent the excessive use of natural resources. This research investigates an in-depth impact of recycling on the mechanical and thermomechanical properties of oak wood flour based PLA composites. Two composite formulations (30 and 50 wt% filler), each with 3 wt% coupling agent (PLA-g-MA), were produced and reprocessed six times by extrusion followed by injection molding. Measurements of fiber length and molecular weight of polymer were, respectively, carried out by gel permeation chromatography (GPC). Scanning electron microscopy (SEM), differential scanning calorimetry (DSC), thermogravimetric analysis (TGA), and Fourier transform infrared spectroscopy (FTIR) tools were used to study morphological and molecular alterations. With consecutive recycling, PLA composites showed a gradual decrease in strength and stiffness properties and an increase in strain properties. The 50% and 30% filler concentration of fibers in the composite showed an abrupt decrease in strength properties after six and two reprocessing cycles, respectively.

1. Introduction

Currently, biodegradable polymers and their natural fiber composites are preferred to nonbiodegradable plastics and their composites all over the world [1–3]. This is because biodegradable polymers and their natural fiber composites are not only environment friendly but also capable of higher physicomechanical properties. Nonbiodegradable polymers can create huge environmental pollution at the end of their life cycle (by disposal in landfills and incineration); in addition, the petroleum based polymers have limited reserves [4–7]. A widely popular biodegradable polymer polylactic acid (PLA) is an aliphatic thermoplastic polyester generated by polymerization or polycondensation of monomers of lactic acid and these monomers can be produced from natural resources such as sugar, beet, wheat, and corn [7, 8]. PLA has higher strength and stiffness and is regarded as the best of all environment friendly polymers [5]. It has a wide range of application such as in biomedical field (e.g., drug delivery

devices, sutures, and scaffolds), electronic field (e.g., portable device housings), automotive industry and construction (e.g., interior panels and indoor furnishing), and disposable items (e.g., packaging, plastic bags, disposable cups and plates, and compostable bottles) [4, 9–13]. However, incorporation of fiber in PLA is desired since the polymer itself is not able to meet all the anticipated properties and is relatively expensive [14]. Wood fiber is an attractive filler due to its cost, availability, biodegradability, and higher specific strength, and stiffness [2, 5, 6, 15]. Therefore, incorporation of wood fiber/filler (WF) in PLA is a logical approach to create an economical and completely biodegradable composite.

Unfortunately, PLA and WF based composites have some minor drawbacks. Even though natural fiber reinforced composites have well established commercial market, PLA as matrix resin is yet to be extensively used in commercial engineering applications because of its lower impact resistance, lower softening temperature, higher brittleness, poor moisture resistance, and lower heat deflection temperature

[2, 8, 16–18]. The manufacturing cost of PLA is comparatively higher, and the profitability of manufacturing has not yet been established strongly enough to be trusted [14]. Similarly, WF also has some shortcomings such as lower bulk density, higher heat sensitivity, and its hygroscopic nature [2]. Nonuniform dispersion of filler can result in less stress transfer from the matrix to fiber and can reduce the load carrying capacity of the composite [15]. Nevertheless, despite these challenges, the use of WF as filler in PLA is increasing globally primarily because of their biodegradability and higher specific properties [5].

With increased demand for WF-PLA composites what concerns is their premature disposal after single use. From a rational point of view, disposal of these WPCs after single use could produce a huge amount of waste material, which is unnecessary and uneconomic. Additionally, this extravagant disposal would create more demand for PLA resulting in greater demand for biobased feedstocks. To overcome this situation, reprocessing or recycling could serve as a competent alternative. Reprocessing would enhance the life cycle (service life) of composites and reduce the consumption of the renewable natural resources. As a result, it will have a positive impact on the environment and could improve the economics of PLA based composites.

Some research has been done to investigate the potential of reprocessing PLA based composites [1, 2, 15]. However, most of the research has focused on understanding the impact of blending recycled and virgin PLA polymers and its impact on physicomechanical properties. The studies found that mechanical properties of the composites initially remained constant and then started decreasing with increase in recycled content or number of reprocessing cycles. The degradation was mainly attributed to fiber length reduction and polymer degradation [1, 15, 19].

The aim of this research is to understand the impact of reprocessing on the mechanical and thermomechanical properties of oak wood flour (WF) filled PLA composites. Two PLA based composite formulations with 30 or 50 wt% wood filler, 3 wt% coupling agent were selected and successively reprocessed six times by extrusion. After each extrusion cycle, test samples were prepared from composite pellets by injection molding.

2. Materials and Methods

2.1. Materials. Oak wood flour was obtained from Southern Wood Services, GA, USA. The particle size distribution of wood flour is shown in Table 1. The initial moisture content of wood flour was 5.5% as measured by using a moisture analyzer (Computrac, Model MAX 4000XL, AZ, USA). PLA (Ingeo Biopolymer 2003D) was obtained from NatureWorks LLC, MN, USA. The polymer has a MFI of 6 g/10 min (210°C, 2.16 Kg), specific gravity of 1.24, and heat deflection temperature of 55°C. Coupling agent PLA grafted maleic anhydride (MA) was manufactured in the laboratory using MA and PLA through extrusion process. Maleic anhydride (2,5-furandione) with molecular weight 98.06 and purity greater than 99% was purchased from Sigma Aldrich (St. Louis, MO, USA).

TABLE 1: Particle size distribution of oak wood flour.

Wood particle size (mm)	Range (%)
0.841	0–5%
0.595	5–15%
0.400	35–42%
0.250	40–50%
0.177	6–9%
0.149	0-1%
<0.149	0–2%

2.2. Methods

2.2.1. Composite Preparation. Two PLA composite formulations with 30 and 50 wt% wood flour each and 3 wt% PLA-g-MA were manufactured and then individually recycled six times by extrusion. Before extrusion, WF, PLA-g-MA, and PLA were dried in an oven at 80°C for 24 hours. The moisture content of dried wood flour and PLA before processing was below 0.5%. The dried materials were manually blended and extruded into pellets using a twin screw corotating extruder (Leistritz Micro 18 GL 40 D, NJ, USA) with seven heating zones. The temperature profile of these zones (from feed section to melting section) was controlled at 157°C, 180°C, 190°C, 200°C, 200°C, 202°C, and 205°C, respectively. Temperature of strand die and the gate adapter was maintained at 205°C and extruder screw speed at 200 rpm. Extruded composite strands were cooled down using a water bath and then pelletized. A portion of composite pellets were dried at 80°C and injection-molded using single-screw injection molder (Model SIM- 5080, Technoplas Inc., OH, USA) to make test samples of "cycle 0" or "virgin" composite material.

The rest of the dried virgin composite pellets were again extruded under the same processing conditions and injection-molded to make test samples of "cycle 1" or "first time recycled" composite. This process was repeated five times to derive test samples representing six reprocessing cycles. All the test samples were conditioned at room temperature (25°C) before testing. The annotations for 30 wt% filler and 50 wt% filler composites are as "WF 30 PLA" and "WF 50 PLA" composites, respectively.

2.2.2. Tensile Testing. Tensile test was performed with a universal testing machine Instron (Model 5567, MA, USA) according to ASTM D 638: standard test method for tensile properties of plastics. A 30 KN load cell and strain rate of 5 mm/min were used in the test.

2.2.3. Flexural Testing. Flexural test was conducted using an universal testing machine (Instron Model 5567, MA, USA) with a load cell of 2 KN, according to ASTM D 790: standard test methods for flexural properties of unreinforced and reinforced plastics and electrical insulating materials. The crosshead speed was set at 1.4 mm/min. The sample dimensions were 75 mm × 12.9 mm × 3.3 mm with a support span of 53 mm.

2.2.4. Coefficient of Thermal Expansion (CTE). The coefficient of thermal expansion was performed by using a dynamic mechanical analyzer (TA Instruments, DMA Q800, DE, USA) with the aid of a tension film clamp. The sample dimensions were 38.1 mm × 12.9 mm × 3.3 mm where the initial length was 12.74 mm. Eight samples, for each cycle, were conditioned at room temperature and then tested from 30°C to 50°C with a ramp rate of 3°C /min. CTE was measured by using the following equation [22]:

$$\alpha = \left(\frac{\Delta L}{\Delta T}\right) \cdot \left(\frac{1}{L}\right). \tag{1}$$

Here, α is the coefficient of thermal expansion, ΔL is the change in length, ΔT is the change in temperature, and L is the initial length (12.74 mm).

2.2.5. Heat Deflection Temperature (HDT). Heat deflection temperature was determined by using a dynamic mechanical analyzer (TA Instruments, DMA Q800, DE, USA) according to ASTM D 648: standard test method for deflection temperature of plastics under flexural load in the edgewise position (pressure δ = 0.455 MPa). Sample dimensions were 65.0 mm × 12.9 mm × 3.3 mm and the supported length was 50 mm. The temperature ramp rate was 3°C/min.

2.2.6. Dynamic Mechanical Analysis. Storage modulus was determined by using the dynamic mechanical analyzer (TA Instruments, DMA Q800, DE, USA) with a dual cantilever beam clamp at 30°C. The soak time was 5 minutes (at 28°C) and the ramp rate was 3°C /min. The frequency was 1 Hz and the amplitude was 15 μm. Test sample dimensions were 65 mm × 12.9 mm × 3.3 mm (supported length 41.6 mm).

2.2.7. Izod Impact Test. Izod impact test (notched) was conducted by using an izod impact tester (Tinius Olsen, Model Impact 104, PA, USA) according to ASTM D 256: standard test method for determining the izod pendulum impact resistance of plastics. No additional load was added to the pendulum. Test sample dimensions were 63.5 mm × 12.9 mm × 3.3 mm with a 2 mm notch being made in the middle of the sample.

2.2.8. Melt Flow Index (MFI). Melt flow index was measured by using an extrusion plastometer (Tinius Olsen, Model MP 600, PA, USA) according to ASTM D 1238: standard test method for melt flow rates of thermoplastics by extrusion plastometer. The applied temperature and load were 190°C and 2.16 Kg, respectively.

2.2.9. Scanning Electron Microscopy (SEM). SEM test samples were obtained from the fracture surfaces of tensile specimens. These fracture surfaces were covered with a conductive gold-palladium layer by using a Balzers SCD 030 sputter coater (BAL-TEC RMC, Tucson, AZ, USA). Coated samples were then tested by using a JEOL JSM-6490LV scanning electron microscope (JEOL USA, Peabody, MA, USA). The accelerating voltage was 15 KeV.

2.2.10. Fiber Length Measurement. Wood fiber length of composites was measured by dissolving the polymer in acetone at 130°C, with constant stirring for two hours. After the polymer dissolved, the wood fibers settled at the bottom of the container. Wood fibers were then separated from the solution and dried in an oven at 105°C for 12 h. A Zeiss microscope (Axiovart 40 Mat) was used to evaluate 100 fiber lengths.

2.2.11. Differential Scanning Calorimetry (DSC). DSC was performed by using a TA instrument (DSC, Q1000, DE, USA) from −10°C to 240°C with a nitrogen flow rate of 50 ml/min. The heating or ramp rate was 10°C/min. Each sample weighed approximately 10 mg.

2.2.12. Thermogravimetric Analysis (TGA). TGA was accomplished by using a TA instrument (TGA, Q500, DE, USA) using temperature range from 25°C to 800°C in air as a sample gas (flow rate 60 ml/min) and nitrogen as a balance gas (flow rate 40 ml/min). The heating rate was controlled at 10°C/min. Each sample weighed approximately 10 mg.

2.2.13. Fourier Transform Infrared Spectroscopy (FTIR). A Thermo Scientific Nicolet 8700 spectrometer was used to perform FTIR. The test was done in photoacoustic mode and in the range of 700–3500 cm^{-1}. All the samples had an approximate thickness of 0.5 mm. FTIR data analysis was done by using OMNIC spectra software.

2.2.14. Gel Permeation Chromatography (GPC). The acetone solution containing dissolved PLA produced in Section 2.2.10 was used for GPC. The acetone solution containing PLA was exposed in the open air at room temperature (25°C) to obtain dried PLA. Dried PLA was then dissolved in tetrahydrofuran (THF) to make another solution (concentration 2 mg/ml). This solution was used in a GPC apparatus (EcoSEC HLC-8320GPC, Tosoh Bioscience, Japan) with an injection volume of 20 μl (at 40°C) and an eluent (THF) flow rate of 0.4 ml/min. Two columns (TSKgel SuperHM-L 6.00 mm ID × 15 cm) and a differential refractometer detector (DRI) were also used with the GPC apparatus, to measure the molecular weight of PLA.

2.2.15. Statistical Analysis. Statistical analysis was conducted on the data of mechanical and thermomechanical tests with a confidence level of 95% by using Minitab 17 (Minitab, PA, USA). The mean comparison of different samples was conducted using Tukey's test. Means with different alphabet show significant difference in the properties. Eight replicates from each cycle were tested for all the mechanical properties and five replicates for melt flow index.

3. Discussion

3.1. Effect of Wood Flour Content on Virgin Composites. Higher WF content (50 wt% filler) increased stiffness properties and tensile and flexural modulus but showed no impact on the tensile and flexural strength. On the contrary, higher filler content decreased composite strain properties. No improvement in the strength properties of composites despite higher filler content can be attributed to poor interfacial adhesion between the fiber and polymer matrix. It is possible that higher filler content could produce fiber agglomeration

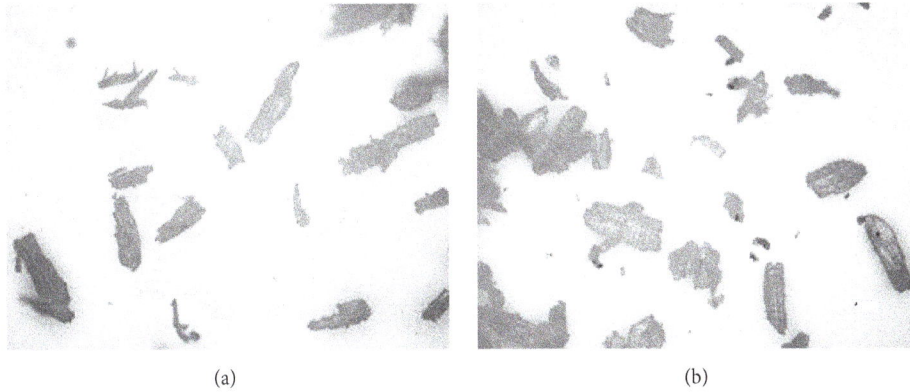

FIGURE 1: Fibers of WF 30 PLA composite at (a) cycle 0 and (b) cycle 6 (magnification 10x).

in the composite matrix resulting in stress concentration, as stress concentration initiates crack propagation which leads to lower composite strength properties [13].

Higher filler composite showed increased tensile, flexural, and storage modulus by 950 MPa, 450 MPa, and 450 MPa, respectively, as compared to lower filler (30 wt% filler) composite. This is because the stiffness properties of biocomposites mainly depend on wood flour content [7, 14]. In higher filler composite, higher constrain to polymer chain movement is created, and hence the composite stiffness is increased [2, 9, 14]. As the composite becomes stiffer, it produces more resistance to mechanical or thermal stresses and thus showed a decrease in strain properties.

Similarly, heat deflection temperature (HDT) properties showed no response to WF loading levels. This is because lower crystalline polymers, such as PLA, show HDT nearly close to their glass transition temperature, while higher crystalline polymers show HDT around their melting temperature [18]. Higher filler content did not play any role in altering the HDT of lower crystalline polymers [23, 24].

3.2. Effect of Recycling on Fiber Length.
The fiber length in WF 30 PLA and WF 50 PLA composites decreased with successive reprocessing. At cycle 0 and cycle 6, the average fiber lengths were, respectively, 253.24 μm and 111.15 μm, which are noticeably lower than the fiber length of original wood flour (Table 1 and Figure 1). This could be due to two probable reasons, lower MFI of PLA (6 g/10 min) and higher screw rpm of the extruder (200 rev/min), both of which can lead to higher shear stress during extrusion which can result in high attrition causing lower average fiber length in the PLA composite. The results are in agreement with previously reported results [1, 22]. Figure 2 shows the normal distribution of fiber lengths of the composite at cycle 0 and cycle 6.

3.3. Effect of Recycling on Molecular Weight of PLA.
Table 2 shows the average molecular weights of WF 30 PLA composite at cycle 0 and cycle 6. Both types of molecular weights, weight averaged (M_w) and number averaged (M_n), showed a decrease with successive recycling. These results agree with the work of other researchers [25]. The polydispersity index (PDI) showed an increase with increased number

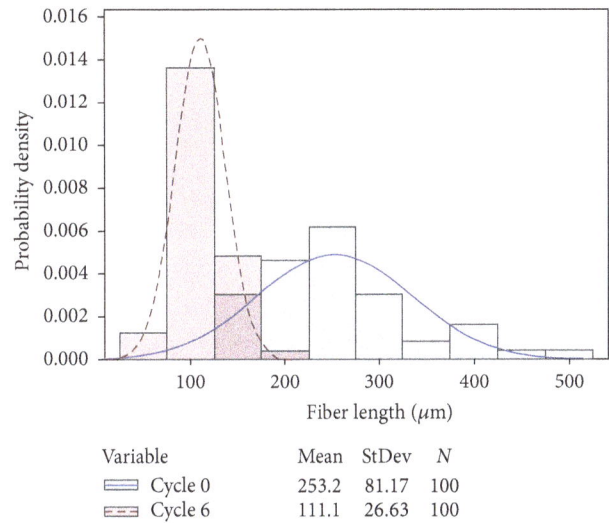

Variable	Mean	StDev	N
Cycle 0	253.2	81.17	100
Cycle 6	111.1	26.63	100

FIGURE 2: Normal distribution plot of fiber length of WF 30 PLA composite at cycle 0 and cycle 6.

TABLE 2: Fiber length and molecular weight of WF 30 PLA composite at cycle 0 and cycle 6.

Property	Cycle 0	Cycle 6
Average fiber length (μm)	253.24	111.15
M_w (g/mole)	151,115	81,958
M_n (g/mole)	133,762	50,742
Polydispersity index (PDI)	1.13	1.62

of reprocessing cycles. This increase in PDI also implies a decrease in molecular weights of PLA [26]. This could be attributed to the breakage of polymer chains due to thermal degradation. Recycling of polymer leads to decrease in M_n resulting in higher PDI.

3.4. Effect of Recycling on Composite Strength Properties.
Strength properties of WF 30 PLA composite (tensile strength, flexural strength, and impact strength) decreased sharply at cycle 6, while they did not exhibit any shift in the previous cycles (Table 3). Strength properties of biocomposites primarily depend on fiber length (or aspect

TABLE 3: Physical and mechanical properties of WF-PLA composites after six reprocessing cycles.

(a) Composite with 30% wood flour

Properties	WF 30 PLA						
	Cycle 0	Cycle 1	Cycle 2	Cycle 3	Cycle 4	Cycle 5	Cycle 6
Tensile strength (MPa)	55.46 (2.96)^{AB}	55.89 (0.73)^A	56.55 (0.68)^A	53.73 (1.53)^{AB}	52.54 (1.49)^{BC}	50.58 (1.52)^C	29.21 (3.23)^D
Flexural strength (MPa)	90.96 (2.54)^A	89.23 (2.21)^A	87.06 (2.72)^A	86.70 (2.70)^A	87.40 (2.08)^A	81.07 (2.94)^B	26.81 (3.93)^C
Impact resistance (J/m)	47.18 (5.48)^A	45.56 (9.62)^A	35.36 (6.09)^B	34.90 (6.05)^B	34.51 (4.45)^B	28.91 (7.17)^B	13.13 (3.73)^C
Tensile modulus (MPa)	6949.95 (332)^A	6361.13 (267.86)^{AB}	6484.70 (490.03)^{AB}	6185.56 (792.19)^{AB}	6533.57 (594.27)^{AB}	5978.74 (541.50)^B	6303.36 (759.82)^{AB}
Flexural modulus (MPa)	5811.89 (273.48)^A	5424.19 (138.78)^B	5368.78 (81.01)^B	5288.02 (205.17)^B	5347.51 (144.87)^B	5256.00 (94.96)^B	4163.17 (74.57)^C
Storage modulus (MPa)	6405.88 (303.48)^A	6179.75 (88.9)^{ABC}	6237.50 (144.73)^A	6365.25 (101.37)^A	5927.38 (302.9)^{BCD}	5844.25 (274.45)^{CD}	5721.88 (271.06)^D
Heat deflection temperature (°C)	56.78 (0.39)^A	56.58 (0.27)^A	56.73 (0.21)^A	56.52 (0.10)^A	56.65 (0.24)^A	56.15 (0.63)^B	54.83 (0.26)^B
Failure strain (%)	1.59 (0.27)^B	2.05 (0.10)^A	2.32 (0.05)^A	2.19 (0.25)^A	2.20 (0.18)^A	2.21 (0.22)^A	0.46 (0.05)^C
Coefficient of thermal expansion (mm/mm/°C) $\times 10^{-5}$	0.68 (0.31)^D	1.22 (0.23)^{BC}	1.31 (0.22)^{BC}	1.19 (0.25)^C	1.63 (0.46)^{ABC}	1.69 (0.24)^{AB}	1.90 (0.33)^A
Melt flow index (g/10 min)	2.85 (0.26)^F	6.45 (0.20)^E	7.84 (0.59)^D	12.00 (0.36)^C	14.26 (0.92)^B	14.62 (0.50)^B	21.09 (0.13)^A

(b) Composite with 50% wood flour

Properties	WF 50 PLA						
	Cycle 0	Cycle 1	Cycle 2	Cycle 3	Cycle 4	Cycle 5	Cycle 6
Tensile strength (MPa)	53.98 (3.61)^A	53.20 (2.90)^A	32.78 (5.17)^B	36.84 (6.19)^B	27.86 (6.08)^B	34.34 (8.15)^B	31.38 (9.56)^B
Flexural strength (MPa)	91.24 (5.81)^A	92.16 (2.71)^A	28.57 (1.45)^C	36.63 (8.83)^B	31.74 (3.73)^{BC}	30.74 (1.87)^{BC}	27.26 (2.16)^C
Impact resistance (J/m)	31.32 (6.13)^A	31.33 (3.65)^A	22.06 (5.46)^B	20.48 (6.19)^B	18.93 (4.90)^B	15.71 (5.45)^B	18.76 (4.15)^B
Tensile modulus (MPa)	8487.95 (698.60)^A	8847.76 (894.24)^A	6925.42 (1428.47)^{AB}	5251.07 (1326)^{BCD}	6156.28 (2333.36)^{BC}	4660.82 (1572.32)^{CD}	3907.93 (725.54)^D
Flexural modulus (MPa)	6685.00 (480.84)^{AB}	5936.28 (547.61)^C	6582.87 (440.04)^{AB}	6850.69 (243.49)^A	6713.83 (104.07)^{AB}	6428.03 (293.98)^{ABC}	6284.67 (135.05)^{BC}
Storage modulus (MPa)	7347.88 (498.78)^A	7286.13 (476.55)^A	7060.13 (825.99)^{AB}	7391 (199.80)^A	7106.13 (653.38)^{AB}	6622.25 (589.05)^{AB}	6317.00 (288.40)^B
Heat deflection temperature (°C)	56.07 (0.33)^A	55.71 (0.59)^{AB}	54.49 (0.43)^D	54.58 (0.33)^D	54.71 (0.33)^{CD}	55.18 (0.28)^{BC}	54.62 (0.20)^{CD}
Failure strain (%)	1.41 (0.15)^{AB}	1.26 (0.18)^B	1.31 (0.52)^B	1.84 (0.43)^{AB}	1.41 (0.24)^{AB}	1.93 (0.45)^A	1.83 (0.55)^{AB}
Coefficient of thermal expansion (mm/mm/°C) $\times 10^{-5}$	1.29 (0.48)^A	0.91 (0.54)^{AB}	0.49 (0.15)^B	0.57 (0.23)^B	0.54 (0.17)^B	0.64 (0.21)^B	0.76 (0.19)^B
Melt flow index (g/10 min)	2.50 (0.21)^F	3.15 (0.12)^{EF}	15.34 (0.99)^E	25.09 (0.55)^D	28.85 (0.59)^C	32.73 (0.56)^B	36.67 (0.58)^A

Values are shown as the mean with standard deviation in parenthesis. Means with different letters are significantly different ($P < 0.05$).

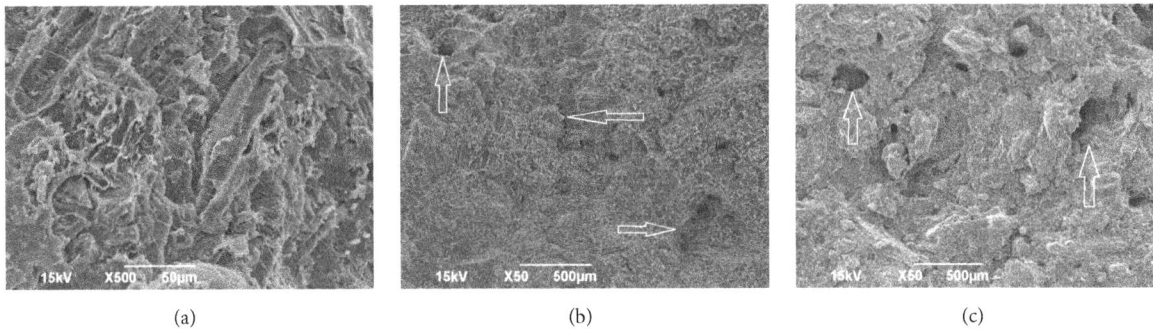

(a)   (b)   (c)

FIGURE 3: SEM of WF-PLA composite at (a) cycle 0 (WF 30 PLA), (b) cycle 6 (WF 30 PLA), and (c) cycle 2 (WF 50 PLA).

TABLE 4: Relative difference in means of properties of WF/PLA composites between cycle 0 and cycle 6.

Properties	WF 30 PLA composite (%)	WF 50 PLA composite (%)
Tensile strength (MPa)	−47.34	−41.86
Tensile modulus (MPa)	−9.30	−53.96
Failure strain (%)	−71.22	29.20
Flexural strength (MPa)	−70.53	−70.12
Flexural modulus (MPa)	−28.37	−5.99
Impact resistance (J/m)	−72.18	−40.10
Heat deflection temperature (°C)	−3.43	−2.58
Storage modulus (MPa)	−10.68	−14.03
Coefficient of thermal expansion (mm/mm/°C)	175.94	−40.93
Melt flow index (g/10 min)	640.37	1364.60

Negative sign shows a decrease and a positive one shows an increase in the value of the property.

ratio), fiber content, void content, interfacial adhesion, and molecular weight of polymer. In this case, fiber length and molecular weight of polymer possibly decreased by a greater extent at cycle 6 than the previous cycles. In addition, crack propagation and fiber pull-outs are observed at cycle 6 that indicate lower interfacial adhesion between the fiber and matrix (Figures 3(a) and 3(b)). This lower interfacial adhesion, attributable to repeated heat and shear stress history due to extrusion, probably results in lower strength properties of the composite [1].

On the contrary, strength properties of WF 50 PLA composite decreased vastly after cycle 2 and then became somewhat constant up to cycle 6 (Tables 3(a) and 3(b)). This could be due to the higher degradation in fiber and polymer, coupled with lower interfacial adhesion due to higher filler content at cycle 2. Microvoids were also observed in the composite matrix at cycle 2 (Figure 3) that could significantly contribute towards lower strength properties. Strength properties of composites were almost constant from cycle 2 to cycle 6, as the number of pores at cycle 2 and cycle 6 remained equivalent.

The sudden change in strength properties for WF 50 PLA at cycle 2 can be related to higher amount of shear stress during extrusion as well as inherent moisture content due to higher amount of wood flour. It is reported that PLA is highly sensitive to shear stress and moisture, which can lead to a rapid decrease in strength properties in highly filled composites [1].

3.5. Effect of Recycling on Composite Stiffness Properties. Composite stiffness properties experimented in this study include tensile modulus, flexural modulus, storage modulus, and heat deflection temperature (HDT). All these properties showed a gradual decrease with increasing number of reprocessing cycles for both composites (Tables 3(a) and 3(b)). This may be due to the less efficient stress transfer from the matrix to fiber, that is, caused by (1) fiber and polymer degradation, (2) increased number of microvoids, (3) less polymer chain entanglement by fibers, and (4) poor fiber-matrix interfacial adhesion with successive recycling [22, 27].

Overall, most stiffness properties showed minor decrease, from cycle 0 to cycle 6 for both formulations of composites, as compared to the corresponding strength properties (Tables 3(a) and 3(b)). For example, after reprocessing six times, flexural modulus for higher and lower filler composite decreased by 6% and 28%, respectively, while both flexural strengths decreased approximately 70% (Table 4). As mentioned earlier (Section 3.1), fiber content is the most important parameter to determine stiffness properties of biocomposites [7, 14]. Since the fiber content was constant for each formulation during reprocessing, the amount of polymer chain entanglement decreased less, which contributed to decrease in composite stiffness properties. These findings are consistent with the findings of other research workers [1, 22].

The tensile modulus of WF 50 PLA composites from cycle 0 and cycle 6 showed a much higher decrease (54%) than the corresponding flexural modulus (6%) (Tables 3(a) and 3(b)).

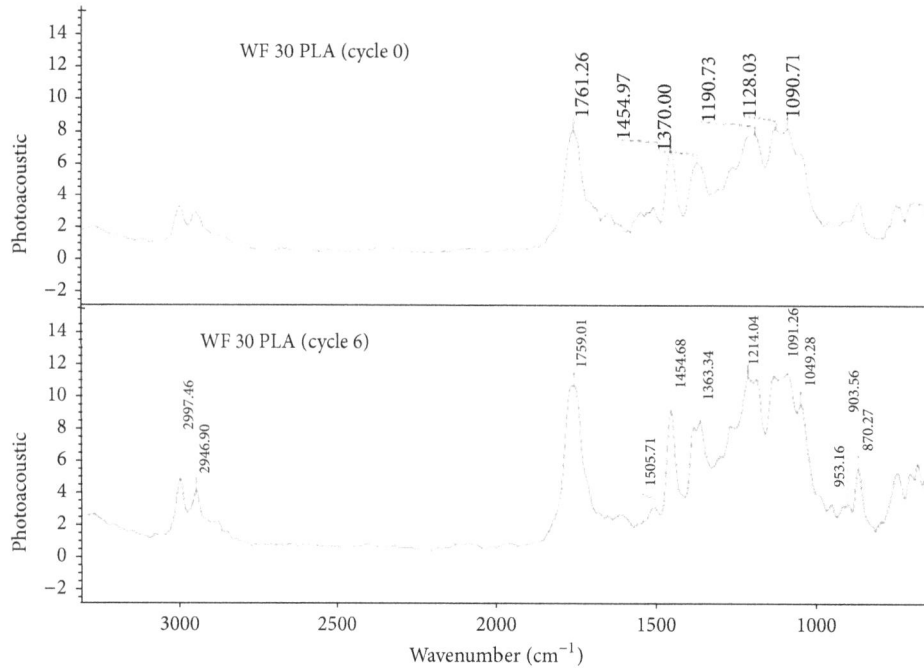

FIGURE 4: FTIR spectra of WF 30 PLA composite at cycle 0 and cycle 6.

TABLE 5: FTIR functional group analysis of WF 30 PLA composite at cycle 0 and cycle 6 [20, 21].

Characteristic absorption (cm^{-1})	Functional group	Indication by functional group
1769	Carboxylic acid	Oxidation
1505	Carboxylic acid	Oxidation
1454	Amine	Weaker bond
1370	Alcohols or phenols	Oxidation
1214	Ether/amine	Oxidation/weaker bond
1091	Ether/amine	Oxidation/weaker bond
1049	Ether/amine	Oxidation/weaker bond
870	Amine	Weaker bond

The difference in the property values could be due to the difference in testing methods since tensile test is thickness dependent and flexural test is surface property dependent [19].

3.6. *Effect of Recycling on Composite Strain Properties.* The impact of reprocessing was investigated on the thermomechanical strain properties of composites such as failure strain, coefficient of thermal expansion (CTE), and melt flow index (MFI). For both composites, most of these strain properties showed an increase with repetitive extrusion cycles (Tables 3(a) and 3(b)). This increase in composite strain is caused by the same factors (or change in factors) that have reduced composite stiffness with successive recycling (Section 3.5).

However, for WF 30 PLA composite, although CTE increased with consecutive recycling, failure strain first increased from cycle 0 to cycle 5 but then sharply decreased at cycle 6 (Table 3(a)). This could be attributed to the fact that failure strain was measured with a tensile load while

CTE was measured with no preload force and is temperature dependent. At cycle 6, the composite may have degraded to the extent that it could not withstand higher load (Table 3(a)) and failed earlier as compared to the previous cycles. As the composite failed earlier, its failure strain decreased instead of increasing. In contrast, CTE showed a higher value at cycle 6 due to the absence of external force that resulted in no failure of the composite.

3.7. *FTIR Analysis.* Figure 4 and Table 5, respectively, show the FTIR spectra and FTIR analysis conducted on WF 30 PLA composite at cycle 0 and cycle 6. This analysis reveals the formation of different functional groups during extrusion that implies composite degradation caused by oxidation and weaker bond formation. Both composites show almost the same characteristic absorptions at cycle 0 and cycle 6. However, because of unequal thickness and heterogeneity of small samples, the relative degradation at cycle 6 as compared to cycle 0 could not be measured.

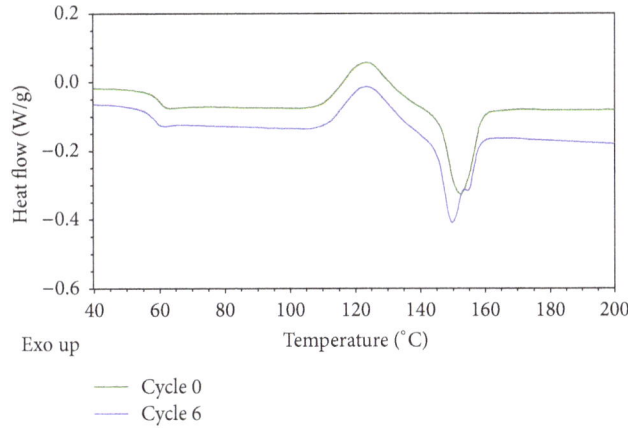

FIGURE 5: DSC analysis curves of WF 50 PLA composite at cycle 0 and cycle 6.

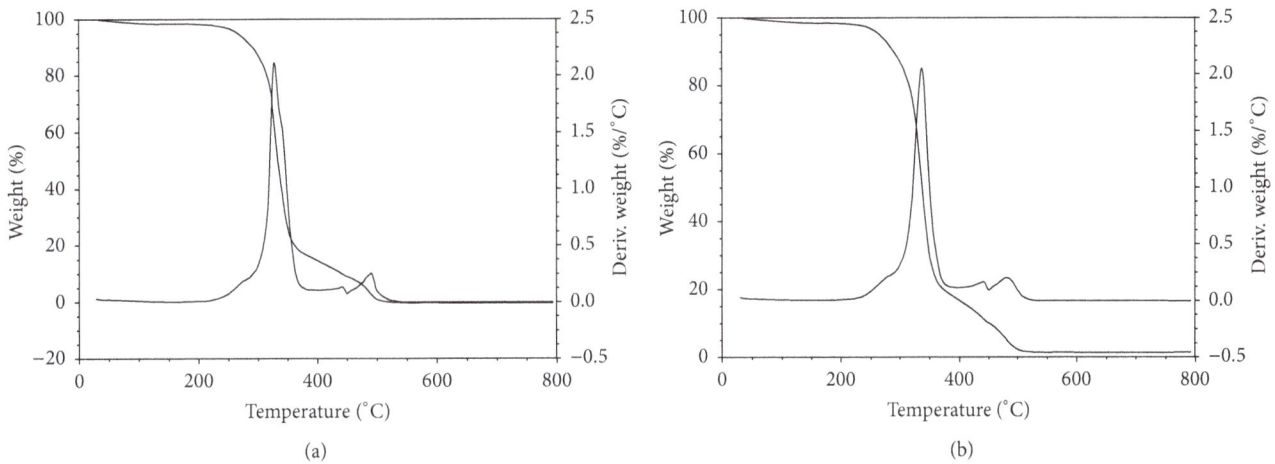

(a)

(b)

FIGURE 6: TGA curve and its 1st derivative of WF 50 PLA composite at (a) cycle 0 and (b) cycle 6.

3.8. Effect of Recycling on Crystallinity and Thermal Stability.

The percentage crystallinity of PLA in the WF 50 PLA composites was measured at cycle 0 and cycle 6 by the following equation [2]:

$$\% \text{ Crystallinity} = \left(\frac{\Delta H_{\text{exp}}}{\Delta H} \right) \times \left(\frac{1}{W} \right) \times 100. \qquad (2)$$

Here, ΔH_{exp} is the experimental heat of fusion provided by DSC, ΔH is the heat of fusion of pure PLA (93.1 J/g), and W is the mass fraction of PLA in the composite. The crystallinity of PLA in composite decreased with increased number of reprocessing cycles (Figure 5 and Table 6). This is because the fibers probably degraded at a relatively higher amount that PLA, at cycle 6. From cycle 0 to cycle 6, the average fiber length decreased approximately 56%, while the weight averaged molecular weight of PLA decreased 45.7% (Table 2). Similarly, the melting temperature of PLA, at cycle 6, decreased with successive reprocessing due to the molecular weight reduction of PLA [22].

However, thermal stability of WF 50 PLA composite increased with consecutive recycling (Table 7 and Figure 6)

TABLE 6: DSC analysis of WF 50 PLA composite at cycle 0 and cycle 6.

Properties	Cycle 0	Cycle 6
Melting point (°C)	152.15	149.64
Crystallization temperature (°C)	145.33	144.24
Heat of fusion (J/g)	16.64	15.42
Crystallinity (%)	38.03	35.24

TABLE 7: TGA of WF 50 PLA composite at cycle 0 and cycle 6.

Property	Cycle 0	Cycle 6
Onset thermal degradation Temperature (°C)	312.42	316.10
Fastest decomposition Temperature (°C)	327.06	337.22
Residue (%)	0	1.35

that was caused by the polymer degradation (molecular weight reduction) at cycle 6 [22]. Lower number of volatiles at cycle 6 could be a probable reason for increased thermal stability of the composites.

4. Summary

Two wood flour/PLA composites were formulated by extrusion, each of which contains 30 or 50 wt% filler, 3 wt% PLA-g-MA, and the rest amount of PLA. Both composites were separately reprocessed six times by extrusion. Samples for mechanical and thermomechanical tests were prepared by injection molding. Both 30 and 50 wt% filler composites showed a sharp decrease in tensile strength, flexural strength, and impact resistance at cycle 6 and cycle 2, respectively. Both composites also showed a decrease in tensile modulus, flexural modulus, and storage modulus with successive recycling, while they exhibited an increase in failure strain, coefficient of thermal expansion, and melt flow index. With consecutive reprocessing, composite thermal stability increased, while polymer crystallinity decreased. To sum up, composite strength and stiffness properties decreased, but strain properties increased with successive recycling due to the fiber length reduction and PLA degradation.

Conflicts of Interest

The authors declare that there are no conflicts of interest regarding the publication of this paper.

Acknowledgments

The authors gratefully acknowledge the financial support from National Science Foundation North Dakota EPSCoR for conducting this research.

References

[1] A. Le Duigou, I. Pillin, A. Bourmaud, P. Davies, and C. Baley, "Effect of recycling on mechanical behaviour of biocompostable flax/poly(l-lactide) composites," *Composites Part A: Applied Science and Manufacturing*, vol. 39, no. 9, pp. 1471–1478, 2008.

[2] S. Pilla, S. Gong, E. O'Neill, L. Yang, and R. M. Rowell, "Polylactide-recycled wood fiber composites," *Journal of Applied Polymer Science*, vol. 111, no. 1, pp. 37–47, 2009.

[3] R. Salehiyan, A. A. Yussuf, N. F. Hanani, A. Hassan, and A. Akbari, "Polylactic acid/polycaprolactone nanocomposite: Influence of montmorillonite and impact modifier on mechanical, thermal, and morphological properties," *Journal of Elastomers and Plastics*, vol. 47, no. 1, pp. 69–87, 2015.

[4] E. Petinakis, L. Yu, G. Edward, K. Dean, H. Liu, and A. D. Scully, "Effect of matrix-particle interfacial adhesion on the mechanical properties of poly(lactic acid)/wood-flour microcomposites," *Journal of Polymers and the Environment*, vol. 17, no. 2, pp. 83–94, 2009.

[5] T. Qiang, D. Yu, and H. Gao, "Wood flour/polylactide biocomposites toughened with polyhydroxyalkanoates," *Journal of Applied Polymer Science*, vol. 124, no. 3, pp. 1831–1839, 2012.

[6] L. Yao, Y. Wang, Y. Li, and J. Duan, "Thermal properties and crystallization behaviors of polylactide/redwood flour or bamboo fiber composites," *Iranian Polymer Journal*, vol. 26, no. 2, pp. 161–168, 2017.

[7] H. Anuar, A. Zuraida, J. G. Kovacs, and T. Tabi, "Improvement of mechanical properties of injection-molded polylactic acid-kenaf fiber biocomposite," *Journal of Thermoplastic Composite Materials*, vol. 25, no. 2, pp. 153–164, 2012.

[8] H. Salmah, S. C. Koay, and O. Hakimah, "Surface modification of coconut shell powder filled polylactic acid biocomposites," *Journal of Thermoplastic Composite Materials*, vol. 26, no. 6, pp. 809–819, 2013.

[9] K. S. Chun, S. Husseinsyah, and H. Osman, "Mechanical and thermal properties of coconut shell powder filled polylactic acid biocomposites: effects of the filler content and silane coupling agent," *Journal of Polymer Research*, vol. 19, no. 5, article 9859, 2012.

[10] K.-W. Kim, B.-H. Lee, H.-J. Kim, K. Sriroth, and J. R. Dorgan, "Thermal and mechanical properties of cassava and pineapple flours-filled PLA bio-composites," *Journal of Thermal Analysis and Calorimetry*, vol. 108, no. 3, pp. 1131–1139, 2012.

[11] K. Hamad, M. Kaseem, H. W. Yang, F. Deri, and Y. G. Ko, "Properties and medical applications of polylactic acid: a review," *Express Polymer Letters*, vol. 9, no. 5, pp. 435–455, 2015.

[12] G. Faludi, G. Dora, K. Renner, J. Móczó, and B. Pukánszky, "Biocomposite from polylactic acid and lignocellulosic fibers: structure-property correlations," *Carbohydrate Polymers*, vol. 92, no. 2, pp. 1767–1775, 2013.

[13] M. Morreale, R. Scaffaro, A. Maio, and F. P. La Mantia, "Effect of adding wood flour to the physical properties of a biodegradable polymer," *Composites Part A: Applied Science and Manufacturing*, vol. 39, no. 3, pp. 503–513, 2008.

[14] A. K. Bledzki, A. Jaszkiewicz, and D. Scherzer, "Mechanical properties of PLA composites with man-made cellulose and abaca fibres," *Composites Part A: Applied Science and Manufacturing*, vol. 40, no. 4, pp. 404–412, 2009.

[15] M. S. Huda, L. T. Drzal, M. Misra, A. K. Mohanty, K. Williams, and D. F. Mielewski, "A study on biocomposites from recycled newspaper fiber and poly(lactic acid)," *Industrial and Engineering Chemistry Research*, vol. 44, no. 15, pp. 5593–5601, 2005.

[16] M. Hrabalova, A. Gregorova, R. Wimmer, V. Sedlarik, M. Machovsky, and N. Mundigler, "Effect of wood flour loading and thermal annealing on viscoelastic properties of poly(lactic acid) composite films," *Journal of Applied Polymer Science*, vol. 118, no. 3, pp. 1534–1540, 2010.

[17] P. Dhar, D. Tarafder, A. Kumar, and V. Katiyar, "Thermally recyclable polylactic acid/cellulose nanocrystal films through reactive extrusion process," *Polymer (United Kingdom)*, vol. 87, pp. 268–282, 2016.

[18] C. Nyambo, A. K. Mohanty, and M. Misra, "Polylactide-based renewable green composites from agricultural residues and their hybrids," *Biomacromolecules*, vol. 11, no. 6, pp. 1654–1660, 2010.

[19] M. S. Huda, A. K. Mohanty, L. T. Drzal, E. Schut, and M. Misra, ""green" composites from recycled cellulose and poly(lactic acid): Physico-mechanical and morphological properties evaluation," *Journal of Materials Science*, vol. 40, no. 16, pp. 4221–4229, 2005.

[20] "Table of ir absorptions," 2016, http://www.Chem.Ucla.Edu/~webspectra/irtable.html.

[21] "Infrared spectroscopy," 2016, https://www2.Chemistry.Msu.Edu/faculty/reusch/virttxtjml/spectrpy/infrared/infrared.htm.

[22] M. D. H. Beg and K. L. Pickering, "Reprocessing of wood fibre reinforced polypropylene composites. Part I: Effects on physical and mechanical properties," *Composites Part A: Applied Science and Manufacturing*, vol. 39, no. 7, pp. 1091–1100, 2008.

[23] V. Nagarajan, K. Zhang, M. Misra, and A. K. Mohanty, "Overcoming the fundamental challenges in improving the impact strength and crystallinity of PLA biocomposites: Influence of

nucleating agent and mold temperature," *ACS Applied Materials and Interfaces*, vol. 7, no. 21, pp. 11203–11214, 2015.

[24] N. Graupner, "Improvement of the mechanical properties of biodegradable hemp fiber reinforced poly(lactic acid) (PLA) composites by the admixture of man-made cellulose fibers," *Journal of Composite Materials*, vol. 43, no. 6, pp. 689–702, 2009.

[25] I. Zembouai, S. Bruzaud, M. Kaci, A. Benhamida, Y.-M. Corre, and Y. Grohens, "Mechanical Recycling of Poly(3-Hydroxybutyrate-co-3-Hydroxyvalerate)/Polylactide Based Blends," *Journal of Polymers and the Environment*, vol. 22, no. 4, pp. 449–459, 2014.

[26] N. Petchwattana, S. Covavisaruch, and J. Sanetuntikul, "Recycling of wood-plastic composites prepared from poly(vinyl chloride) and wood flour," *Construction and Building Materials*, vol. 28, no. 1, pp. 557–560, 2012.

[27] K. B. Adhikary, S. Pang, and M. P. Staiger, "Dimensional stability and mechanical behaviour of wood-plastic composites based on recycled and virgin high-density polyethylene (HDPE)," *Composites Part B: Engineering*, vol. 39, no. 5, pp. 807–815, 2008.

7

Evaluating Mechanical Properties of Few Layers MoS$_2$ Nanosheets-Polymer Composites

Muhammad Bilal Khan, Rahim Jan, Amir Habib, and Ahmad Nawaz Khan

School of Chemical and Materials Engineering (SCME), National University of Sciences and Technology (NUST), H-12 Campus, Islamabad 44000, Pakistan

Correspondence should be addressed to Rahim Jan; rahimjan@scme.nust.edu.pk

Academic Editor: Mikhael Bechelany

The reinforcement effects of liquid exfoliated molybdenum disulphide (MoS$_2$) nanosheets, dispersed in polystyrene (PS) matrix, are evaluated here. The range of composites (0~0.002 volume fraction (V_f) MoS$_2$-PS) is prepared via solution casting. Size selected MoS$_2$ nanosheets (3~4 layers), with a lateral dimension ⟨L⟩ 0.5~1 μm, have improved Young's modulus up to 0.8 GPa for 0.0002 V_f MoS$_2$-PS as compared to 0.2 GPa observed for PS only. The ultimate tensile strength (UTS) is improved considerably (~ ×3) with a minute addition of MoS$_2$ nanosheets (0.00002 V_f). The MoS$_2$ nanosheets lateral dimension and number of layers are approximated using atomic force microscopy (AFM). The composites formation is confirmed using X-ray diffraction (XRD) and scanning electron microscopy (SEM). Theoretical predicted results (Halpin-Tsai model) are well below the experimental findings, especially at lower concentrations. Only at maximum concentrations, the experimental and theoretical results coincide. The high aspect ratio of MoS$_2$ nanosheets, homogeneous dispersion inside polymer, and their probable planar orientation are the possible reasons for the effective stress transfer, resulting in enhanced mechanical characteristics. Moreover, the micro-Vickers hardness (H_V) of the MoS$_2$-PS is also improved from 19 (PS) to 23 (0.002 V_f MoS$_2$-PS) as MoS$_2$ nanosheets inclusion may hinder the deformation more effectively.

1. Introduction

The experimental discovery of graphene, due to its superior mechanical, electrical, thermal, and optical properties, has generated enormous interest in the field of polymer nanocomposites (PNCs) [1]. Graphene holds the best achieved mechanical characteristics with Young's modulus (Y) in the range of ~1 TPa and ultimate tensile strength (UTS) around 130 GPa [2]. The electrical properties are no less amazing with zero band gap, very high carrier mobility (~2×10^5 cm^2/Vs at room temp.), and electrical conductivity (~10^4 – 10^8 S/cm) [3]. With graphene, a surge in 2D materials followed a diverse range of materials with hexagonal boron nitride (hBN) and molybdenum disulphide (MoS$_2$) to be the other most followed layered materials. While hBN is electrically insulating, MoS$_2$ is semiconducting. The direct band gap of around 1.8 eV along with carrier mobility of ~200 cm^2/Vs for monolayer MoS$_2$ makes it go to material for switching and optoelectronics applications [4]. The in-plane stable

structure of 2D materials is common among all, whether graphene, hBN, or MoS$_2$. The in-plane stability is mostly responsible for the extraordinary mechanical characteristics of graphene which are utilized extensively in PNCs. Other 2D materials like hBN and MoS$_2$ have been underutilized for reinforcement purpose; especially MoS$_2$, which has not been specifically assessed for its reinforcements effects in PNCs. Various research groups have tried to simulate as well as experimentally find Young's modulus and strength of single layer MoS$_2$. By using molecular dynamic simulations, Jiang et al. [5] predicted that the in-plane "Y" ~230 GPa which is compromised considering the "Y" for bulk MoS$_2$ is ~240 GPa. The thickness of single layer MoS$_2$ is around 6.092197 Å. Castellanos-Gomez et al. have experimentally studied the few layers (5–25) MoS$_2$ elastic deformations by utilizing atomic force microscope (AFM) bend tests. The "Y" ~330 ± 70 GPa, measured here, is very high and only one-third to "Y" of graphene (1 TPa) while considerably higher than graphene oxide, hBN, and bulk MoS$_2$ [6]. Bertolazzi et al. found the in-plane stiffness of single layer MoS$_2$ ~270 ± 100 GPa along with the average breaking strength to be ~23 GPa [7]. These results show that few layer MoS$_2$ can be employed effectively for the reinforcement purpose as well as an alternating option to graphene in flexible electronics applications [8, 9].

Recently, the mechanical properties of the polyimide (PI) were improved considerably by a slight addition of MoS$_2$ nanosheets. Both the strength and "Y" were enhanced by 43% and 47%, respectively, at 0.75 wt% MoS$_2$ [10]. In another work, chemically modified 4 wt% MoS$_2$-polyurethane composites showed an increment of 140% and 85% in strength and "Y," respectively [11]. Same reinforcement effects have been observed for chitosan with strength being improved to 200% for 0.5 wt% MoS$_2$ [12]. The trend for utilizing MoS$_2$ as reinforcing filler is very compromised in comparison to graphene and graphene oxide. One possible reason may be the lack of high yield production of 2D nanosheets required for PNCs. Coleman group has been able to produce large quantities of defect-free 2D nanosheets with their famous liquid phase exfoliation method [13]. Here in this work we have prepared few layer MoS$_2$ via Coleman method of liquid phase exfoliation and dispersed in polystyrene matrix at various volume fractions (V_f) ranging from 0 to 0.002 V_f. The tensile properties of MoS$_2$-PS composites are measured as a function of MoS$_2$ V_f. The enhanced Young's modulus is evaluated based on two different theoretical models. We report the highest degree of reinforcement (dY/dV_f ~ 320) GPa for MoS$_2$ based polymer composites. The ultimate tensile strength and micro-Vickers hardness of MoS$_2$-PS composites are also enhanced.

2. Experimental Methods

As supplied MoS$_2$ (avg. grain size 6 μm), tetrahydrofuran (THF ≥ 99.9%), 1-methyl-2-pyrrolidinone (≥99.7%), and polystyrene (avg. molecular weight (MW): 35000) used in the work were purchased from Aldrich. MoS$_2$ was exfoliated in 1-methyl-2-pyrrolidinone solvent (20 mg/ml) by using ultrasonication. The probe sonicator (Model: UP50H, 50 watts, 30 kHz) was run at 0.3 cycle and 40% amplitude for 60 hrs at 2°C. After sonication, centrifugation was carried out (500 rpm for 90 min) to size select the exfoliated MoS$_2$ (by taking out the supernatant) and to remove any unexfoliated material. The supernatant was filtered out using nylon membrane (pore size ~ 0.4 μm) and dried at 120°C for 24 hrs. Solution processing method was used for the fabrication of polymer nanocomposites. Both MoS$_2$ (exfoliated) and PS were dispersed in THF with concentrations of 5 mg/ml and 50 mg/ml, respectively. A range of composites were prepared by taking dispersions from both filler and polymer solutions. The composite solutions were bath sonicated for 4 hrs to ensure the homogenous dispersions. The composite dispersions were poured in to Teflon molds. The solvent is evaporated at room temperature for 24 hrs followed by the vacuum drying at 65°C for 72 hrs. Tensile testing was performed using a Shimadzu tensile tester at a strain rate of 5 mm/min. The dimensions of samples (3–5) for each

concentration were ~20 mm in length and ~10 mm in width while the thickness was around 0.05~0.06 mm.

3. Results and Discussion

The main aspect of liquid phase exfoliation of 2D nanosheets lies in the size selection. Ultracentrifugation at various rpm plays a vital role with a basic principal of continually changing the centrifugation speed from higher to lower range and getting the subsequent lower to higher aspect ratio nanosheets [14–17]. The lateral dimension of nanosheets (L) and number of layers per nanosheet (N) can be measured by utilizing Raman spectroscopy, transmission electron microscopy (TEM), and atomic force microscopy (AFM) [18]. Here in this work we have used AFM to estimate the "L" and "N" of MoS$_2$ nanosheets. Few drops of the MoS$_2$/THF were dropped onto the silicon substrate and scanned by utilizing AFM (Jeol SPM 5200) in tapping mode [19]. The average lateral dimension is around 0.5~1 micron while the number of layers is around 3~ 5. These are the average values based on the number of micrographs; a representative AFM image is shown as Figure 1(a). These values are highly approximated but give an indication about the exfoliation state of MoS$_2$. For lateral dimension consideration, scanning electron micrograph is shown in Figure 1(b), supporting the AFM approximations particularly for lateral dimension. The formation of MoS$_2$-PS composites and their crystalline phases were determined with the help of well-established X-ray diffraction (XRD) technique. Powder X-ray diffractometer is used having CuKα (1.54060 Å) as a source of radiation operated (40 mA and 40 kV) at room temperature and 2-theta (θ) ~ 5°–25°. The amorphous peak of PS appears around 2θ ~ 21.3°, while for MoS$_2$ the main peak is clearly visible at 2θ ~ 14.4° shown in Figure 2. The strong 2θ ~ 14.4°[002] peak indicates a well-stacked layered structure for MoS$_2$ [20]. In composite state both polymer and filler peaks are retained at the same positions. Tensile testing results; Young's modulus (Y); and ultimate tensile strength (UTS) are shown in Figures 3(a)-3(b). Strain at break shows an improvement, but considering the error bar it remains constant up to 0.002 V_f MoS$_2$. Young's modulus, on the other hand, increases as a function of MoS$_2$ concentration and reaches up to maximum (0.84 GPa) at 0.0002 V_f MoS$_2$ as compared to the PS (0.21 GPa) shown in Figure 3(a). Beyond this loading (0.0002 V_f MoS$_2$), "Y" decreases a bit but remains constant ~0.6 GPa up to maximum loading (0.002 V_f MoS$_2$). The degree of reinforcement, dY/dV_f, is an indicator utilized for the filler effect on the outcome of PNCs mechanical characteristics [21]. The slope seems to be peaking at 0.0002 V_f MoS$_2$, but the error bar is large enough to consider the whole range of MoS$_2$ concentration for degree of reinforcement. Theoretical explanation for reinforcement can be done based on modified rule of mixtures (MRoM), expressed as [22]

$$Y_C = \left(\eta\eta_0 Y_F - Y_M\right)V_f + Y_M. \qquad (1)$$

In the above equation, Y_C, Y_F, Y_M, and V_f represent Young's modulus of the composite, filler, polymer matrix, and volume fraction of filler, respectively. The length efficiency factor is

(a)

(b)

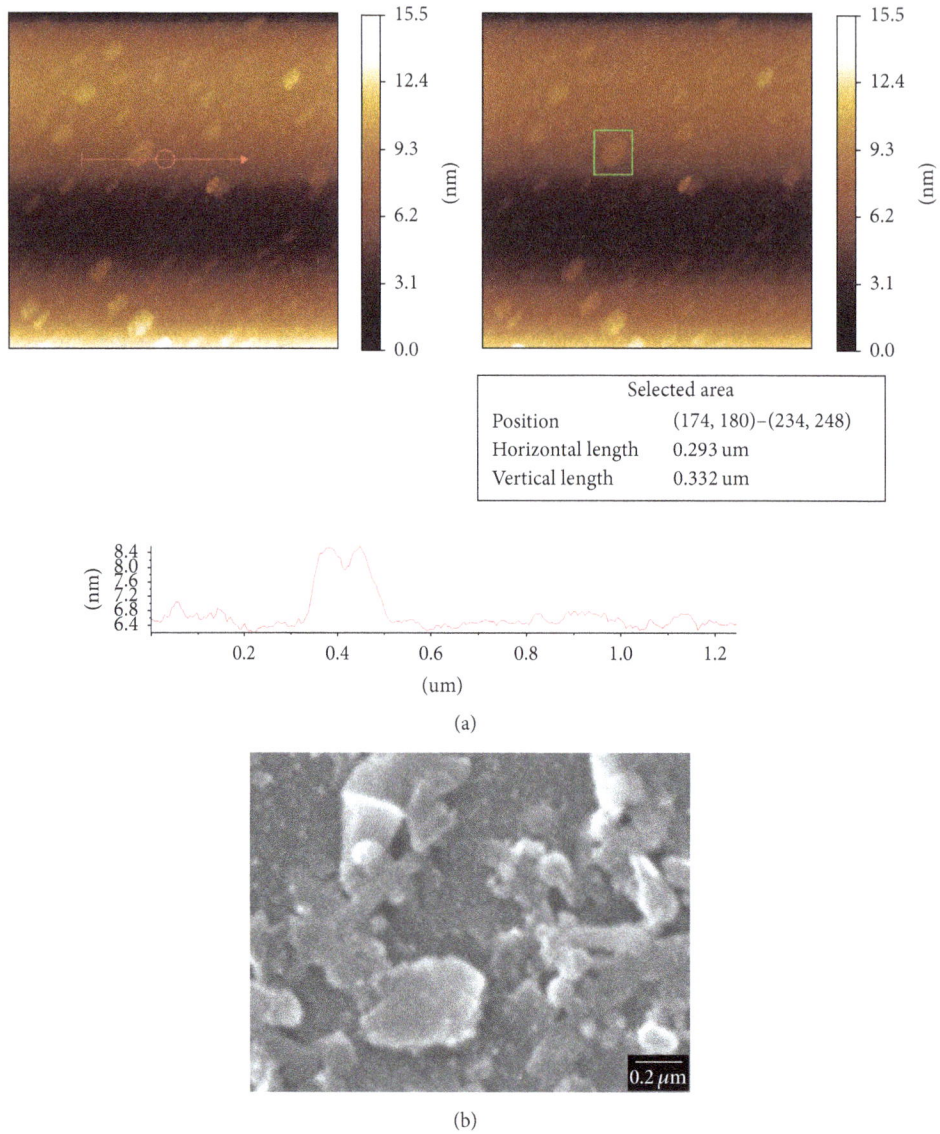

FIGURE 1: (a) Representative AFM image and dimensional analysis for MoS_2 nanosheets length (L) and number of layers (N). (b) Scanning electron micrograph of exfoliated MoS_2 for lateral dimension approximation.

depicted by η, evaluating the matrix-filler stress transfer effect [23].

$$\eta\,(\mathrm{MRoM}) = 1 - \frac{\tanh\,(nL/t)}{nL/t} \quad \therefore n = \sqrt{\frac{G_P V_f}{Y_F\left(1 - V_f\right)}}. \quad (2)$$

The length efficiency factor depends on the filler aspect ratio (L/t) and Young's modulus (Y_F) along with the shear modulus of polymer matrix (G_P). The shear modulus of PS is calculated to be 0.078 by using the relationship $Y_P = 2G_P(1 + \upsilon)$, where υ is the Poisson ratio of PS. Another term in the above-mentioned MRoM expression is η_0, termed as orientation efficiency factor. Generally, 2D materials are randomly oriented in the polymer matrices, having orientation parameter ~0.38 [21]. The filler aspect ratio is an important parameter which can be utilized for estimation Y_C as shown above (see (2)). Both length and thickness of MoS_2 are estimated with AFM. The length is considered ~1 μm while thickness is ~2 nm. Analyzing the present scenario based on MRoM model, the experimentally found Young's modulus value of 0.213 GPa for PS is used. Young's modulus value for MoS_2 is 330 GPa [6, 8]. Overestimating both orientation parameter and length of MoS_2 to be in the range of 0.5~1 and ~2 μm, respectively, still the estimated theoretical Y_C from the MRoM model underestimates Young's modulus of these composites as shown in Figure 3(a). Another approach to predict Young's modulus as a function of filler volume fraction is Halpin-Tsai model. This model can particularly be

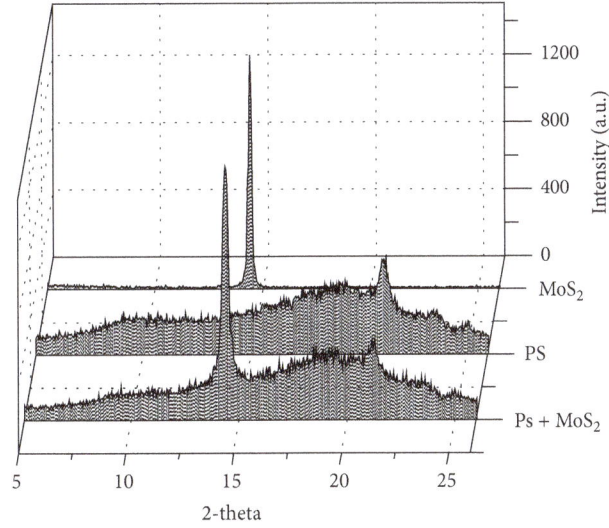

FIGURE 2: XRD results of PS, MoS_2, and MoS_2-PS composite.

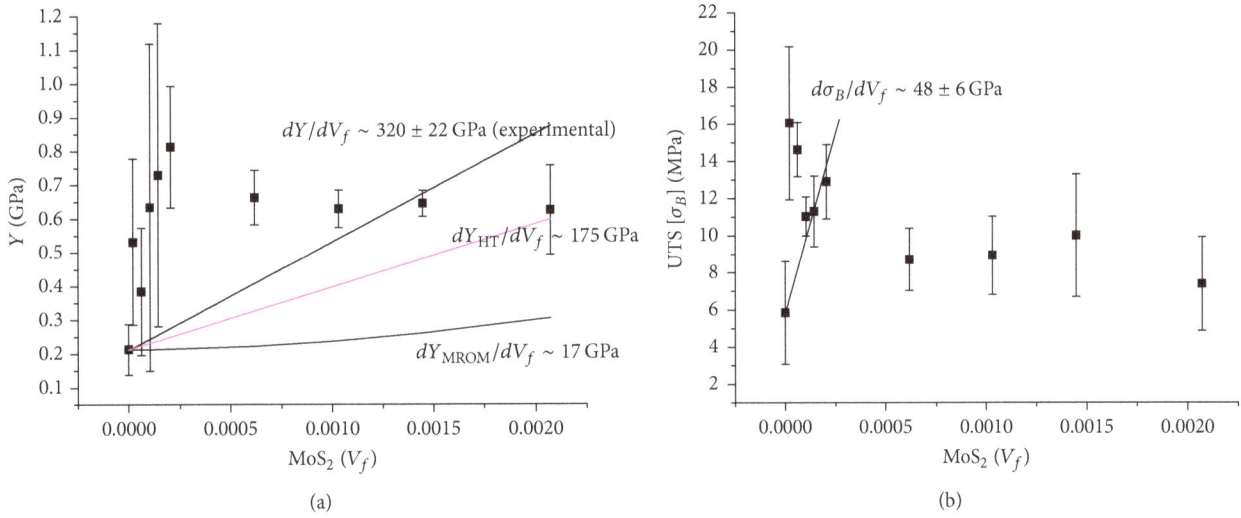

(a)

(b)

FIGURE 3: Modulus and ultimate tensile strength of MoS_2-PS composites as a function of MoS_2 volume fraction.

used for the well dispersed and aligned systems, expressed as under [21–23]

$$Y_C = Y_M \left[\frac{1 + 2V_f \eta L/t}{1 - V_f \eta} \right], \qquad (3)$$

where

$$\eta\,(\mathrm{HT}) = \frac{Y_F/Y_M + 1}{Y_F/Y_M + 2L/t}. \qquad (4)$$

Applying (3) to MoS_2-PS composites, the Halpin-Tsai modelled reinforcement is shown in Figure 3(a). The estimated theoretical results come close to the experimental results, especially when the overestimated MoS_2 length (2 μm) is used. But still these values are compromised as compared to the experimental findings. The rate of increase of Young's modulus (dY/dV_f) is a clear indicator of the reinforcement

trend. The slope (dY/dV_f) of the H-T model is in the range 175 GPa while the experimental dY/dV_f ~ 320 ± 22 GPa (red line) shown in Figure 3(a). The Halpin-Tsai model predicts relatively better than MRoM as the main difference is reported to be in the dissimilar scaling behavior of the length efficiency factor. The experimental dY/dV_f ~320 GPa is almost equal to the few layer (5~25) MoS_2 in-plane stiffness (~330 ± 70 GPa) measured by Castellanos-Gomez et al. [6, 8]. The reason for such reinforcement (dY/dV_f ~ 320 GPa) of MoS_2-PS composites may be the improved dispersion of filler inside polymer, filler alignment, and strong interface between filler and polymer. It would be of great interest if these composites are drawn and then characterized for degree of reinforcement as was done for graphene and hBN based PNCs [19, 21]. UTS of the PS is enhanced considerably with a slight addition of MoS_2 as shown in Figure 3(b). The polymer only sample, when tested for UTS, reached around 4 MPa.

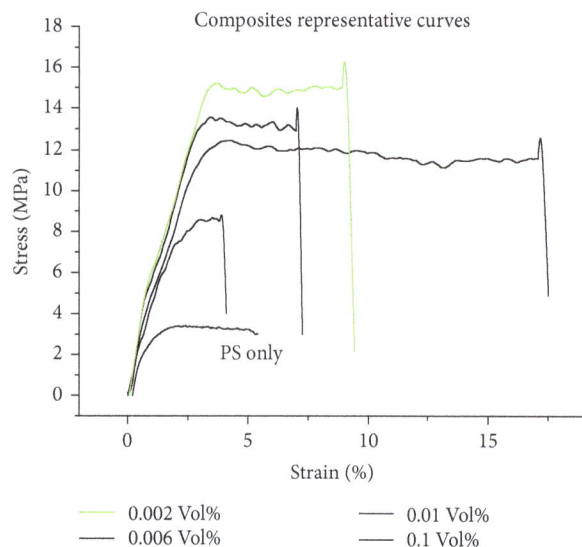

FIGURE 4: Representative stress-strain curves for the MoS_2-PS composites.

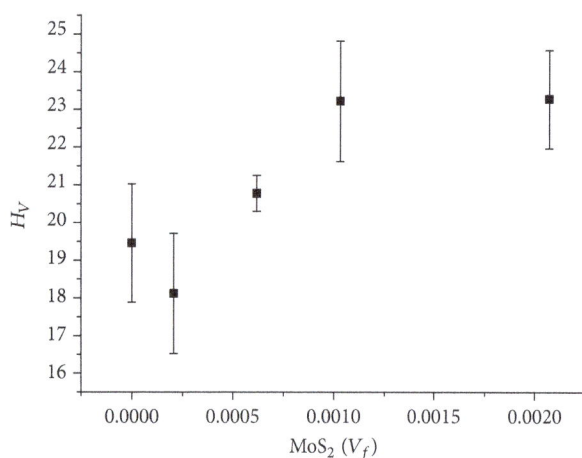

FIGURE 5: Hardness results for MoS_2-PS composites as a function of MoS_2 volume fraction.

By increasing MoS_2 concentration up to 0.0002 V_f, the UTS is well above 10 MPa and reaches up to ~16 MPa. Beyond 0.0002 V_f MoS_2 loading, the UTS decreases but remains around 8~10, well above the PS strength. Agglomeration may be adding up to the slight decrease in UTS at higher loading but seems that these aggregates are dispersed well for the overall effect. Interestingly $d(\sigma_B)/dV_f$ for MoS_2-PS composites is ~48 ± 6 GPa, almost doubled to the measured value of breaking strength of MoS_2 which is 23 GPa found by Bertolazzi et al. [7]. The representative stress-strain curves for the MoS_2-PS composites in various concentrations are shown in Figure 4. Hardness test was carried out with the help of digital Brinell hardness tester (Model: SHB3000C). Maximum increase in hardness value was up to 23.28 H_V at 0.001 V_f MoS_2 concentration as compared to 19.46 H_V of PS matrix as shown in Figure 5. The reason of increase in hardness value is presence of 2D layered structure present inside in composite phase. The reinforcement properties will be of great use for the further utilization of these free standing composite films/membranes. These mechanically robust composite films can be readily utilized for various applications like gas barrier properties and dielectric spectroscopy measurements and for filtration.

4. Conclusion

Liquid exfoliated, few layered, high aspect ratio MoS_2 nanosheets are dispersed in polystyrene matrix for reinforcement purpose. The degree of reinforcement is ~320 GPa, reaching up to the filler's Young's modulus (~330 GPa). The strength at break is also improved considerably as the rate of UTS as a function of MoS_2 volume fraction is around 48 GPa, twice the UTS of MoS_2 nanosheets. This work may lead to more utilization of MoS_2 nanosheets for the plastic reinforcement. Uniaxial drawing of MoS_2-polymer composites may also provide some more insight into the effect of aspect ratio and alignment inside composites.

Conflicts of Interest

The authors declare that they have no conflicts of interest.

References

[1] R. Jan, A. Habib, and I. H. Gul, "Stiff, strong, yet tough free-standing dielectric films of graphene nanosheets-polyurethane nanocomposites with very high dielectric constant and loss," *Electronic Materials Letters*, vol. 12, no. 1, pp. 91–99, 2016.

[2] C. Lee, X. Wei, J. W. Kysar, and J. Hone, "Measurement of the elastic properties and intrinsic strength of monolayer graphene," *Science*, vol. 321, no. 5887, pp. 385–388, 2008.

[3] X. Huang, X. Qi, F. Boey, and H. Zhang, *Chem. Soc. Rev*, vol. 41, pp. 666–686, 2012.

[4] X. Li and H. Zhu, *Journal of Materiomics*, vol. 1, pp. 33–44, 2015.

[5] J. W. Jiang, H. S. Park, and T. Rabczuk, *Journal of Applied Physics*, vol. 114, Article ID 064307, 2016.

[6] A. Castellanos-Gomez, M. Poot, G. A. Steele, H. S. J. Van Der Zant, N. Agraït, and G. Rubio-Bollinger, "Elastic properties of freely suspended MoS 2 nanosheets," *Advanced Materials*, vol. 24, no. 6, pp. 772–775, 2012.

[7] S. Bertolazzi, J. Brivio, and A. Kis, "Stretching and breaking of ultrathin MoS 2," *ACS Nano*, vol. 5, no. 12, pp. 9703–9709, 2011.

[8] A. Castellanos-Gomez, M. Poot, G. A. Steele, H. S. J. van der Zant, N. Agraït, and G. Rubio-Bollinger, "Mechanical properties of freely suspended semiconducting graphene-like layers based on MoS2," *Nanoscale Research Letters*, vol. 7, pp. 1–7, 2012.

[9] S. Bertolazzi, J. Brivio, A. Radenovic et al., *Microscopy and Analysis*, vol. 27, p. 21, 2013.

[10] H. Yuan, X. Liu, L. Ma et al., "Application of two-dimensional MoS2 nanosheets in the property improvement of polyimide matrix: Mechanical and thermal aspects," *Composites Part A: Applied Science and Manufacturing*, vol. 95, pp. 220–228, 2017.

[11] X. Wang, W. Xing, X. Feng et al., "Enhanced mechanical and barrier properties of polyurethane nanocomposite films with randomly distributed molybdenum disulfide nanosheets," *Composites Science and Technology*, vol. 127, pp. 142–148, 2016.

[12] X. Feng, X. Wang, W. Xing, K. Zhou, L. Song, and Y. Hu, "Liquid-exfoliated MoS2 by chitosan and enhanced mechanical and thermal properties of chitosan/MoS2 composites," *Composites Science and Technology*, vol. 93, pp. 76–82, 2014.

[13] K. R. Paton and etal., *Nature Materials*, vol. 13, pp. 623–630, 2014.

[14] U. Khan, A. O'Neill, H. Porwal, P. May, K. Nawaz, and J. N. Coleman, "Size selection of dispersed, exfoliated graphene flakes by controlled centrifugation," *Carbon*, vol. 50, no. 2, pp. 470–475, 2012.

[15] M. Lotya, P. J. King, U. Khan, S. De, and J. N. Coleman, "High-concentration, surfactant-stabilized graphene dispersions," *ACS Nano*, vol. 4, no. 6, pp. 3155–3162, 2010.

[16] A. O'Neill, U. Khan, and J. N. Coleman, "Preparation of high concentration dispersions of exfoliated MoS 2 with increased flake size," *Chemistry of Materials*, vol. 24, no. 12, pp. 2414–2421, 2012.

[17] P. May, U. Khan, A. O'Neill, J. N. Coleman, and J. Mater, *Chem*, vol. 22, pp. 1278–1282, 2012.

[18] C. Backes, K. R. Paton, D. Hanlon et al., "Spectroscopic metrics allow in situ measurement of mean size and thickness of liquid-exfoliated few-layer graphene nanosheets," *Nanoscale*, vol. 8, no. 7, pp. 4311–4323, 2016.

[19] R. Jan, A. Habib, M. A. Akram, T.-U. Zia, and A. N. Khan, "Uniaxial Drawing of Graphene-PVA nanocomposites: improvement in mechanical characteristics via strain-induced exfoliation of graphene," *Nanoscale Research Letters*, vol. 11, no. 1, article 377, pp. 1–9, 2016.

[20] G. Du, Z. Guo, S. Wang, R. Zeng, Z. Chen, and H. Liu, *Chem. Commun*, vol. 46, pp. 1106–1108, 2010.

[21] R. Jan, P. May, A. P. Bell, A. Habib, U. Khan, and J. N. Coleman, "Enhancing the mechanical properties of BN nanosheet-polymer composites by uniaxial drawing," *Nanoscale*, vol. 6, no. 9, pp. 4889–4895, 2014.

[22] Q. Waheed, A. N. Khan, and R. Jan, "Investigating the reinforcement effect of few layer graphene and multi-walled carbon nanotubes in acrylonitrile-butadiene-styrene," *Polymer (United Kingdom)*, vol. 97, pp. 496–503, 2016.

[23] A. N. Khan, Q. Waheed, R. Jan, K. Yaqoob, Z. Ali, and I. H. Gul, "Experimental and theoretical correlation of reinforcement trends in acrylonitrile butadiene styrene/single-walled carbon nanotubes hybrid composites," *Polymer Composites*, 2017.

8

Preparation and Characterization of Nano-Dy$_2$O$_3$-Doped PVA + Na$_3$C$_6$H$_5$O$_7$ Polymer Electrolyte Films for Battery Applications

J. Ramesh Babu ⓘ,[1] K. Ravindhranath ⓘ,[2] and K. Vijaya Kumar[1]

[1]*Department of Physics, K L University, Vaddeswarram, 522 502 Guntur, India*
[2]*Department of Chemistry, K L University, Vaddeswarram, 522 502 Guntur, India*

Correspondence should be addressed to J. Ramesh Babu; jallirameshura@gmail.com

Academic Editor: Frederic Dumur

Composite polymer electrolyte films containing various concentrations of nano-Dy$_2$O$_3$ (1.0 to 4.0%) in PVA + sodium citrate (90 : 10) are synthesized adopting solution cast method and are characterized using FTIR, XRD, SEM, and DSC techniques. The investigations indicate that all components are homogenously dispersed. Films containing 3% of nano-Dy$_2$O$_3$ are more homogenous and less crystalline, and the same is supported by DSC studies indicating the friendly nature to ionic conductivity. Transference number studies reveal that the major charge carriers are ions. With the increase in % of nano-Dy$_2$O$_3$, the conductivity increases and reaches maximum in 3% film with a value of 1.06×10^{-4} S/cm (at 303 K). Further, the conductivity of the film increases with raise in temperature due to the hopping of interchain and intrachain ion movements and fall in microscopic viscosity at the matrix interface of the film. Electrochemical cells are fabricated using these films with the configuration "anode (Mg + MgSO4)/[PVA (90%) + Na$_3$C$_6$H$_5$O$_7$ (10%) + (1–4% nano-Dy$_2$O$_3$)]/cathode (I$_2$ + C + electrolyte)," and various discharge characteristics are evaluated. With 3% nano-Dy$_2$O$_3$ film, the maximum discharge time of 118 hrs with open-circuit voltage of 2.68 V, power density of 0.91 W/kg, and energy density of 107.5 Wh/kg are observed. These findings reflect the successful adoption of the developed polymer electrolyte films in electrochemical cells.

1. Introduction

Solid polymer electrolytes are endowed with characteristics that are intermediate between the solid inorganic electrolytes and liquid electrolytes. These electrolytes by virtue of possessing good ionic conductivity, mechanical strength, and chemical, thermal, and electrochemical stabilities and furthermore good compatibility with the electrode materials are proving to be promising ingredients in electrochemical devices, fuel cells, supercapacitors, solar cells, electrochromic displays, and so on [1–5]. Many investigations are concentrating on improving the conductivity of the solid polymer electrolytes by chemically modifying the surface morphology to promote the formation of protecting layers of low resistivity at the electrolyte-electrode interface. But such surface modified polymer electrolytes are costly. Hence, investigations are being undertaken using multicomponent composite polymer films loaded with salts and inorganic compounds with an aim

to enhance the conductivity by modifying the physicochemical properties of the blended films.

PVA-based polymer films are finding more importance in this regard as they form good films with high mechanical strength, and they are nontoxic. The surface functional groups such as –OH offer easy surface modifications [6–8].

Doping of nanoparticles in the PVA-based films with an aim to enhance the mobility of ions/electrons in the films besides stabilizing the films, is a relevant and recent development [9]. The nanoparticle-studded films are showing improved properties because the less surface area/volume ratio, quantum confinements, and other morphological changes attributed to the nanoparticles are influencing the conductivity of the films. Investigations are reported to the literature by doping nano-Al$_2$O$_3$, TiO$_2$, SiO$_2$, and Fe$_2$O$_3$ in PVA films with respect to different electrical properties [10–24].

In this present investigation, nano-Dy$_2$O$_3$-doped PVA + sodium citrate composite films are synthesized and characterized

TABLE 1: DSC DATA pertaining to T_g, M_p, and % of crystallinity of composite films.

Number	Film composition	T_g	M_p	% of crystallinity
1	Pure PVA	92.98	224.05	100
2	PVA + Na$_3$C$_6$H$_5$O$_7$ + Dy$_2$O$_3$ (90:10:1%)	89.26	223	78.27
3	PVA + Na$_3$C$_6$H$_5$O$_7$ + Dy$_2$O$_3$ (90:10:2%)	87	222	72.50
4	PVA + Na$_3$C$_6$H$_5$O$_7$ + Dy$_2$O$_3$ (90:10:3%)	85	221	65.26
5	PVA + Na$_3$C$_6$H$_5$O$_7$ + Dy$_2$O$_3$ (90:10:4%)	90	223	79.91

TABLE 2: Ionic conductivity, activation energy, and transference numbers of PVA + sodium citrate + nano-Dy$_2$O$_3$ polymer electrolyte films at different compositions.

Number	Polymer electrolyte	Conductivity at 303 K (RT) (S/cm)	Activation energy (E_a)	Transference number	
				t_{ion}	t_{ele}
1	Pure PVA	5.59×10^{-10}	0.42	—	—
2	PVA + Na$_3$C$_6$H$_5$O$_7$ + Dy$_2$O$_3$ (90:10:1%)	5.31×10^{-5}	0.36	0.94	0.06
3	PVA + Na$_3$C$_6$H$_5$O$_7$ + Dy$_2$O$_3$ (90:10:2%)	4.48×10^{-5}	0.28	0.97	0.03
4	PVA + Na$_3$C$_6$H$_5$O$_7$ + Dy$_2$O$_3$ (90:10:3%)	1.06×10^{-4}	0.19	0.98	0.02
5	PVA + Na$_3$C$_6$H$_5$O$_7$ + Dy$_2$O$_3$ (90:10:4%)	3.26×10^{-6}	0.26	0.93	0.07

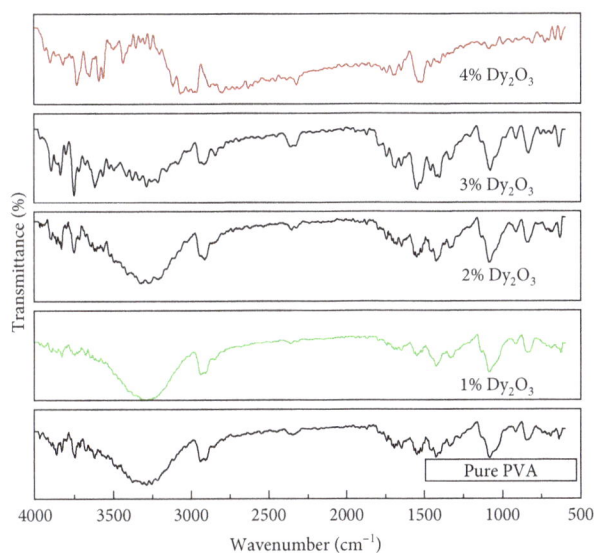

FIGURE 1: FTIR spectra of PVA (90%) + Na$_3$C$_6$H$_5$O$_7$ (10%) + nano-Dy$_2$O$_3$ (1–4%) films.

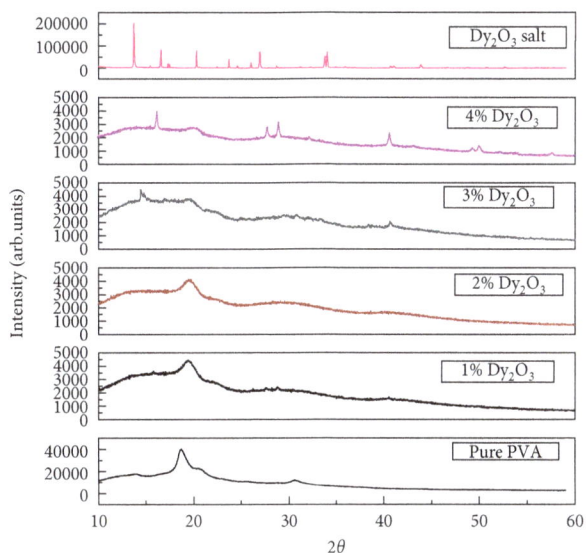

FIGURE 2: XRD pattern of films of PVA (90%) + Na$_3$C$_6$H$_5$O$_7$ (10%) + nano-Dy$_2$O$_3$ PVA (1–4%) along with pure PVA and Na$_3$C$_6$H$_5$O$_7$.

using XRD, SEM, DSC, FTIR, and AC impedance spectroscopy, and their utility in electrochemical cell fabrication is probed.

2. Materials and Methods

All chemicals, namely, PVA (MW 85,000), Na$_3$C$_6$H$_5$O$_7$, and Dy$_2$O$_3$ used in the present work were of AR Grade. Dy$_2$O$_3$ was grinded for 8 hrs in ball milling machine until the size was between 30 and 60 nm. These particles were used in the preset investigation.

The films were casted using solution cast technique [25]. Different proportions of PVA, Na$_3$C$_6$H$_5$O$_7$, and nano-Dy$_2$O$_3$ as detailed in Tables 1 and 2 were added to triple distilled water and stirred for 48 hours to get a homogenous solution.

Then, the solution was poured in Petri dishes, dried at 40°C for 24 hrs, and then vacuum-dried for 24 hrs. The dried films were peeled and characterized with XRD, SEM, FTIR, and DSC, and their electrical properties were measured and presented in Figures 1–7 and Tables 1–3. Further, by using these films as solid polymer electrolytes, electrochemical cells were fabricated, and their characteristics were assessed and presented in Table 4.

FTIR spectra were recorded using Perkin Elmer FTIR Spectrophotometer in the range 4000 to 500 cm^{-1} adopting KBr pellet method (with the resolution of 0.5 cm^{-1}). The spectra obtained are shown in Figure 1. XRD Bruker D8 instrument with Cu Kα radiation for 2θ angles between 10° and 60° (with scan rate 2°/min and step size 0.02°) was used to record spectra of the films, and the obtained spectra are presented in Figure 2.

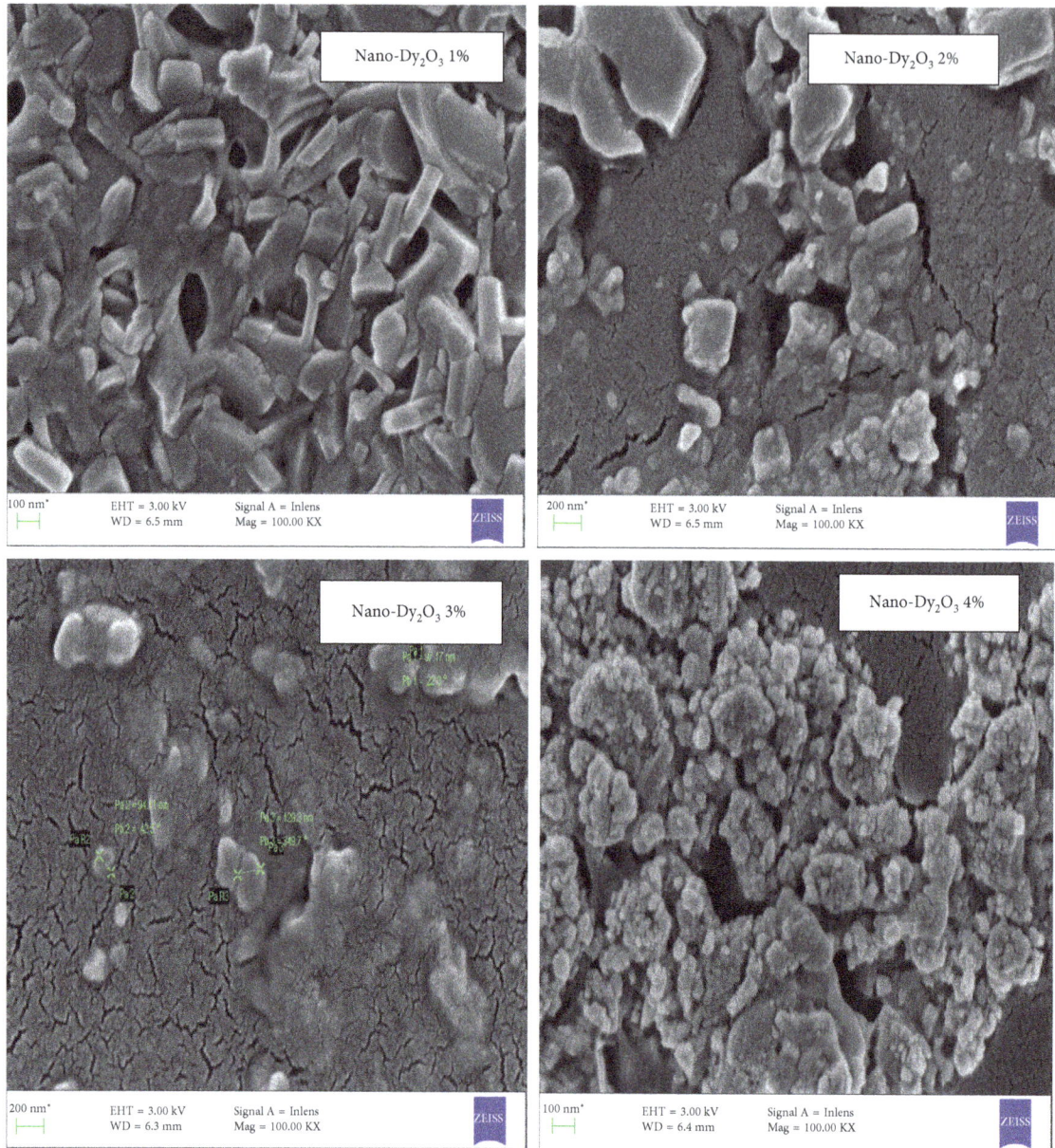

FIGURE 3: SEM images of composite films of PVA (90%) + $Na_3C_6H_5O_7$ (10%) + nano-Dy_2O_3 (1–4%).

The scanning electron microscope (SEM) images were recorded using FE-SEM (Carl Zeiss, Ultra 55 model) and presented in Figure 3.

Differential scanning calorimeter (DSC) thermograms were recorded using DSC Q20 V24.11 Build 124 for the composite films of PVA (90%) + $Na_3C_6H_5O_7$ (10%) + nano-Dy_2O_3 (1–4%) films in determining the glass transition and melting temperatures and % of crystallinity. The results are presented in Figure 4 and Table 1.

The conductivity measurements were carried out using a HIOKI3532-50 impedance analyzer in the frequency range 50 Hz to 1 MHz at various temperatures of 303 K to 333 K. The obtained results are presented in Tables 2 and 3 and Figures 5(a) and 5(b).

The transference number measurements were made using Wagner's polarization technique [26] and Watanabe technique [27], and the results are presented in Table 2 and Figure 6. Electrochemical cells were fabricated with the configuration of "Mg/$MgSO_4$ (anode)/polymer electrolyte/ (I_2 + C + electrolyte) (cathode)." The discharge characteristics of the cell like open-circuit voltage (OCV), short-circuit current (SCC), power density, energy density, current capacity, and other parameters were measured under a constant load of 100 kΩ, and the obtained values are presented in Figure 7 and Tables 4 and 5.

3. Results and Discussion

Various films synthesized were characterized using various surface morphological techniques and are discussed below.

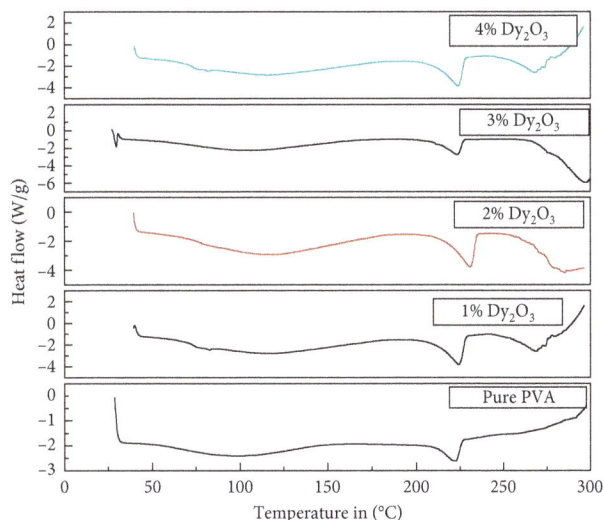

FIGURE 4: DSC curves of pure PVA and composite films of PVA (90%) + Na$_3$C$_6$H$_5$O$_7$ (10%) + nano-Dy$_2$O$_3$ (1–4%).

TABLE 3: Ionic conductivity values of PVA films at different temperatures.

Number	Composition of film: $90:10:3\%$	Conductivity (S/cm)
1	303 K	1.06×10^{-4}
2	313 K	3.86×10^{-4}
3	323 K	6.30×10^{-4}
4	333 K	9.72×10^{-4}

TABLE 4: Cell parameters using the polymer electrolye PVA (90%) + Na$_3$C$_6$H$_5$O$_7$ (10%) + nano-Dy$_2$O$_3$ (1–4%) at constant load of 100 kΩ.

Number	Cell parameters	PVA + Na$_3$C$_6$H$_5$O$_7$ + nano-Dy$_2$O$_3$ $(90:10:1\%)$	PVA + Na$_3$C$_6$H$_5$O$_7$ + nano-Dy$_2$O$_3$ $(90:10:2\%)$	PVA + Na$_3$C$_6$H$_5$O$_7$ + nano-Dy$_2$O$_3$ $(90:10:3\%)$	PVA + Na$_3$C$_6$H$_5$O$_7$ + nano-Dy$_2$O$_3$ $(90:10:4\%)$
1	Open-circuit voltage (V)	1.96	2.37	2.68	2.14
2	Short-circuit current (μA)	230	372	340	260
3	Area of the cell (cm^2)	1.34	1.34	1.34	1.34
4	Weight of the cell (g)	1.40	1.40	1.40	1.40
5	Discharge time (h)	96	108	118	96
6	Power density (W/kg)	0.32	0.62	0.91	0.39
7	Energy density (Wh/kg)	31.55	68	107.5	38.15
8	Current density (μA/cm^2)	171.6	277	253	194
9	Discharge capacity (mA-hr)	22.54	40.17	40.12	24.96

3.1. FTIR Analysis.

The FTIR spectra of the composite films of various compositions: PVA (90%) + Na$_3$C$_6$H$_5$O$_7$ (10%) + nano-Dy$_2$O$_3$ PVA (1–4%), were noted and presented in Figure 1.

The IR data reveal the presence of functional groups like hydroxyl, carbonyl, and carboxylates. The broad peak in pure PVA at 3581–3055 cm^{-1} with center at 3312 cm^{-1} pertains to the stretching frequencies of–OH. The broadness indicates the hydrogen bonding between the surface –OH groups of PVA. With the increase in the % of nano-Dy$_2$O$_3$ in the films, the broadness of the film is decreased, and the position of frequencies are shifted to 3514–3088 cm^{-1} (with center at 3289 cm^{-1}) and 3480–3110 cm^{-1} (with center at 3267 cm^{-1}) with 1 and 2% of nano-Dy$_2$O$_3$, respectively. With films having 3 and 4% of nano-Dy$_2$O$_3$, sharp peaks are

appeared, respectively, at 3602 and 3748 cm^{-1} for 3% film and 3580 and 3737 cm^{-1} for 4% film. The conversion of broad peak to sharp peaks with change in frequencies as the % of nano-Dy$_2$O$_3$ increases may be attributed to the complex formation between the –OH and nano-Dy$_2$O$_3$.

Further, there is a shift of –CH$_2$ frequencies from 2944 cm^{-1} (pure PVA) to 2920 cm^{-1}, 2909 cm^{-1}, 2887 cm^{-1}, and 2954 cm^{-1} as the concentration of Dy$_2$O$_3$ in the film increases from 1 to 4%, respectively. The –C–O– stretching frequencies are also shifted from 1098 cm^{-1} (for pure PVA) to 1073 cm^{-1}, 1062 cm^{-1}, 1051 cm^{-1}, and 1040 cm^{-1} as the % of Dy$_2$O$_3$ is increased from 1 to 4%, respectively.

The frequencies pertaining to carboxylate ions, namely, 1466 and 1544 cm^{-1} (in citrates) are shifted to 1308, 1420, and

(a)

(b)

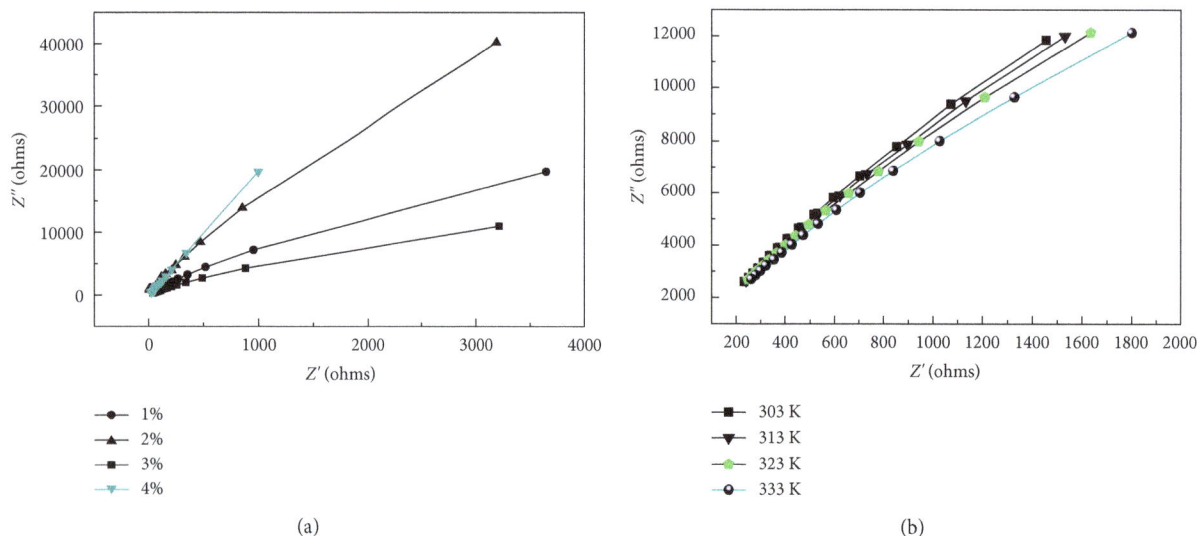

FIGURE 5: (a) Impedance plots of PVA:Na$_3$C$_6$H$_5$O$_7$:nano-Dy$_2$O$_3$ (90:10:1–4%) polymer electrolyte films at room temperature. (b) Conductivity plots for PVA (90%) + Na$_3$C$_6$H$_5$O$_7$ (10%) + nano-Dy$_2$O$_3$ (3%) at different temperatures.

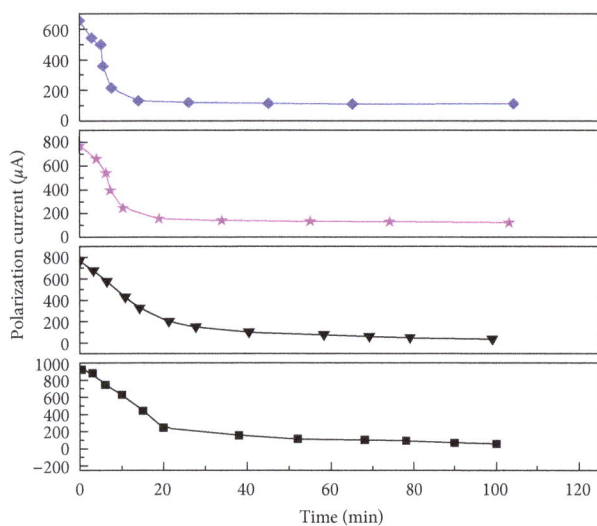

FIGURE 6: Transference number measurements for the composite films PVA (90%) + Na$_3$C$_6$H$_5$O$_7$ (10%) + nano-Dy$_2$O$_3$ (1–4%).

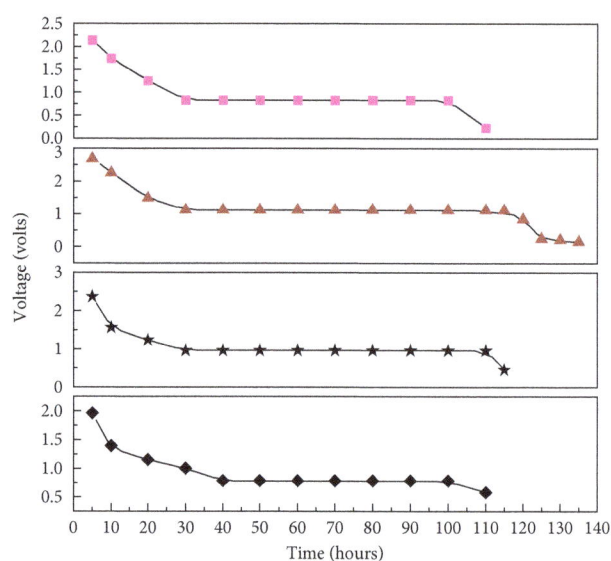

FIGURE 7: Discharge characteristic plots of PVA (90%) + Na$_3$C$_6$H$_5$O$_7$ (10%) + nano-Dy$_2$O$_3$ (1–4%) electrochemical cell for constant load of 100 kΩ.

1555 cm^{-1} for 1%; 1331, 1421, and 1577 cm^{-1} for 2%; 1419, 1432, and 1544 cm^{-1} for 3%; and 1398, 1465, and 1544 cm^{-1} for 4% nano-Dy$_2$O$_3$-impregnated films. Further, carbonyl group frequencies are also progressively increased towards longer wave side: 1655 cm^{-1} for 1%, 1689 cm^{-1} for 2%, 1667 cm^{-1} for 3%, and 1700 cm^{-1} for 4% nano-Dy$_2$O$_3$ films. These changes in the frequency positions and their nature, indicate some kind of complex formation between the groups like –OH, C=O, and –COO$^-$ with the Dy$_2$O$_3$ and thereby results in the uniform and plasticized films.

3.2. XRD.

XRD data of films of different composite are presented in Figure 2.

Pure PVA has broadened the characteristic peak from 18 to 22°, in the 2θ range (110), while pure nano-Dy$_2$O$_3$ has well-defined peaks reflecting its crystalline nature (Figure 2). With the addition of 10% of sodium citrate salt in the PVA film, the peak is broadened and shifted to 19.0° with decrease in intensity [34]. Further, in the binary blended films of PVA + sodium citrate, different nano-Dy$_2$O$_3$ concentrations are doped, and the peaks are shifted with further broadening until the % of nanoparticles is 3%. With 4% Dy$_2$O$_3$ films, again sharp peaks pertaining to Dy$_2$O$_3$ are noted. This indicates that the film has reached saturation at 3% of Dy$_2$O$_3$, and above this concentration, the nanoparticles are precipitating. Further, the broadening of peaks indicate the generation of more amorphous region in the film, which is conducive to the ionic mobility. These changes in the XRD spectrum reflect that the ingredients in the film are homogenously dispersed in the film

TABLE 5: Comparison of present cell parameters with the data of other cells reported earlier.

Number	Solid-state electrochemical cell configuration	Open-circuit voltage (V)	Discharge time (hrs)	Reference number
1	Ag/(PVP + AgNO$_3$)/(I$_2$ + C + electrolyte)	0.46	82	[18]
2	Ag/(PEO + AgNO$_3$)/(I$_2$ + C + electrolyte)	0.61	48	[28]
3	Mg/PEO + Mg(NO$_3$)$_2$/(I$_2$ + C + electrolyte)	1.85	142	[29]
4	K/(PVP + PVA + KBrO$_3$)/(I$_2$ + C + electrolyte)	2.30	72	[30]
5	Mg/PVA + Mg(CH$_3$COO)$_2$/(I$_2$ + C + electrolyte)	1.84	87	[31]
6	K/(PEO + KYF$_4$)/(I$_2$ + C + electrolyte)	2.40	51	[32]
7	K/(PVP + PVA + KClO$_3$)/(I$_2$ + C + electrolyte)	2.00	52	[33]
8	Mg + MgSO$_4$/[PVA (90%) + Na$_3$C$_6$H$_5$O$_7$ (10%) + nano-Dy$_2$O$_3$ (3%)]/(I$_2$ + C + electrolyte)	2.68	118	Present

and that the film containing the 3% of nano-Dy$_2$O$_3$ possesses more amorphous regions and more homogenous nature, indicating the suitability of it for conductivity studies.

3.3. SEM Analysis. The SEM images of the films at different concentrations of nano-Dy$_2$O$_3$ are presented in Figure 3.

It is seen from the images that as the concentration of nano-Dy$_2$O$_3$ is increased from 1 to 3% the edges, corners, gaps and so on are decreasing and thereby the homogeneity, so also the amorphous nature, of the film is increased. The more homogenous film is found with 3% Dy$_2$O$_3$ film. This increasing homogeneousness is attributed to the bonding between the functional groups of the film, namely, –OH and –COO$^-$ with nano-Dy$_2$O$_3$. This bonding may be through the occupation of vacant coordinating sites of Dy$_2$O$_3$ or due to some sort of hydrogen bonding such as "Dy=O . . . H–O" or "Dy–OH . . . O=C–." When the % of Dy$_2$O$_3$ is more than 3%, again edges, boundaries, and holes are appeared, indicating the precipitation of undissolved nano-Dy$_2$O$_3$ in the film. These inferences are supported by XRD data.

3.4. DSC Studies. For assessing the degree of crystallinity, differential scanning calorimeter (DSC) investigations were undertaken, and the obtained results are presented in Figure 4 and Table 1. T_g and M_p were evaluated from graphs (Figure 4).

Percentage of crystallinities of the films were calculated using the equation: % $\chi_c = \{(\Delta H_m)/(\Delta H_m^0)\} \times 100$, where ΔH_m^0 is the melting enthalpy of PVA : sodium citrate (90 : 10) film and ΔH_m is the melting enthalpy of related composite films containing nano-Dy$_2$O$_3$ at varied percentages [35]. The glass transition temperature (T_g), melting temperature (T_m), and relative percentage of crystallinity (% χ_c) values are presented in Table 1.

At 3% of nano-Dy$_2$O$_3$, the crystallinity is lowest, and the film is more homogenous and plasticized with more amorphous nature. This kind of structure provides more paths to proton transport in the film and hence more conductivity.

3.5. Impedance Analysis. The conductivity of the different films was measured by sandwiching PVA (90%) + Na$_3$C$_6$H$_5$O$_7$ (10%) + nano-Dy$_2$O$_3$ (1–4%) films between thin stainless steel plates using HIOKI3532-50 impedance analyzer at temperature 303 K. The conductivity was calculated using the formula: $\sigma = l = R_b/A$ in S/cm, where σ = ionic conductivity, l = thickness

of the polymer electrolyte film, R_b = bulk resistance, and A = area of the stainless steel electrode, contacting the polymer electrolyte film.

Findings are presented in Figures 5(a) and 5(b) and Table 2. Further, the conductivities at various temperatures were measured (using HIOKI3532-50 impedance analyzer) for the PVA (90%) + Na$_3$C$_6$H$_5$O$_7$ (10%) + nano-Dy$_2$O$_3$ PVA (3%) film in view of the fact that the said film was showing higher conductivity.

It is observed from Figure 5(a) that the absence of high-frequency semicircle portion reflects the charge carriers are ions. The presence of nano-Dy$_2$O$_3$ in the films has remarkably increased the conductivity of the films. The conductivity of pure PVA is of the order 10^{-10} S/cm [28]. Increase in conductivity is found with increase in concentration of nano-Dy$_2$O$_3$, and the conductivity reached maximum at 3% and any further increase causes the decrease of conductivity: 5.31×10^{-5} with 1%; 4.48×10^{-5} with 2%; 1.06×10^{-4} with 3%, and 3.26×10^{-6} with 4%.

The nano-Dy$_2$O$_3$ by virtue of possessing high surface area, quantum confinements, paramagnetic nature, and bonding tendencies of accepting electrons (Lewis acidic nature) from donor atoms of the surface functions groups influences the conductivity of the composite films [36]. Moreover, the bonding results in cross-linking PVA + Na$_3$C$_6$H$_5$O$_7$ segments through the bonding tendencies of nano-Dy$_2$O$_3$ and thereby generates additional pathways in the film that complements the ionic movement [22–24]. Thus, acquired homogeneity in more plasticized films brings some kind of order in the films, and this facilitates easy movement of the charge carriers that results in the enhancement of conductivity. It seems that PVA + Na$_3$C$_6$H$_5$O$_7$ (90 : 10) film is statured with nano-Dy$_2$O$_3$ at 3% and any further increase results in the precipitation of the nanoparticles in the pathways, thereby blocking short ways in the polymer matrix for the movement of ions. This results in low conductivity. It may be noted that the activation energy is minimum in the composite film having 3% of nano-Dy$_2$O$_3$ with a value of 0.19 V (Table 1).

Further, the bulk resistance decreases with increase in temperature, and this results in the enhancement of the conductivity. At 303 K, the conductivity is 1.06×10^{-4} S/cm, while at 333 K, it is 9.72×10^{-4} S/cm (vide Table 3). This increase in conductivity may be attributed to the hopping of interchain and intrachain ion movements and also due to decrease in microscopic viscosity at interface in solid polymer electrolyte matrixes.

3.6. Transference Numbers.

For the classification of polymer electrolyte films, ionic transference number is considered to be one of the most significant parameters. By using Wagner's polarizing technique, the transference numbers for the films were measured by constructing a system of Mg/(PVA $+ \text{Na}_3\text{C}_6\text{H}_5\text{O}_7 + \text{Dy}_2\text{O}_3$)/C and polarizing it at 303 K at a constant dc potential of 1.5 V in order to assess the contributions of ions and electrons to the entire conductivity of the polymer electrolyte films. The equations used were

$$t_{\text{ion}} = \frac{I_{\text{initial}} - I_{\text{final}}}{I_{\text{initial}}} = \frac{I_{\text{total}} - I_{\text{electronic}}}{I_{\text{total}}} = \frac{I_{\text{ionic}}}{I_{\text{total}}}, \quad (1)$$

where I_{initial} is the initial current and I_{final} is the final current. The ionic transference number (t_{ion}) values were in the range 0.93–0.98. The results are presented in Figure 6 and Table 2.

It is seen from Table 2 that the charge carriers are predominantly ions in all the composite films, and the contribution of electrons is very minute. For the composite films containing 3.0% of Dy_2O_3, more conductivity is observed and in fact, at that composition, the film has low activation energy and crystallinity. Further, the ionic transference numbers (t_{ion}) of the films are near to unity, indicating their suitability as a polymer electrolyte for solid-state electrochemical cells [25–27].

3.7. Discharge Studies.

The solid-state electrochemical cells were fabricated with the configuration "anode (Mg $+ \text{MgSO}_4$)/[PVA (90%) $+ \text{Na}_3\text{C}_6\text{H}_5\text{O}_7$ (10%) + nano-Dy_2O_3 (1–4%)]/cathode ($\text{I}_2 + \text{C} + \text{electrolyte}$)." The thickness of both the electrodes was 1 mm, while the surface area and thicknesses of the "PVA $+ \text{Na}_3\text{C}_6\text{H}_5\text{O}_7$ + nano-Dy_2O_3" polymer electrolyte were 1.34 cm^2 and 150 μm, respectively. The presence of carbon in the cathode increases the conductivity, and polyelectrolyte reduces the resistance by allowing more interfacial contact between the cathode and electrolyte [35]. The discharge characteristics, namely, open-circuit voltage (OCV), short-circuit current (SCC), current density, power density, energy density, discharge time, and discharge capacity of the cell for a constant load of 100 kΩ were evaluated at room temperature and are shown in Figure 7 and Table 4.

It is seen from the data that the film with 3% of nano-Dy_2O_3 shows better discharge characters: discharge time: 118 h; energy density: 107.5 Wh/kg; power density: 0.91 W/kg; and open-circuit voltage: 2.68 V. These indicate that these composite films loaded with optimum amounts of nano-Dy_2O_3 can be successfully adopted as polyelectrolyte films in the battery applications.

4. Comparison with Previous Work

The electrochemical cell developed in this work is compared with the works available in the literature, and a comparative statement is presented in Table 5.

It is inferred from Table 5 that the present developed solid-state electrolyte system is more efficient than many reported in the literature with respect to simplicity, efficiency, reliability, and good discharge times besides being economical and environment friendly. These systems find applications as cost-effective electrolytes in high-density solid-state electrochemical cells.

5. Conclusions

Nano-Dy_2O_3-doped polymer electrolyte films of composition "PVA (90%) + sodium citrate (10%) + nano-Dy_2O_3 (1 to 4%)" are synthesized by solution cast technique. The films are found to be good, stable, and endowed with excellent plasticity.

The films surface morphological characteristics and other physicochemical features are assessed by using FTIR, XRD, and SEM methods. The studies reveal that the different components in the film are homogenously and completely dispersed. It is attributed to the bonding tendencies of various functional groups present in PVA and sodium citrate such as –OH and –COO$^-$ with the Dy_2O_3. This binding nature helps to form a kind of surface complex, which is conducive for the movement of ions. The optimum % of Dy_2O_3 for forming more homogenous and less crystalline film is found to be 3%. Even DSC investigations reveal that at 3.0% of nano-Dy_2O_3, the amorphous region of the film is more, and this enables the ions to penetrate more in the film, thereby increasing the conductivity of the films.

The presence of nano-Dy_2O_3 remarkably enhances the conductivity of these PVA-based films. The conductivity for pure PVA film is 5.59×10^{-10} S/cm at 303 K. But the conductivity is increased to 5.31×10^{-5}, 4.48×10^{-5}, 1.06×10^{-4}, and 3.26×10^{-6} S/cm (at 303 K) with the films containing 1%, 2%, 3%, and 4% of nano-Dy_2O_3, respectively. With the increase in temperature, the conductivity is increased: 1.06×10^{-4} S/cm at 303 K; 3.86×10^{-4} S/cm at 313 K; 6.30×10^{-4} S/cm; at 323 K; and 9.72×10^{-4} S/cm at 333 K. This enhancement in conductivity is due to the hopping of interchain and intrachain ion movements and falling of microscopic viscosity at the matrix interface of the film. The studies on transference numbers reveal that the charge carriers are ions, and the contribution of electrons is almost negligible.

These polymer electrolyte films are investigated for their utility by constructing electrochemical cell with the configuration "anode (Mg + MgSO_4)/[PVA (90%) + $\text{Na}_3\text{C}_6\text{H}_5\text{O}_7$ (10%) + (1–4% nano-Dy_2O_3)]/cathode ($\text{I}_2 + \text{C} + \text{electrolyte}$)." The various discharge characteristics are evaluated and found that the films with 3% nano-Dy_2O_3 have maximum discharge time of 118 hrs, open-circuit voltage of 2.68 V, power density of 0.91 W/kg, and energy density of 107.5 Wh/Kg. These good findings emphasize the successful adoption of the developed polymer electrolyte films. These films may find good utility in solid-state battery applications.

Conflicts of Interest

The authors declare that there are no conflicts of interest regarding the publication of this paper.

Authors' Contributions

All authors have contributed significantly to this work.

Acknowledgments

The authors thank the K L University authorities for providing the necessary facilities and financial help to carry out this research work.

References

[1] J. C. Lasseques and P. Colombon, *Proton Conductors: Solids, Membranes and Gels*, University Press, Cambridge, UK, 1992.

[2] C. S. Ramya, S. Selvasekarapandian, T. Savitha et al., "Conductivity and thermal behavior of proton conducting polymer electrolyte based on poly (N-vinyl pyrrolidone)," *European Polymer Journal*, vol. 42, no. 10, pp. 2672–2677, 2006.

[3] B. Smitha, S. Sridhar, and A. A. Khan, "Chitosan–sodium alginate polyion complexes as fuel cell membranes," *European Polymer Journal*, vol. 41, no. 8, pp. 1859–1866, 2005.

[4] N. Srivastava and S. Chandra, "Studies on a new proton conducting polymer system: poly(ethylene succinate) + NH_4ClO_4," *European Polymer Journal*, vol. 36, no. 2, pp. 421–433, 2000.

[5] H. T. Pu and D. Wang, "Studies on proton conductivity of polyimide/H3PO4/imidazole blends," *Electrochimica Acta*, vol. 51, no. 26, pp. 5612–5617, 2006.

[6] M. M. Coleman and P. C. Painter, "Hydrogen bonded polymer blends," *Progress in Polymer Science*, vol. 20, no. 1, pp. 1–59, 1995.

[7] T. Kanbara, M. Inami, and T. Yamamoto, "New solid-state electric double-layer capacitor using poly(vinyl alcohol)-based polymer solid electrolyte," *Journal of Power Sources*, vol. 36, no. 1, pp. 87–93, 1991.

[8] H. A. Every, F. Zhou, M. Forsyth, and D. R. MacFarlane, "Lithium ion mobility in poly(vinyl alcohol) based polymer electrolytes as determined by 7Li NMR spectroscopy," *Electrochimica Acta*, vol. 43, no. 10-11, pp. 1465–1469, 1998.

[9] D. Kumar and S. A. Hashmi, "Ion transport and ion–filler-polymer interaction in poly(methyl methacrylate)-based, sodium ion conducting, gel polymer electrolytes dispersed with silica nanoparticles," *Journal of Power Sources*, vol. 195, no. 15, pp. 5101–5108, 2010.

[10] F. Croce, L. Persi, B. Scrosati, F. Serraino-Fiory, E. Plichta, and M. A. Hen-drickson, "Role of the ceramic fillers in enhancing the transport properties of composite polymer electrolytes," *Electrochimica Acta*, vol. 46, no. 16, pp. 2457–2461, 2001.

[11] Z. Li, G. Su, D. Gao, X. Wang, and X. Li, "Effect of Al_2O_3 nanoparticles on the electrochemical characteristics of P(VDF-HFP)-based polymer electrolyte," *Electrochimica Acta*, vol. 49, no. 26, pp. 4633–4639, 2004.

[12] C. C. Tambelli, A. C. Bloise, A. V. Rosario, E. C. Pereira, C. J. Magon, and J. P. Donosa, "Characterisation of PEO–Al_2O_3 composite polymer electrolytes," *Electrochimica Acta*, vol. 47, no. 11, pp. 1677–1682, 2002.

[13] P. A. R. D. Jayathilaka, M. A. K. L. Dissanayake, I. Albinsson, and B.-E. Mel-lander, "Effect of nano-porous Al_2O_3 on thermal, dielectric and transport properties of the (PEO)9LiTFSI polymer electrolyte system," *Electrochimica Acta*, vol. 47, no. 20, pp. 3257–3268, 2002.

[14] M. A. K. L. Dissanayake, P. A. R. D. Jayathilaka, R. S. P. Bokalawala, I. Albinsson, and B.-E. Mellander, "Effect of concentration and grain size of alumina filler on the ionic

[15] conductivity enhancement of the (PEO)9LiCF3SO3:Al_2O_3 composite polymer electrolyte," *Journal of Power Sources*, vol. 119–121, pp. 409–414, 2003.

[15] C. H. Park, D. W. Kim, J. Prakash, and Y.-K. Sun, "Electrochemical stability and conductivity enhancement of composite polymer electrolytes," *Solid State Ionics*, vol. 159, no. 1-2, pp. 111–119, 2003.

[16] B. Kumar, S. J. Rodrigues, and S. Koka, "The crystalline to amorphous transition in PEO-based composite electrolytes: role of lithium salts," *Electrochimica Acta*, vol. 47, no. 25, pp. 4125–4131, 2002.

[17] Y. Liu, J. Y. Lee, and L. Hong, "Functionalized SiO_2 in poly (ethylene oxide)-based polymer electrolytes," *Journal of Power Sources*, vol. 109, no. 2, pp. 507–514, 2002.

[18] L. Fan, C. W. Nan, and S. Zhao, "Effect of modified SiO_2 on the properties of PEO-based polymer electrolytes," *Solid State Ionics*, vol. 164, no. 1-2, pp. 81–86, 2003.

[19] H. M. Xiong, K. K. Zhao, X. Zhao, Y. W. Wang, and J. S. Chen, "Elucidating the conductivity enhancement effect of nano-sized SnO_2 fillers in the hybrid polymer electrolyte PEO–SnO_2–$LiClO_4$," *Solid State Ionics*, vol. 159, no. 1-2, pp. 89–95, 2003.

[20] L. Fan, Z. Dang, G. Wei, C. W. Nan, and M. Li, "Effect of nanosized ZnO on the electrical properties of (PEO)16$LiClO_4$ electrolytes," *Materials Science and Engineering: B*, vol. 99, no. 1–3, pp. 340–343, 2003.

[21] J. Adebahr, N. Byrne, M. Forsyth, D. R. MacFarlane, and P. Jacobsson, "Enhancement of ion dynamics in PMMA-based gels with addition of TiO_2 nano-particles," *Electrochimica Acta*, vol. 48, no. 14–16, pp. 2099–2103, 2003.

[22] B. Kumar, S. J. Rodrigues, and L. G. Scanlon, "Poly(ethylene oxide)-based composite electrolytes: crystalline $\rightleftharpoons$ amorphous transition," *Journal of The Electrochemical Society*, vol. 148, no. 12, p. A1191, 2001.

[23] J. D. Kim and I. Honma, "Proton conducting polydimethylsiloxane/zirconium oxide hybrid membranes added with phosphotungstic acid," *Electrochimica Acta*, vol. 48, no. 24, pp. 3633–3638, 2003.

[24] A. D'epifanio, F. S Fiory, S. Licoccia, E. Traversa, and B. Scrosati, "Metallic-lithium, LiFePO4-based polymer battery using PEO–ZrO_2 nanocomposite polymer electrolyte," *Journal of Applied Electrochemistry*, vol. 34, pp. 403–408, 2004.

[25] M. White, "Thin polymer films," *Thin Solid Films*, vol. 18, no. 2, pp. 157–172, 1973.

[26] J. B. Wagner and C. J. Wagner, "Electrical conductivity measurements on cuprous halides," *Journal of Chemical Physics*, vol. 20, p. 1597, 1957.

[27] M. Watanabe, S. Nagano, K. Sanui, and N. Ogata, "Estimation of Li+ transport number in polymer electrolytes by the combination of complex impedance and potentiostatic polarization measurements," *Solid State Ionics*, vol. 28–30, pp. 911–917, 1988.

[28] A. Lewandowski, M. Zajder, E. Frackowiak, and F. Beguin, "Supercapacitor based on activated carbon and polyethylene oxide–KOH–H_2O polymer electrolyte," *Electrochimica Acta*, vol. 46, no. 18, pp. 2777–2780, 2001.

[29] M. J. Reddy and U. V. SubbaRao, "Transport studies of poly (ethylene oxide)-based polymer electrolyte complexed with sodium yttrium fluoride," *Journal of Materials Science Letters*, vol. 17, no. 19, pp. 1613–1615, 1998.

[30] P. Balaji Bhargav, V. Madhu Mohan, A. K. Sharma, and V. V. R. N. Rao, "Investigations on electrical properties of (PVA: NaF) polymer electrolytes for electrochemical cell applications," *Current Applied Physics*, vol. 9, no. 1, pp. 165–171, 2009.

[31] R. M. Hodge, G. H. Edward, and G. P. Simon, "Water absorption and states of water in semicrystalline poly(vinyl alcohol) films," *Polymer*, vol. 37, no. 8, pp. 1371–1376, 1996.

[32] P. Anji Reddy and R. Kumar, "Ionic conductivity and discharge characteristic studies of PVA-Mg(CH$_3$COO)$_2$ solid polymer electrolytes," *International Journal of Polymeric Materials*, vol. 62, no. 2, pp. 76–80, 2012.

[33] C. V. Subba Reddy, A. K. Sharma, and V. V. R Narasimha Rao, "Effect of plasticizer on electrical conductivity and cell parameters of PVP+PVA+KClO$_3$ blend polymer electrolyte system," *Journal of Power Sources*, vol. 111, no. 2, pp. 357–360, 2002.

[34] J. Ramesh Babu, K. Ravindhranath, and K. Vijaya Kumar, "Structural and electrical properties of sodium citrate doped poly(vinyl alcohol) films for electrochemical cell applications," *Asian Journal of Chemistry*, vol. 29, no. 5, pp. 1049–1055, 2017.

[35] M. Hema, S. Selvasekerapandian, G. Hirankumar, A. Sakunthala, D. Arunkumar, and H. J. Nithya, "Structural and thermal studies of PVA:NH$_4$I," *Journal of Physics and Chemistry of Solids*, vol. 70, no. 7, pp. 1098–1103, 2009.

[36] K. Kattel, J. Y. Park, W. Xu et al., "Paramagnetic dysprosium oxide nanoparticles and dysprosium hydroxide nanorods as T2 MRI contrast agents," *Biomaterials*, vol. 33, no. 11, pp. 3254–3261, 2012.

9

Rake Angle Effect on a Machined Surface in Orthogonal Cutting of Graphite/Polymer Composites

Dayong Yang ⓘ, **Zhenping Wan** ⓘ, **Peijie Xu, and Longsheng Lu**

School of Mechanical and Automotive Engineering, South China University of Technology, Guangzhou 510640, China

Correspondence should be addressed to Dayong Yang; ydy425@126.com

Academic Editor: Massimiliano Barletta

Graphite and its composites have been widely used in various industrial fields. It has been generally accepted that, for positive rake angles, there is a significant increase in tension stress at the cutting zone during the machining of brittle materials, and cracks occur and spread easily, degrading the quality of the machined surface quality. However, it is found in this study that positive rake angles can improve the machined surface finish during the orthogonal cutting of graphite/polymer composites. Better machined surface finish is obtained for a larger rake angle. A finite element model is developed to reveal the mechanism of influence of the positive rake angle on the machined surface. Based on the effective stress field obtained from finite element analysis, it can be predicted that the crack initiates at the tool tip, subsequently propagates downward and forward, and later spreads gradually toward the free surface of the workpiece. A larger rake angle can promote crack propagation far from the machined surface. The crack initiation and propagation laws are validated by the edge-indentation experiments. In addition, the cutting force at various rake angles is investigated.

1. Introduction

Graphite and its composites (G/GCs) have been increasingly used in the fabrication of various precision parts such as biomedical implants [1], thermal sinks [2], bipolar plates of fuel cells [3], electrical discharge machining (EDM) electrodes [4], semiconductor jigs, pile cores, and mechanical seals [5]. G/GCs have abrasive and brittle characters, and the material in cutting zone undergoes localized fractures rather than plastic deformation during machining [6]. Moreover, the crack initiation and propagation laws of G/GCs, such as graphite/polymer composites, are unique and distinct from other brittle materials [7]. Hence, the machining mechanism of G/GCs is different from that of the other brittle materials, for example, structural ceramics [8] and other types of composite materials, such as carbon/epoxy composites [9] and carbon fiber-reinforced polymer/plastic [10]. Therefore, the machining of G/GCs has attracted considerable attention.

To date, the investigations on the machining of G/GCs mainly focus on the tool wear and machined surface quality. Graphite is extremely abrasive due to the bond strength between individual carbon molecules [11]. Therefore, in graphite machining, randomly oriented graphite aggregates lead to severe tool wear due to the abrasive character of graphite [12], and different process parameters will result in different tool wear modes. When the hot-filament chemical vapor-deposited and time-modulated chemical vapor-deposited diamond-coated tool inserts were used for the machining of graphite, the main wear modes were crater wear and notching resulting from the action of the graphite powder coming into contact with and sliding against the tool surface during machining [6]. During the high-speed milling of graphite using an AlTiN-coated carbide micro-end mill, the flank wear is the dominant wear pattern in the steady wear stage [13]. When the graphite electrodes were turned by microcrystalline and nanocrystalline diamond-coated Si_3N_4 ceramic inserts, the main tool wear mode was abrasion-induced by the powdery graphite [12]. According to the studies of Lei et al. [14] and Hashimoto et al. [15], the deposition of microcrystalline diamond films on cocemented tungsten carbide micro-drills and the introduction of nitrogen into the cutting area are effective methods for extending tool life. On the other hand, the investigations conducted by Wan et al. [7] indicated that there were a huge number of tiny concavities on the machined surface of graphite/polymer composites. Hence, it is difficult to obtain a good surface finish because graphite is a brittle

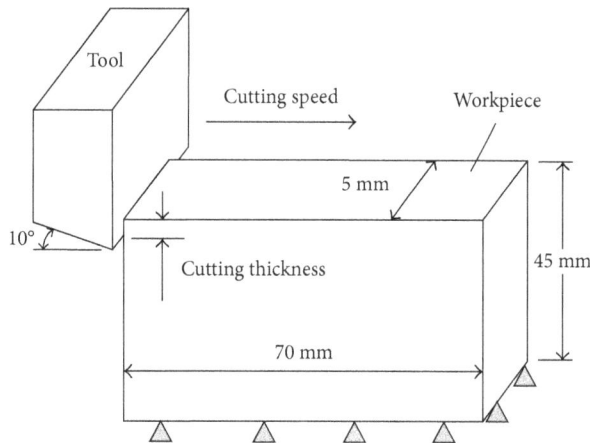

FIGURE 1: Schematic diagram of orthogonal cutting experiments.

material, and brittle fracture is the main characteristic during graphite machining [16]. To obtain a good surface finish, great efforts have been conducted. Wang et al. [17] found that a smoother surface could be achieved with a small feed per tooth in the high-speed milling of a graphite electrode. The experimental results presented by Zhou et al. [18] showed that the surface roughness of the machined surface increased clearly with the depth of cut increasing in the orthogonal cutting of graphite. Huo et al. [11] found that the feed rate has the most significant influence on surface roughness, and the surface roughness is not sensitive to cutting speed on the micro-milling of fine-grained graphite. The statistical analysis also showed that the surface roughness decreased with a low feed rate [19]. Experimental design methods [20], artificial neural networks [21], and the gray relational analysis method [22] were used to optimize the machining parameters for high-purity graphite in the end milling process. Additionally, Bajpai and Singh investigated the orthogonal micro-grooving of anisotropic pyrolytic carbon [23] and established a finite element model to understand the mechanics of material removal in the plane of transverse isotropy of pyrolytic carbon [24].

From the above literature survey, little work has been reported on the effect of the tool rake angle on the machined surface quality during graphite/polymer composite machining. This paper investigates the influence of the tool rake angle on the machined surface finish during the orthogonal cutting of graphite/polymer composites. To reveal the mechanism of influence of a positive tool rake angle on the machined surface finish, the effective stress field of the cutting zone was determined out by finite element analysis, and then, the crack initiation and propagation path in the cutting zone were predicted based on the effective stress field. The laws of crack initiation and propagation were verified by edge-indentation experiments, which are effective methods for the study of the material removal mechanism during brittle material cutting [6, 25, 26]. In addition, the effect of the tool rake angle on the cutting force was investigated.

2. Experimental Methodologies

2.1. Experimental Conditions of Orthogonal Cutting. Dry orthogonal cutting experiments were conducted on planer

BC6063B. The dimensions of the workpiece are 70 mm × 45 mm × 5 mm. A superhard high-speed steel tool with the clearance angle of 10° and different rake angles is used to machine the workpiece at the cutting speed of 6 m/min. The cutting thicknesses are 0.2 mm and 0.4 mm. A schematic diagram of the orthogonal cutting experiments is shown in Figure 1.

The graphite/polymer composites used in this study are a powder mixture of coke and natural graphite added to a binder. The paste is first homogenized and placed in a mold and then is sufficiently compacted. The material is then baked slowly at high temperature. The physical and mechanical properties of the graphite/polymer composites are shown in Table 1.

2.2. Edge-Indentation Experimental Setup. The edge-indentation experiments are conducted using a universal material testing machine (CMT5105). A schematic drawing of the edge-indentation experiments is shown in Figure 2, with the indenter positioned on the edge of the graphite/polymer composite specimen. The edge-indentation surface of the specimen is polished until no cracks can be observed in order to eliminate the influence of tiny cracks on crack initiation and propagation. The indenter can rotate adaptively to ensure well-distributed pressure on the indentation surface. The dimensions of the graphite/polymer composite specimen are 70 mm × 45 mm × 5 mm. All specimens are indented at the constant speed of 0.5 mm/min.

It can be seen from Figure 2 that the edge indentation is similar to orthogonal cutting. In Figure 2, γ_s is the rake angle of the indenter, h_D is the edge-indentation thickness, and P is the normal load imposed on the indenter. The indenter is similar to a cutting tool with the rake angle of γ_s and clearance angle of 0°, and h_D is the cutting thickness. Therefore, the laws of crack initiation and propagation obtained from the edge-indentation experiments can be applied to validate the predicted crack initiation and propagation rules for the graphite/polymer composite orthogonal cutting process.

3. Effect of Rake Angle on Machined Surface Finish

Figure 3 shows the machined surface roughness Ra obtained at different rake angles for the cutting thicknesses of 0.2 mm and 0.4 mm. It can be seen from Figure 3 that the surface roughness Ra is largest when the rake angles are 0° and 5°, and the surface roughness Ra decreases gradually with either an increase or decrease in the rake angle. The entire curve can be fitted by a parabola. The fact that a negative rake angle can reduce the surface roughness conforms to the long-held notion that the compressive stresses induced by a negative rake angle can weaken the breaking of the material, resulting in improved surface qualities. However, the fact that a positive rake angle can also reduce the surface roughness and improve the machined surface qualities conflicts with the long-held notion that flaking due to tensile stresses induced by a positive rake angle damages the machined surface in the machining of brittle materials.

Figure 4 shows the surface morphologies machined at different rake angles for the cutting thickness of 0.2 mm. Due to the

TABLE 1: Mechanical properties of graphite/polymer composites.

Properties	Density (g/cm^3)	Hardness (shores)	Tensile strength (MPa)	Compressive strength (MPa)	Modulus of elasticity (GPa)	Porosity (%)
Parameter	1.9	75	17.3	107.2	15.9	0.5

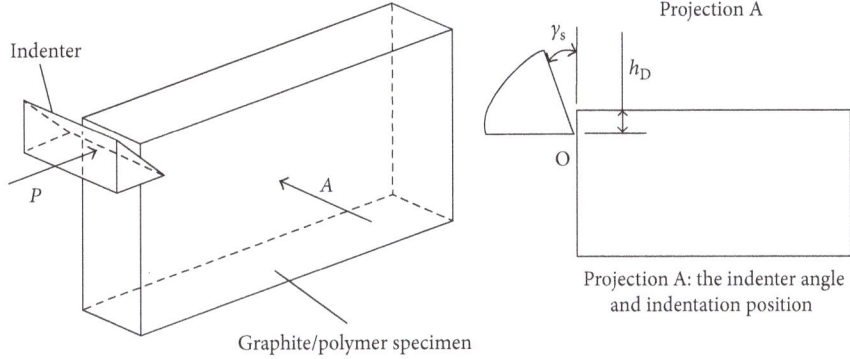

FIGURE 2: Schematic drawing of edge-indentation experiment.

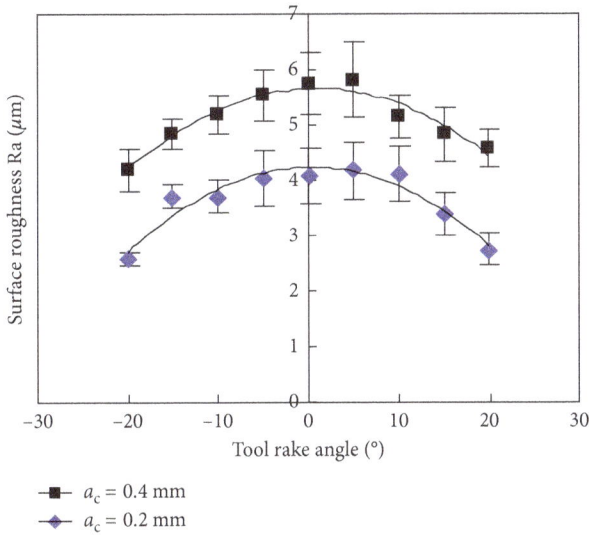

FIGURE 3: Machined surface roughness of graphite/polymer composites at different rake angles.

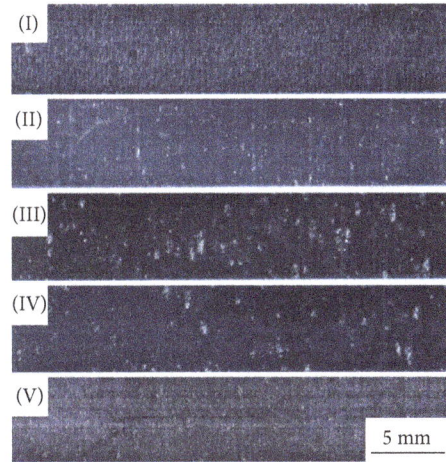

FIGURE 4: Surface morphologies machined at cutting thickness of 0.2 mm and different rake angles: (I) −20°, (II) −10°, (III) 0°, (IV) 10°, and (V) 20°.

breaking of the surface material, many concavities are observed on the machined surface. Furthermore, the largest concavities are formed on the surface machined at the rake angle of 0°. With increased or decreased of rake angles, the concavities gradually decrease. It is suggested that, with increasing rake angle, the machined surface qualities are improved for positive rake angles during graphite/polymer composite orthogonal cutting.

4. Mechanism of Influence of a Positive Rake Angle on a Machined Surface

The discussion in this section focuses only on the mechanism of influence of a positive rake angle on the machined surface finish because the mechanism of influence of a negative rake angle on machined surface qualities has been studied thoroughly. The crack initiation and propagation at the cutting zone are the main processes during graphite machining that result in large amounts of crack and concavity formation on the machined surface. Hence, the determination of the rules that govern crack initiation and propagation is highly important for understanding the mechanism of influence of a positive rake angle on the machined surface.

4.1. Effective Stress Field of the Cutting Zone.
During graphite/polymer cutting, surface materials are removed through the brittle fracture of the material. The brittle fracture of material is determined by the effective stress field of the cutting zone. Hence, the analysis of the effective stress field of the cutting zone is helpful for understanding the material removal mechanism.

4.1.1. Fracture Criterion of Material.
According to Paul and Mirandy's theory [27], if $\sigma_{III}/\sigma_b < (1 - N^2)$, the fracture criterion for the three-dimensional stress state is given by

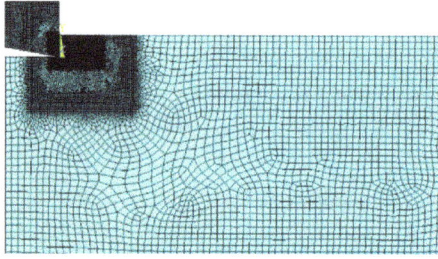

FIGURE 5: Mesh models of graphite composite orthogonal cutting.

$$\sigma_{\mathrm{eff}} = \frac{N\left(\sigma_{\mathrm{I}} + \sigma_{\mathrm{III}}\right) + 2\sqrt{N^2 \sigma_{\mathrm{I}} \sigma_{\mathrm{III}} + \left(\sigma_{\mathrm{I}} + \sigma_{\mathrm{III}}\right)^2}}{N\left(4 - N^2\right)}. \qquad (1)$$

If $\sigma_{\mathrm{III}}/\sigma_{\mathrm{b}} \geq (1 - N^2)$, the fracture criterion for the three-dimensional stress state is

$$\sigma_{\mathrm{eff}} = \sigma_{\mathrm{I}}, \qquad (2)$$

where σ_{eff} is the effective stress; σ_{I} and σ_{III} are the maximal principal stress and minimum principal stress, respectively; and σ_{b} is the tension strength. N is a constant determined by the following equation:

$$N = \sqrt{2 + \frac{\sigma_{\mathrm{bc}}}{\sigma_{\mathrm{b}}} - 2\sqrt{1 + \frac{\sigma_{\mathrm{bc}}}{\sigma_{\mathrm{b}}}}}, \qquad (3)$$

where σ_{bc} is the compressive strength.

4.1.2. Finite Element Modeling. ANSYS explicit dynamics are used to calculate the effective stress field of the cutting zone. The sketch of the graphite/polymer composite orthogonal cutting is shown in Figure 1. In finite element modeling, the dimensions of the workpiece are $50\,\mathrm{mm} \times 25\,\mathrm{mm} \times 5\,\mathrm{mm}$, and the properties of the workpiece material are described in Table 1. The clearance angle of the tool is $10°$, and the cutting thickness is $2\,\mathrm{mm}$. The workpiece is fixed in all directions, and the cutting tool is modeled as a rigid body that moves forward at the speed of $6\,\mathrm{m/min}$. The interaction between the chip and the tool can be modeled as sliding frictional behavior. The friction coefficient is defined as $\mu = F_{\mathrm{t}}/F_{\mathrm{n}}$, where F_{t} is the tangential force acting on the rake face and F_{n} is the normal force acting on the rake face. Based on the cutting force measurements, the friction coefficient is assumed to be 0.5. In the graphite cutting process, high temperature is almost non-existent. The orthogonal cutting is simplified as two-dimensional cutting, and a total of 28,300 quadrilateral meshes are used to model the workpiece, as shown in Figure 5. A very fine mesh density is defined at the contact zone of the tool tip and the workpiece in order to obtain fine process output distributions.

4.1.3. Effective Stress Field and Crack Initiation and Propagation Prediction. The effective stress contours of the cutting zone during the cutting process are shown in Figure 6 when the rake angles of cutting tool are $0°$, $10°$, and $20°$. Figures 6(a)–6(c) show snapshots when the effective stresses in the vicinity of the tool tip just reach or approach the tension strength. In the effective stress field, the crack

initiates at the point where the stress reaches the tension strength and propagates along the direction of the minimum stress gradient. In Figure 6, the curve c denotes the direction of the minimum stress gradient. Thus, the curve c also indicates the crack propagation path in the cutting zone. It can be seen from Figure 6 that the crack initiates at the tool tip and then propagates downward and forward, subsequently spreading gradually upward until it intersects the free surface of the workpiece. Therefore, the block of the workpiece material surrounded by the crack is removed by brittle fracture, and a concavity forms on the machined surface. By comparing Figures 6(a) and 6(b), it can be easily seen that when the rake angle of the tool is $10°$, the crack c spreads toward the free surface of the workpiece soon after its first appearance, and a small concavity remains. When the rake angle is $0°$, the crack propagation path is long, and a large concavity forms. As shown in Figure 6(c), the crack hardly propagates downward, and in this case, a good surface finish can be obtained. Therefore, a large positive rake angle can facilitate crack spreading toward the free surface of the workpiece and improves the machined surface qualities.

4.2. Validation of Crack Initiation and Propagation Rules. The rules governing crack initiation and propagation predicted from the effective stress field of the cutting zone can elucidate the mechanism of influence of the positive rake angle on the machined surface qualities. However, the predictions should be verified by experiments. This section investigates the rules governing crack initiation and propagation during graphite/polymer orthogonal cutting through edge-indentation experiments.

4.2.1. Crack Initiation and Propagation Path. Figure 7 shows the crack initiation and propagation path under the action of an indenter with rake angles (γ_{s}) of $5°$, $10°$, and $20°$ when h_{D} is $2\,\mathrm{mm}$. With the increase in the normal load P, no crack occurs at the beginning of loading. At this point, a notch forms at the contact of the indenter and graphite/polymer specimen, as shown in Figure 7(a). The crack initiates suddenly and propagates rapidly when the normal load increases to a critical value. In this case, the propagation of the crack will terminate if the load can be removed quickly. However, there is usually not enough time to remove the normal load before the crack reaches the free surface. Therefore, a large block of graphite surrounded by the crack sheds, as shown in Figure 7, where the crack spreads downward in the direction of the indentation in the initial stage and then propagates gradually toward the free surface of specimen. The propagation path of the crack is close to an arc shape. The crack spreads farther from the inside of the specimen with the increase of rake angle of the indenter. That is, a large rake angle can facilitate the crack spreading further toward the free surface of the workpiece if the indenter is considered as a cutting tool. When γ_{s} is $20°$, the formed crack hardly damages the machined surface, as shown in Figure 7(c). This outcome suggests that the experimental results are in good agreement with the predictions of crack initiation and propagation. Furthermore,

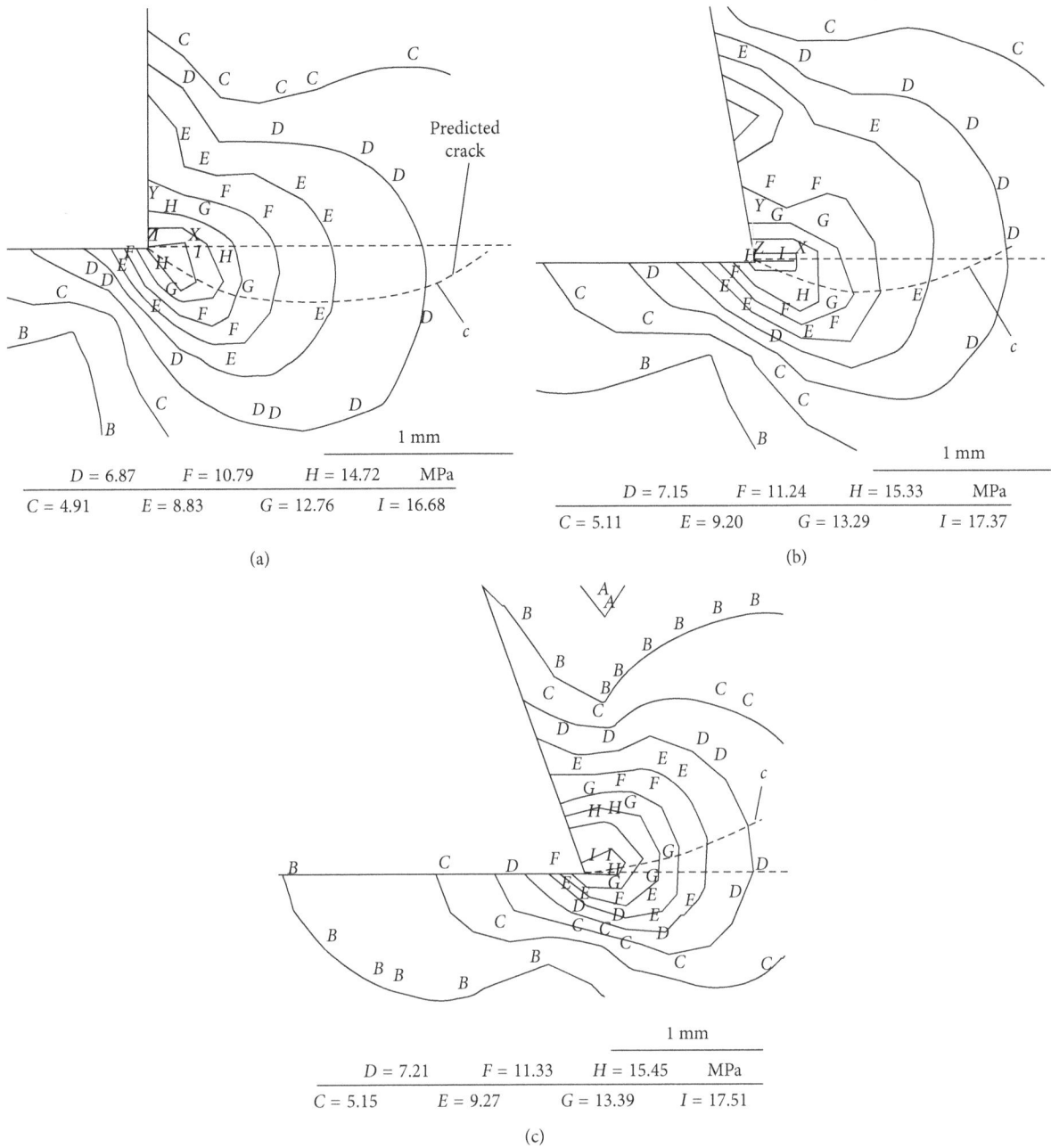

		D = 6.87		F = 10.79		H = 14.72		MPa
C = 4.91		E = 8.83		G = 12.76		I = 16.68		

(a)

		D = 7.15		F = 11.24		H = 15.33		MPa
C = 5.11		E = 9.20		G = 13.29		I = 17.37		

(b)

		D = 7.21		F = 11.33		H = 15.45		MPa
C = 5.15		E = 9.27		G = 13.39		I = 17.51		

(c)

FIGURE 6: Effective stress contours in cutting zone and predicted crack propagation paths (the stresses in the vicinity of the tool tip just reach tension strength) at rakes angles of (a) 0°, (b) 10°, and (c) 20°.

the crack propagation path in the graphite/polymer composite is different from the glass crack propagation path, which shows a rectilinear configuration in edge-indentation experiments. More importantly, a smaller rake angle leads to the glass crack propagating farther away from the machined surface. Hence, the rules governing the crack propagation of graphite/polymer composites are in contrast to the laws of glass crack propagation. Thus, graphite/polymer composite machining has its own unique laws that are unlike those of glass cutting.

4.2.2. Relationship of the Rake Angle of the Indenter to Crack Propagation Path. According to the above experimental

results, the crack initiates at the contact of the indenter and specimen and propagates in an arc-shaped path. As a result, a concavity forms. Thus, the initial angle of crack propagation θ, concavity depth c_d, and concavity width c_w are used to describe the crack propagation path, as shown in Figure 8. Figure 9 depicts the changing trend of the width and depth of concavities with the rake angle of the indenter at the edge-indentation thickness of 2 mm. The width and depth of the concavities decrease as the rake angle of the indenter increases. Larger rake angle of indentation leads to smaller size of the concavities. Furthermore, the concavity depth decreases linearly as the rake angle of the indenter increases,

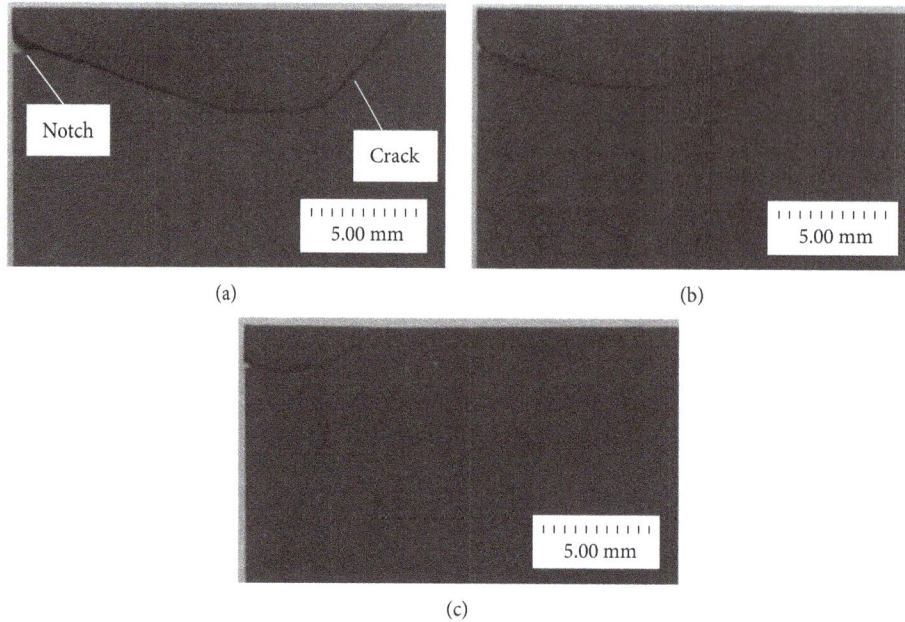

FIGURE 7: Crack initiation and propagation paths at different rake angles of indenter ($h_D = 2$ mm): (a) $\gamma_s = 5°$, (b) $\gamma_s = 10°$, and (c) $\gamma_s = 20°$.

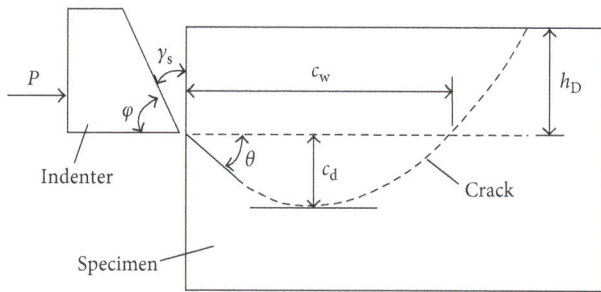

FIGURE 8: Parameters of crack initiation and propagation path.

while the largest width of the concavity formed by the action of the indenter is obtained for the rake angle of 5°. This outcome occurs because the indented specimen is vulnerable to collapse (as shown in Figure 10) when the rake angle of the indenter is 0°, and once the material of the indentation surface is disintegrated, the real edge-indentation thickness will be reduced. Thus, the concavity width produced at indenter rake angle of 0° is even smaller than that produced at indenter rake angle of 5°, as shown in Figure 9. Consequently, the surface roughness obtained at rake angle of 5° is largest as shown in Figure 3.

Figure 11 shows the dependence of the initial angle of crack propagation on the rake angle of the indenter. It can be seen from Figure 11 that the initial angles of crack propagation decrease linearly with increasing rake angle of the indenter. That is, the larger the rake angle of the indenter is, the smaller the initial angle of crack propagation. Thus, a large rake angle of the indenter can make the crack propagate farther away from the inside of the specimen. This outcome further confirms that a large rake angle of the cutting tool can improve the machined surface finish of graphite/polymer composites.

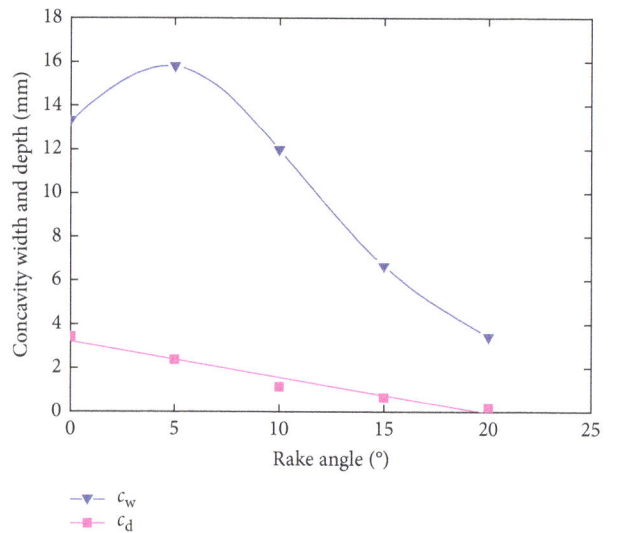

FIGURE 9: Width and depth of concavities at different rake angles of indenter ($h_D = 2$ mm).

FIGURE 10: Crack initiation and propagation path formed at indenter rake angle of 0° ($h_D = 2$ mm).

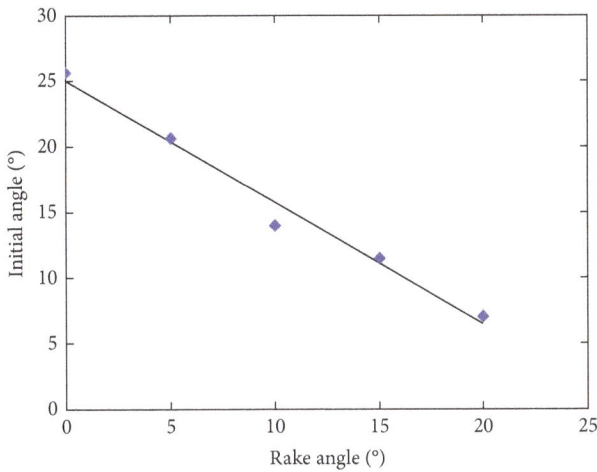

FIGURE 11: Initial angle of crack propagation at different indenter rake angles.

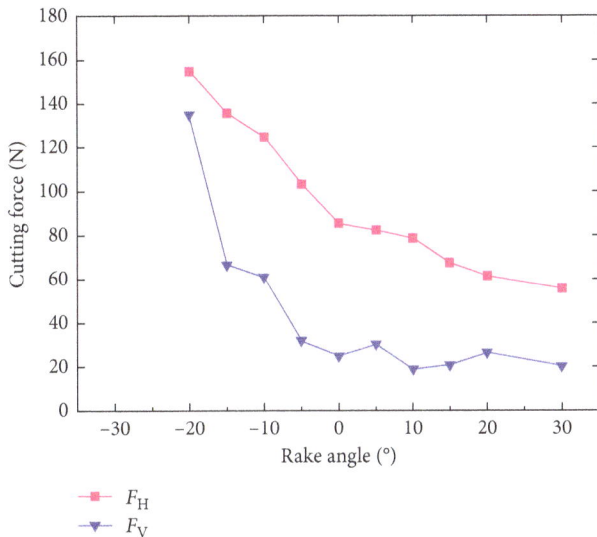

FIGURE 12: Cutting forces at different rake angles ($v = 6$ m/min, $a_c = 0.4$ mm).

5. Cutting Force at a Different Rake Angle

The cutting force was measured by a Kistler dynamometer. The horizontal cutting forces F_H and vertical cutting forces F_V at different tool rake angles are shown in Figure 12 for the cutting speed v of 6 m/min and the depth of cut a_c of 0.4 mm. It can be observed from Figure 12 that the cutting force decreases with increasing rake angle regardless of whether the rake angle is negative or positive. Furthermore, the cutting force decreases dramatically with increasing rake angle when the rake angle is negative. When the rake angle is positive, the horizontal cutting force decreases with a small slope, and the vertical cutting force is almost invariable while the rake angle is increasing. Thus, it can be observed that there is a turning point in the variation curve of cutting force when the rake angle is 0°. Considering the influence of the rake angle on the machined surface finish and cutting force, it can be concluded that a positive rake angle that is higher than 5° should be selected and used for the machining of graphite/polymer composites.

6. Conclusions

(1) When the rake angle during the graphite/polymer composite orthogonal cutting is between 0° and 5°, the machined surface finish is the poorest, and the machined surface finish is improved with decreasing or increasing rake angle.

(2) The crack initiates at the tool tip, subsequently propagates downward and forward, and then spreads gradually toward the free surface of the workpiece, resulting in the removal of a block of graphite surrounded by the crack and the formation of a concavity on the machined surface. A large positive rake angle can facilitate crack propagation far away from the machined surface. Therefore, the concavities formed on the machined surface decrease. Thus, a large positive rake angle can improve the machined surface finish.

(3) The cutting force decreases with increasing rake angle. Furthermore, when the rake angle is negative, the cutting force decreases rapidly with increasing rake angle. When the rake angle is positive, the horizontal cutting force decreases slowly, while the vertical cutting force remains almost constant with increasing rake angle.

(4) Considering the machined surface finish and cutting force, it is better to use a positive rake angle greater than 5° to machine graphite/polymer composites.

Notations

a_c: Cutting depth (mm)
c_d: Concavity depth (mm)
c_w: Concavity width (mm)
F_H: Horizontal cutting force (N)
F_n: Normal force (N)
F_t: Tangential force (N)
F_V: Vertical cutting force (N)
h_D: Edge-indentation thickness (mm)
N: Constant
P: Normal load imposed on indenter (N)
γ_s: Rake angle of indenter (degree)
θ: Initial angle of crack propagation (degree)
μ: Friction coefficient
v: Cutting speed (m/min)
σ_b: Tension strength (MPa)
σ_{bc}: Compressive strength (MPa)
σ_{eff}: Effective stress (MPa)
σ_I: Maximal principal stress (MPa)
σ_{III}: Minimum principal stress (MPa).

Conflicts of Interest

The authors declare that there are no conflicts of interest regarding the publication of this paper.

Acknowledgments

This work is supported by the National Natural Science Foundation of China (no. 51775198) and the Science and Technology Project of Guangzhou, China (201804010182).

References

[1] V. Starý, L. Bačáková, J. Horník, and V. Chmelík, "Biocompatibility of the surface layer of pyrolytic graphite," *Thin Solid Films*, vol. 433, no. 1-2, pp. 191–198, 2003.

[2] T. C. Chang, S. Lee, Y. K. Fuh, Y.-C. Peng, and Z.-Y. Lin, "PCM based heat sinks of paraffin/nanoplatelet graphite composite for thermal management of IGBT," *Applied Thermal Engineering*, vol. 112, pp. 1129–1136, 2017.

[3] K. Kang, S. Park, A. Jo, K. Lee, and H. Ju, "Development of ultralight and thin bipolar plates using epoxy-carbon fiber prepregs and graphite composites," *International Journal of Hydrogen Energy International Journal of Hydrogen Energy*, vol. 42, no. 3, pp. 1691–1697, 2017.

[4] M. Zeis, "Deformation of thin graphite electrodes with high aspect ratio during sinking electrical discharge machining," *CIRP Annals*, vol. 66, no. 1, pp. 185–188, 2017.

[5] M. Masuda, Y. Kuroshima, and Y. Chujo, "The machinability of sintered carbons based on the correlation between tool wear rate and physical and mechanical properties," *Wear*, vol. 195, pp. 178–185, 1996.

[6] G. Cabral, P. Reis, R. Polini et al., "Cutting performance of time-modulated chemical vapour deposited diamond coated tool inserts during machining graphite," *Diamond and Related Materials*, vol. 15, no. 10, pp. 1753–1758, 2006.

[7] Z. Wan, D. Yang, L. Lu, J. Wu, and Y. Tang, "Mechanism of material removal during orthogonal cutting of graphite/polymer composites," *International Journal of Advanced Manufacturing Technology*, vol. 82, no. 9–12, pp. 1815–1821, 2016.

[8] C. Nath, G. C. Lim, and H. Y. Zheng, "Influence of the material removal mechanisms on hole integrity in ultrasonic machining of structural ceramics," *Ultrasonics*, vol. 52, no. 5, pp. 605–613, 2012.

[9] H. Gao, Y. J. Bao, and Z. M. Feng, "A study of drilling unidirectional carbon/epoxy composites," *International Journal of Abrasive Technology*, vol. 4, no. 1, pp. 1–13, 2011.

[10] X. Wang, X. Shen, G. Yang, and F. Sun, "Evaluation of boron-doped-microcrystalline/nanocrystal-line diamond composite coatings in drilling of CFRP," *Surface and Coatings Technology*, vol. 330, pp. 149–162, 2017.

[11] D. Huo, C. Lin, and K. Dalgarno, "An experimental investigation on micro machining of fine-grained graphite," *International Journal of Advanced Manufacturing Technology*, vol. 72, no. 5–8, pp. 943–953, 2014.

[12] F. A. Almeida, J. Sacramento, F. J. Oliveira, and R. F. Silva, "Micro- and nano-crystalline CVD diamond coated tools in the turning of EDM graphite," *Surface and Coatings Technology*, vol. 203, no. 3-4, pp. 271–276, 2008.

[13] L. Zhou, C. Y. Wang, and Z. Qin, "Tool wear characteristics in high-speed milling of graphite using a coated carbide micro endmill," *Proceedings of the Institution of Mechanical Engineers, Part B: Journal of Engineering Manufacture*, vol. 223, no. 3, pp. 267–277, 2009.

[14] X. L. Lei, L. Wang, B. Shen, F. Sun, and Z. Zhang, "Effect of boron-doped diamond interlayer on cutting performance of diamond coated micro drills for graphite machining," *Materials*, vol. 6, no. 8, pp. 3128–3138, 2013.

[15] M. Hashimoto, K. Kanda, and T. Tsubokawa, "Reduction of diamond-coated cutting tool wear during graphite cutting," *Precision Engineering*, vol. 51, pp. 186–189, 2018.

[16] R. B. Schroeter, R. Kratochvil, and J. D. Gomes, "High-speed finishing milling of industrial graphite electrodes," *Journal of Materials Processing Technology*, vol. 179, no. 1–3, pp. 128–132, 2006.

[17] C. Y. Wang, L. Zhou, H. Fu et al., "High speed milling of graphite electrode with endmill of small diameter," *Chinese Journal of Mechanical Engineering*, vol. 20, no. 4, pp. 27–31, 2007.

[18] L. Zhou, C. Y. Wang, and Z. Qin, "Investigation of chip formation characteristics in orthogonal cutting of graphite," *Materials and Manufacturing Processes*, vol. 24, no. 12, pp. 1365–1372, 2009.

[19] O. S. López, A. R. González, and I. H. Castillo, "Statistical analysis of surface roughness of machined graphite by means of CNC milling," *Ingeniería e Investigación*, vol. 36, no. 3, pp. 89–94, 2016.

[20] Y. K. Yang, M. T. Chuang, and S. S. Lin, "Optimization of dry machining parameters for high-purity graphite in end milling process via design of experiments methods," *Journal of Materials Processing Technology*, vol. 209, no. 9, pp. 4395–4400, 2009.

[21] J. R. Shie, "Optimization of dry machining parameters for high-purity graphite in end-milling process by artificial neural networks: a case study," *Materials and Manufacturing Processes*, vol. 21, no. 8, pp. 838–845, 2006.

[22] Y. K. Yang, J. R. Shie, and C. H. Huang, "Optimization of dry machining parameters for high-purity graphite in end-milling process," *Materials and Manufacturing Processes*, vol. 21, no. 8, pp. 832–837, 2006.

[23] V. Bajpai and R. K. Singh, "Orthogonal micro-grooving of anisotropic pyrolytic carbon," *Materials and Manufacturing Processes*, vol. 26, no. 12, pp. 1481–1493, 2011.

[24] V. Bajpai and R. K. Singh, "Brittle damage and interlaminar decohesion in orthogonal micromachining of pyrolytic carbon," *International Journal of Machine Tools and Manufacture*, vol. 64, pp. 20–30, 2013.

[25] B. Lawn and R. Wilshaw, "Review indentation fracture: principles and applications," *Journal of Materials Science*, vol. 10, no. 6, pp. 1049–1081, 1975.

[26] Z. P. Wan and Y. Tang, "Brittle–ductile mode cutting of glass based on controlling cracks initiation and propagation," *International Journal of Advanced Manufacturing Technology*, vol. 43, no. 11-12, pp. 1051–1059, 2009.

[27] B. Paul and L. Mirandy, "Improved fracture criterion for three-dimensional stress states," *Journal of Engineering Materials and Technology*, vol. 98, no. 2, pp. 159–163, 1976.

10

Molecular Association Studies on Polyvinyl Alcohol at Different Concentrations

Richa Saxena(iD)[1,2] **and S. C. Bhatt**[1]

[1]*Ultrasonic and Dielectric Laboratory, Department of Physics, H.N.B. Garhwal University, Srinagar, Uttarakhand, India*
[2]*IFTM University, Lodhipur Rajput, Delhi Road, Moradabad 244001, India*

Correspondence should be addressed to Richa Saxena; saxena.richa23@gmail.com

Academic Editor: Antonio Facchetti

Ultrasonic velocities, densities, and viscosities have been measured for the solution of polyvinyl alcohol in water at concentration range of 0.3% to 1% at temperature 35°C. Ultrasonic velocities have been measured using variable path ultrasonic interferometer at 1 MHz frequency. The acoustical parameters like adiabatic compressibility, acoustic impedance, intermolecular free length, and relaxation time have been calculated by using above-mentioned values of ultrasonic velocities, densities, and viscosities. The variation of these acoustical parameters is explained in terms of solute-solvent interaction in a polymer solution.

1. Introduction

For many years ultrasound has been used in a variety of fields such as biology, biochemistry, dentistry, engineering, geology, industry, medicine, and polymers [1]. The study of molecular interaction in binary and ternary liquid mixtures plays an important role in molecular sciences. In recent years ultrasound has become a powerful tool in providing information about the physiochemical properties of liquid system [2–5]. The study of molecular interactions in the polymer and solvent throws light on the processes involving polymer production and their uses. The deviation of ultrasonic sound velocity and several other thermodynamic properties of electrolytic solutions and binary and ternary mixtures with various concentrations has been investigated by various researchers [6–8].

The ion-dipole interaction depends mainly on ion size and polarity of the solvent. The strength of ion-dipole attraction is directly proportional to size of the ion, charge, and the magnitude of the dipole, but inversely proportional to the distance between the ions and the dipolar molecules [9–13] of polymer solutions and has shown that ultrasonic velocity and its allied parameters provide more information on molecular interactions which are of the utmost importance for process involving polymer production and their

uses [14]. In the present paper an attempt has been made to calculate adiabatic compressibility, acoustic impedance, intermolecular free length, and relaxation time by using measured values of ultrasonic velocities, densities, and viscosities at concentration range of 1.0%, 0.8%, 0.6%, 0.5%, 0.4%, and 0.3% at 35°C temperature.

2. Experimental Details

In the present investigation polyvinyl alcohol in solid form of molecular weight approximately (140,000 Da) is used. The solutions were prepared by adding known volume of polyvinyl alcohol to fixed volume of water and stirring under reflex, until a clear solution was obtained. The concentration range studied in the solution is 1.0%, 0.8%, 0.6%, 0.5%, 0.4%, and 0.3% (m/v). Different acoustical parameters like intermolecular free length and relaxation time were calculated at different concentration like 1.0%, 0.8%, 0.6%, 0.5%, 0.4%, and 0.3% and at 35°C temperatures at 1 MHz frequency by using variable path ultrasonic interferometer with reproducibility of ±0.4 m/s at 35°C. The temperature of the solution has been kept constant by circulating water from the thermostatically controlled (±0.1°C) water bath. The densities at different temperature were measured using 10 ml specific gravity bottle and single pan macrobalance. The

TABLE 1: Density ($\times 10^3$ kg/m^3) of polyvinyl alcohol (PVA) at 35°C temperature and concentration at 1 MHz frequency.

Concentration (m/v)	Density (at 35°C)
1.0%	0.996
0.8%	0.983
0.6%	0.978
0.5%	0.975
0.4%	0.973
0.3%	0.970

TABLE 2: Viscosity ($\times 10^{-1}$ Pa·s) of polyvinyl alcohol (PVA) at 35°C temperature and concentration at 1 MHz frequency.

Concentration (m/v)	Viscosity at 35°C
1.0%	0.089
0.8%	0.086
0.6%	0.077
0.5%	0.074
0.4%	0.073
0.3%	0.072

TABLE 3: Ultrasonic velocity (m/s) of polyvinyl alcohol (PVA) at different concentration at 1 MHz frequency.

Concentration (m/v)	Ultrasonic velocity at 35°C
1.0%	1509.9
0.8%	1503.2
0.6%	1501.3
0.5%	1495.9
0.4%	1483.8
0.3%	1482.5

TABLE 4: Adiabatic compressibility ($\times 10^{-10}$ kg^{-1} ms^2) at different concentration at 1 MHz for polyvinyl alcohol (PVA).

Concentration (m/v)	Adiabatic compressibility At 35°C
1.0%	4.404
0.8%	4.502
0.6%	4.537
0.5%	4.583
0.4%	4.668
0.3%	4.691

TABLE 5: Acoustic impedance ($\times 10^3$ kg m^2 s^{-1}) at different concentration at 1 MHz for polyvinyl alcohol (PVA).

Concentration (m/v)	Acoustic impedance at 35°C
1.0%	1503.9
0.8%	1477.6
0.6%	1468.3
0.5%	1458.5
0.4%	1443.7
0.3%	1438

TABLE 6: Intermolecular free length ($\times 10^{-13}$ m) at different concentration at 1 MHz for polyvinyl alcohol (PVA).

Concentration (m/v)	Intermolecular free length at 35°C
1.0%	2.84
0.8%	2.872
0.6%	2.883
0.5%	2.898
0.4%	2.924
0.3%	2.931

TABLE 7: Relaxation time ($\times 10^{-12}$ s) at different concentration at 1 MHz for polyvinyl alcohol (PVA).

Concentration (m/v)	Relaxation time at 35°C
1.0%	5.23
0.8%	5.16
0.6%	4.66
0.5%	4.61
0.4%	4.54
0.3%	4.5

uncertainty in density measurements was found to be about 0.5 kg/m^3. The viscosity of the mixtures was determined by using Ostwald's viscometer, which was kept inside a double-walled jacket, in which water from thermostat water bath was circulated. The inner cylinder of this double-wall-glass jacket was filled with water of desired temperature so as to establish and maintain the thermal equilibrium. The accuracy in the viscosity measurements is within ±0.5%. These parameters are calculated by using standard relations [15–17].

3. Result and Discussion

In the present work density, viscosity, and ultrasonic velocity have been measured at different concentration of polyvinyl alcohol, at 35°C which is shown in Tables 1, 2, and 3, respectively. By using these values for PVA, adiabatic compressibility, acoustic impedance, intermolecular free length, relaxation time, and ultrasonic attenuation have been calculated by using well known relations [18–20] and the results have been presented in Tables 4, 5, 6, 7, and 8, respectively.

The variations of these parameters with temperature and concentration have been shown in Figures 1–8, respectively.

Polyvinyl alcohol in solid form of molecular weight approximately 140,000 is taken. Solution were prepared by adding known weight of polyvinyl alcohol of molecular weight approximately 140,000 to fixed volume of water and

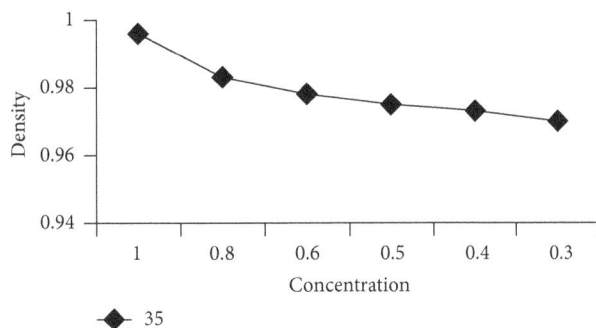

FIGURE 1: Variation of density with concentration at 35°C temperature.

TABLE 8: Ultrasonic absorption ($\times 10^{-15}$ s^2 m^{-1}) at different concentration at 1 MHz for polyvinyl alcohol (PVA).

Concentration (m/v)	Ultrasonic absorption at 35°C
1.0%	6.84
0.8%	6.79
0.6%	6.13
0.5%	6.21
0.4%	6.05
0.3%	6.00

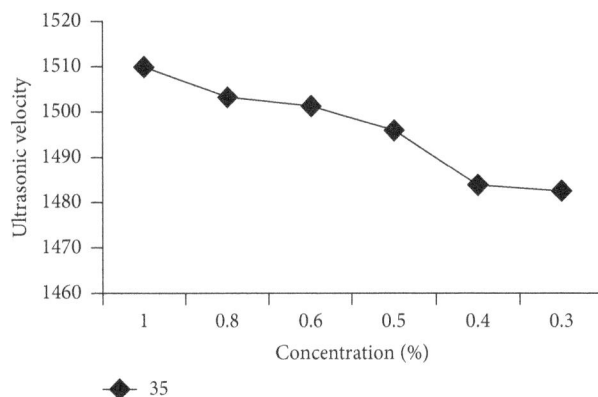

FIGURE 3: Variation of ultrasonic velocity with concentration at v35°C temperature.

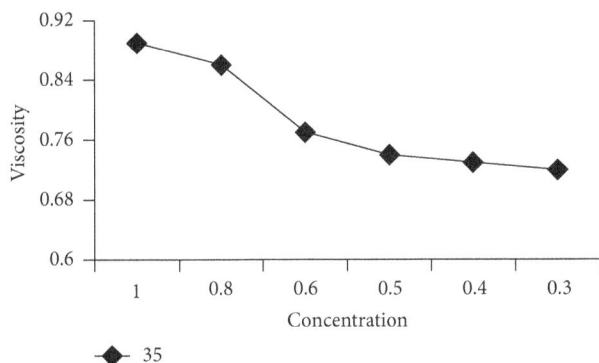

FIGURE 2: Variation of viscosity with concentration at 35°C temperature.

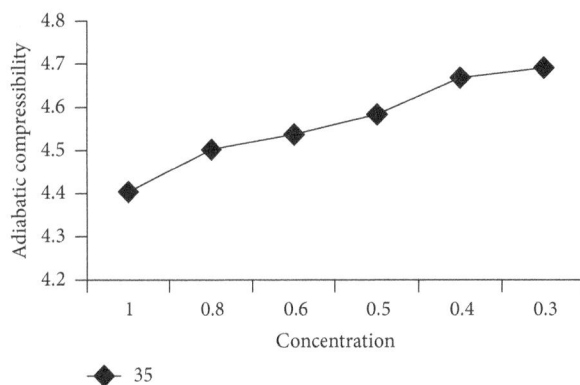

FIGURE 4: Variation of adiabatic compressibility with concentration at 35°C temperature.

stirring under reflex, until a clear solution was obtained. Figure 1 represents the variation of density with concentration at 35°C. Density increases with increase in concentration. These are in agreement with earlier workers [21]. It may be due to electro striction in that solution. This electrostriction decreases the volume and hence increases the density as a number of solute molecules increase the electrostriction and density. It is evident from Figure 2 that viscosity increases with increase in concentration of PVA. This is showing similar trend as reported by earlier workers [22]. The variations of ultrasonic velocity with concentration have been shown in Figure 3. Ultrasonic velocity increases with increase in concentration of PVA. The increase in ultrasonic velocity with concentration indicates increase in cohesion forces due too powerful-solvent interactions. The results

are in good agreement with earlier worker. Variation of adiabatic compressibility with concentration is shown in Figure 4. It is evident that adiabatic compressibility decreases with increase in concentration of polyvinyl alcohol in solution. This decrease in adiabatic compressibility indicates the enhancement of the bond strength at this concentration. Figure 5 depicts the variation of acoustic impedance with concentration. It is seen that it increases with increase in concentration of polyvinyl alcohol in the solution. This is in agreement with the requirement as both ultrasonic velocity

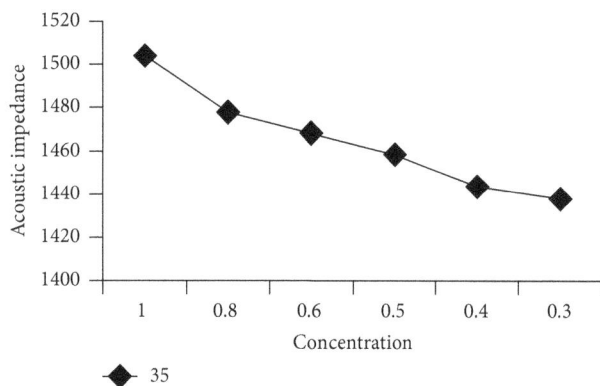

FIGURE 5: Variation of acoustic impedance with concentration at 35°C temperature.

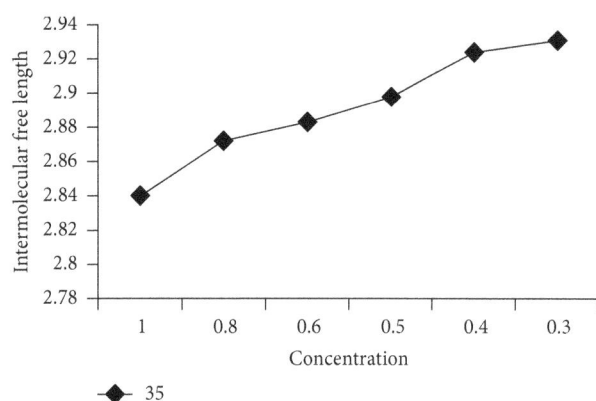

FIGURE 6: Variation of intermolecular free length with concentration at 35°C temperature.

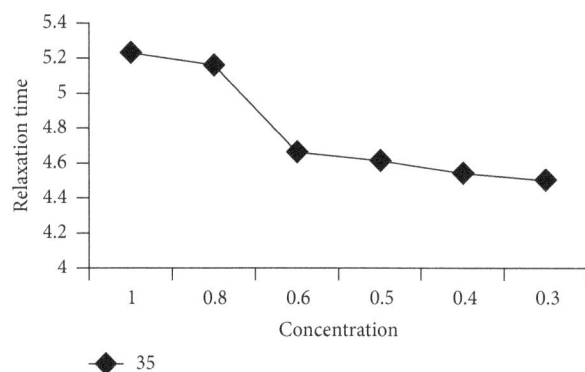

FIGURE 7: Variation of relaxation time with concentration at 35°C temperature.

and density increase with increase in concentration of the solute and also effective due to solute-solvent interactions. Variation of intermolecular free length with concentration is presented in Figure 6 which shows that it decreases with increase in concentration of PVA in solution. Similar results are obtained by earlier researchers. It is evident from Figure 7 that relaxation time increases with increase in concentration of polyvinyl alcohol in solution. Variation of ultrasonic

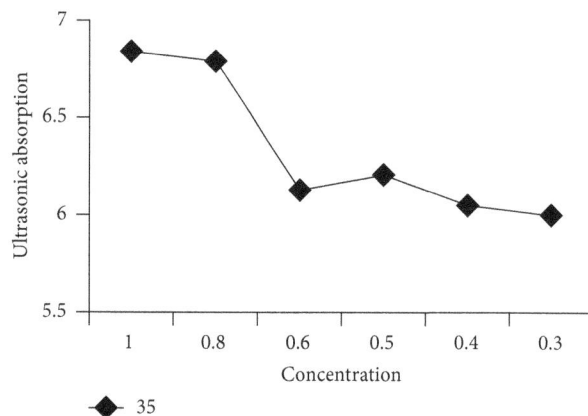

FIGURE 8: Variation of ultrasonic absorption with concentration at 35°C temperature.

absorption with concentration is shown in Figure 8. It is observed that ultrasonic absorption increases with increase in concentration of PVA in the solution. Increase in the value of relaxation time and ultrasonic absorption with concentration can be explained in terms of the motion of the molecular interaction forces.

Conflicts of Interest

The authors declare that they have no conflicts of interest.

References

[1] T. J. Mason, Ed., *Sonochemistry: The Uses of Ultrasound in Chemistry*, Royal Society of Chemistry, 1990.

[2] A. K. Dash and R. Paikaray, "Acoustical study in binary liquid mixture containing dimethyl acetamide using ultrasonic and viscosity probes," *Der. Chem. Sinica*, vol. 5, pp. 81–88, 2014.

[3] R. Palani, S. Saravanan, and R. Kumar, "Ultrasonic studies on some ternary organic liquid mixtures at 303, 308 and 313k," *RASĀYAN Journal of Chemistry*, vol. 2, no. 3, pp. 622–629, 2009.

[4] T. S. Antony, M. P. John Peter, and J. Y. Raj, "Phytochemical Analysis of Stylosanthes fruticosa using UV-VIS, FTIR and GC-MS," *Research Journal of Chemical Sciences*, vol. 3, no. 11, pp. 14–23, 2013.

[5] A. A. Mistry, V. D. bhandakkar, and O. P. Chimankar, *J. of Chem and Pharma Res*, vol. 4, no. 1, pp. 170–174, 2012.

[6] S. Oswal and A. T. Patel, "Speeds of sound, isentropic compressibilities, and excess volumes of binary mixtures. 2. Mono-n-alkylamines with cyclohexane and benzene," *Journal of Chemical & Engineering Data*, vol. 40, no. 1, pp. 194–198, 1995.

[7] A. Awasthi and J. P. Shukla, "Ultrasonic and IR study of intermolecular association through hydrogen bonding in ternary liquid mixtures," *Ultrasonics*, vol. 41, no. 6, pp. 477–486, 2003.

[8] Y. V. Patel and P. H. Parsania, "Ultrasonic velocity study of poly(R,R/,4,4/-cyclohexylidene diphenylene diphenyl ether 4,4/-disulfonate) solutions at 30, 35 and 40 ∘C," *European Polymer Journal*, vol. 38, no. 10, pp. 1971–1977, 2002.

[9] S. Kalyana Sundaram and B. Sundaram, *Journal of Acoustical Society of India*, vol. 27, p. 375, 1999.

[10] S. Kalyana Sundaram and B. Sundaram, *Journal of Acoustical Society of India*, vol. 27, p. 363, 1999.

[11] B. Sundaresan and A. Srinivasa Rao, "Molecular interaction studies on poly(vinyl chloride) in chlorobenzene by acoustic method," *Polymer Journal*, vol. 26, no. 11, pp. 1286–1290, 1994.

[12] S. Kalyanasundaram, A. Manuel Stephen, and A. Gopalan, "Ultrasonic studies of poly (methyl acrylate) in dimethyl formamide," *J Polym. Matt*, vol. 12, p. 177, 1995.

[13] T. H. Polym, "Solution Butterworth, Scientific," London, 1956.

[14] I. Johnson, M. Kalidoss, and R. Srinivasamoorthy, *Journal of Pure Applied Ultrasonics*, vol. 25, p. 136, 2003.

[15] R. Saxena and S. C. Bhatt, "Molecular Interactions in Binary Mixture of Polymethylmethacrylate with Acetic Acid," *International Journal of Chemistry*, vol. 2, no. 2, 2010.

[16] S. Bhatt C, V. Lingwal, K. Singh, and B. Semwal S, "Acoustical parameters of some molecular liquids," *Journal of the Acoustical Society of India*, vol. 28, pp. 293–296, 2000.

[17] S. Ravichandran and K. Ramanathan, "Ultrasonic investigations of MnSO4, NiSO4 and CuSO4 aqueous in polyvinyl alcohol solution at 303K," *RASĀYAN Journal of Chemistry*, vol. 3, no. 2, pp. 375–384, 2010.

[18] A. R. Shah and P. H. Parsania, "Ultrasonic Velocity Studies on Poly(4,4/-Cyclohexylidene-2,2/-Dimethyldiphenylene/ Diphenylene-3, 3/-Benzophenone Disulfonates) in Chlorinated and Aprotic Solvents," *Journal of Polymer Materials*, vol. 14, no. 1, pp. 33–41, 1997.

[19] J. Desai A and P. H Parsania, "Ultrasonic studies of poly(1, 1-cyclohexlidonyl-4, 4 diphenylene2, 4-toulene) disulfonate at 300 C," *Journal of Pure Applied Ultrasonic*, vol. 19, pp. 65–69, 1997.

[20] V. Kagathara, M. Sanariya, and P. Parsania, "Sound velocity and molecular interaction studies on chloro epoxy resins solutions at 30∘C," *European Polymer Journal*, vol. 36, no. 11, pp. 2371–2374, 2000.

[21] J. Vidyanand, "Quality circle implementation in India organization: an alternative," *DECISION*, vol. 24, pp. 1–4, 1997.

[22] M. Umadevi, R. Kesavasamy, V. Ponnusamy, N. S. Priya, and K. Ratina, "Intermolecular interactions in ternary liquid mixtures by ultrasonic measurements," *Indian Journal of Pure & Applied Physics*, vol. 53, no. 12, pp. 796–803, 2015.

11

Effect of Molecular Chain Structure on Fracture Mechanical Properties of Aeronautical Polymethyl Methacrylate Using Extended Digital Image Correlation Method

Wei Shang,[1] Xiaojing Yang,[2] Xinhua Ji,[3] and Zhongxian Liu[1]

[1]*Key Laboratory of Protection and Retrofitting for Civil Building and Construction of Tianjin,
Tianjin Chengjian University, Tianjin 300384, China*
[2]*School of Materials Science and Engineering, Hebei University of Technology, Tianjin 300130, China*
[3]*Department of Mechanics, Tianjin University, Tianjin 300072, China*

Correspondence should be addressed to Wei Shang; wshang@tju.edu.cn and Xiaojing Yang; yxjing@mail.nankai.edu.cn

Academic Editor: Luciano Lamberti

The main purpose of this work is to investigate the fracture mechanical properties of aeronautical polymethyl methacrylate, which has been treated with directional tensile technology. Because of the special processing of directional polymethyl methacrylate, the molecular chain structures are different in different directions. The mechanical properties depend on the specific molecular chain structures. We use extended digital image correlation to measure the displacement field near the tip of the crack when the cracks grow in different directions in directional polymethyl methacrylate. We then tested the critical load for different specimens and analyzed the fracture morphology of the different specimens. Thanks to the experimental results, a molecular chain model of directional polymethyl methacrylate could be established. The analysis results using the molecular chain model are consistent with the experiments, which confirms the reliability of the molecular chain model.

1. Introduction

Polymethyl methacrylate (PMMA) is widely used in the field of aviation, due to its excellent properties, such as lightweight, high temperature resistance, high light transmittance, and good mechanical properties. The mechanical properties of PMMA have attracted additional attention after several plane crashes were caused by cracks in a hatch made of PMMA.

Research on fatigue and fracture properties of polymer materials started in the 1960s because of high requirements for strength and reliability. Berry [1] confirmed that Griffith strength theory could be used to analyze the brittle fracture of PMMA. Recently, this theory and related experiments have become important means to analyze the fracture of polymer material in general. Mukherjee and Burns [2] proposed that fatigue of PMMA was determined by three parameters: stress intensity factor amplitude, average stress intensity factor, and frequency. Woo and Chow [3] unified the fatigue crack propagation formula for metal aluminum and the nonmetallic

PMMA. They proposed that the strain energy release rate amplitude should be used to analyze crack propagation but not the stress intensity factor amplitude. Cheng et al. [4, 5] studied the influence of temperature and loading rate on the tensile strength and fracture toughness of PMMA. Kim et al. [6, 7] proposed that the fatigue crack growth rate of most polymers increases with increasing temperature and decreases with increasing loading frequency. Ramsteiner and Armbrust [8] solved some fatigue practical problems with polymers, such as the measurement of crack propagation, the influence of the specimen shape, the applied frequency, the measurement with constant or increasing stress intensity amplitude, and the propagating crack as a signal for transitions in the internal deformation process. Yao et al. [9] investigated the dynamic fracture behavior of thin PMMA plates with three- and four-parallel edge cracks using the method of caustics and a high-speed Schardin camera. Yao et al. [10] investigated dynamic fracture behavior of a thin PMMA sheet with two overlapping offset-parallel cracks under tensile loading using the optical method of caustics in combination with a Cranz-Schardin high-speed camera. Xu et al. also [11] studied the fracture characterization of a V-notch tip in PMMA material using an optical caustics method. Sahraoui et al. [12] measured the dynamic fracture toughness of notched PMMA at high impact velocities, where the classical method is limited by the inertial effects. The direct measurements of the specimen deflection are successfully used for toughness evaluation. Zhang et al. [13] studied the fracture characteristics of PMMA with different offset cracks under three-point bending using the digital gradient sensing method. Ayatollahi et al. [14] studied the brittle fracture characteristics of PMMA under compressive loading using the theoretical and experimental methods. Berto et al. [15, 16] studied the fracture characteristics of PMMA using notched specimens tested under torsion at room temperature and at −60 degrees C.

Sauer and Hsiao began to investigate the craze phenomenon of polymers in 1949. Kies and his coworkers were inspired because the top of PMMA hatch has improved craze resistance. They studied biaxial and multiaxial tension oriented PMMA. Oriented PMMA is manufactured as follows: a PMMA plate is pulled under directional stresses following a preselected temperature profile that includes heating, keeping, and cooling. Oriented PMMA has a higher pull strength and elasticity module than normal PMMA. Some important components (e.g., hatches) of airplanes are often made of oriented PMMA plates.

There have been extensive studies of the mechanical properties of PMMA. However, most of the previous studies use the isotropic mechanical model. Because of the special processing of oriented PMMA, the molecular chain structures vary in different directions. The fracture mechanical properties depend on the molecular chain structures. Therefore, the isotropic mechanical model is unable to describe the mechanical properties of oriented PMMA [17].

Extended digital image correlation was used to measure the displacement field near the tip of the crack when cracks grow in different directions. In addition, the critical load for different samples was tested. The fracture morphology represents the historical record of the material fracture. The mechanism of the fracture can be found via analysis of the fracture morphology. The fracture morphology of different specimens was also analyzed. Based on the experimental results, the molecular chain model of oriented PMMA was established. The molecular chain model can be used to analyze the fracture mechanical properties of oriented PMMA. The analysis results of the molecular chain model are consistent with the experimental results, which also confirm the reliability of the molecular chain model.

2. Extended Digital Image Correlation

Digital image correlation (DIC) [18–20] is an established method to measure mechanical parameters. However, this method is not applicable for displacement measurement of a discontinuous interface and cannot solve fracture problems. Recently, an algorithm named extended digital image correlation (X-DIC) [21–25] has been used to solve this problem. The method tracks the gray value pattern in small neighborhoods called subsets during deformation. Two digital images (the reference image and the deformed image) are used to record the surface changes of the specimen. Here, $f(x, y)$ is the gray level value at coordinate (x, y) of the reference image, and $g(x^*, y^*)$ is the gray level value at coordinate (x^*, y^*) of the deformed image. Any point with coordinates (x, y) in the reference image relates to the deformed image via the deformation

$$x^* = x + u$$
$$y^* = y + v. \tag{1}$$

In this equation, u and v are the horizontal and vertical displacements of the point (x, y), respectively.

According to the gray level values of the reference image and the deformed image, the correlation factor is expressed as

$$S = 1 - \frac{\sum f(x, y) \cdot g(x + u^{\text{test}}, y + v^{\text{test}})}{\sqrt{\sum f^2(x, y) \cdot \sum g^2(x + u^{\text{test}}, y + v^{\text{test}})}}. \tag{2}$$

In (2), when the correlation factor S has the minimum value, the real displacement is $u = u^{\text{test}}$ and $v = v^{\text{test}}$.

In this step, a correlation algorithm that is directly linked with finite element simulations is developed. We chose the Bilinear Quadrilateral (Q4) finite elements as the simplest basis. The element is considered a subset of the correlation algorithm. Within the subset, the displacement field is [26]

$$\vec{u}(\vec{x})$$

$$= \sum_{I \in K} N_I(\vec{x}) \left[\vec{a}_I + \underbrace{H(\vec{x})\,\vec{b}_I}_{I \in K_\Gamma} + \underbrace{\sum_{l=1}^{4} B_I^l(\vec{x})\,\vec{c}_I^l}_{I \in K_\Lambda} \right]. \tag{3}$$

In (3), $\vec{u}(\vec{x})$ is the displacement vector; N_I is the Q4 shape function; $\vec{a}_I$ are the displacement components of the nodes; $\vec{b}_I$ are the improved degrees of freedom of the nodes on the unit cut by the crack; $\vec{c}_I^l$ are the improved degree of freedom of the nodes on the unit that includes the tip of the crack; K is the set of all nodes; K_Γ is the set of nodes cut by crack; K_Λ is the set of nodes on the unit that includes the tip of the crack; $H(\vec{x})$ is called Heaviside Function and expressed as

$$H(\vec{x}) = \begin{cases} 1 & \text{if } (x - x^*) \cdot n > 0, \\ -1 & \text{otherwise,} \end{cases} \tag{4}$$

where x is the sample point, x^* is the closest point to x that lies on the crack, and n is the direction vector perpendicular to the crack face at x^*. $B_I^l(\vec{x})$ is the function of the tip of the crack and expressed in the local coordinate system as follows:

$$\left[B_I^l(r, \theta) \right] = \left[\sqrt{r} \sin\left(\frac{\theta}{2}\right),\ \sqrt{r} \cos\left(\frac{\theta}{2}\right),\ \sqrt{r} \sin\left(\frac{\theta}{2}\right) \right.$$
$$\left. \cdot \sin(\theta),\ \sqrt{r} \cos\left(\frac{\theta}{2}\right) \sin(\theta) \right]. \tag{5}$$

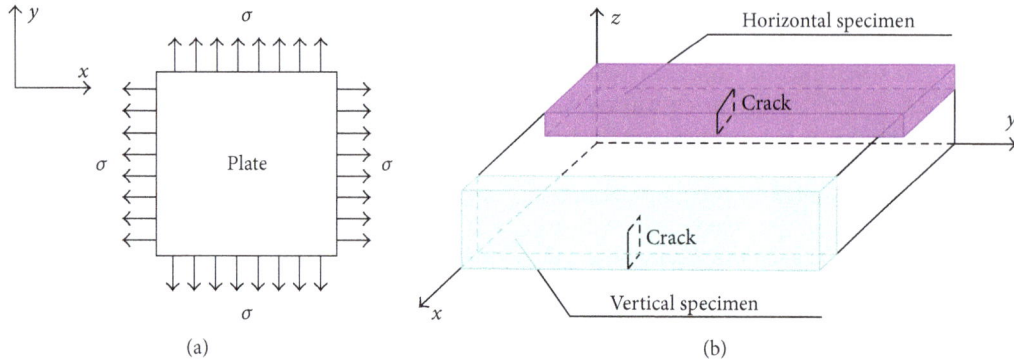

FIGURE 1: Two types of specimens: (a) biaxial tension PMMA plate; (b) two specimens in different directions.

r and θ are the polar coordinates in the local tip of the crack coordinate system.

3. Experiment and Analysis

3.1. Displacement Near the Tip of the Crack. In order to investigate the mechanical properties in different directions of the oriented PMMA plate, two specimens were cut from an oriented PMMA plate in different directions shown in Figure 1. Pulling loads during the process are in x-y plane.

The fracture mechanical properties of oriented PMMA are studied with the three-point bending beams shown in Figure 2. The length of the horizontal specimen is 120 mm; the width is 30 mm; the crack length is 14 mm. The crack direction is x, which is also the oriented direction. The length of the vertical specimen is 30 mm; the width is 7.5 mm; the crack length is 2.2 mm. The crack direction is z, which is perpendicular to the oriented direction.

Usually the front of a crack is subjected to three-dimensional effects that should be considered. Three-dimensional effects near the front were subjected to many analytical, numerical, and experimental studies in the past 50 years [27–29]. In this paper, the direction of crack propagation is studied using the displacement field of the tip of the crack. The three-dimensional effects do not affect the conclusions of this paper. Extended digital image correlation was used to test the displacement field near the tip of the crack of specimens. Considering the experimental results, the fracture mechanical properties of oriented PMMA were analyzed. The surfaces of both the horizontal and vertical specimen have the prefabricated spot shown in Figures 3(a) and 3(c). According to the surface images of loads 355 N and 817 N, extended digital image correlation was used to measure the y direction displacement field near the tip of the crack of the horizontal specimen shown in Figure 3(b). In addition, according to the surface images of loads 114 N

and 187 N, extended digital image correlation is also used to measure the displacement field in y direction near the tip of the crack of the vertical specimen shown in Figure 3(d). The displacement field in y direction near the tip of the crack of the horizontal specimen is symmetrical. Therefore, the crack of the horizontal specimen grows in the crack direction, while the displacement field in y direction near the tip of the crack of the vertical specimen is not symmetric. The crack of the vertical specimen grew in an oblique direction.

The crack propagation directions of different specimens were analyzed. The maximum circumferential normal stress theory is a type of crack propagation criterion. When the circumferential normal stress reaches the limit values, the crack begins to grow. This crack propagation criterion is the simplest and often used. According to maximum circumferential normal stress theory, the crack of the three-point bending beam grows in the crack direction. The experiments show that the crack of the horizontal specimen grows in the crack direction. Therefore, maximum circumferential normal stress theory is consistent with our experimental results. However, the theory is not suitable to be applied to the vertical specimen. The crack of the vertical specimen grows in an oblique direction. The reason will be analyzed. Because the molecular chains vertical to the oriented direction are not oriented, the mechanical properties are inferior compared with the oriented direction. The nonoriented molecular chains of the vertical specimen are damaged first, so the crack of the vertical specimen grows in an oblique direction.

3.2. Critical Load. Fracture toughness is important for the fracture mechanical properties of oriented PMMA. Fracture toughness is tested in two directions for the horizontal and the vertical specimen [13]. One is in the direction of the directional tension; the other is perpendicular to it. Equation (6) is the formula for the stress intensity factor.

$$K_I = \frac{PS}{BW^{3/2}} f\left(\frac{a}{W}\right) \tag{6}$$

$$f\left(\frac{a}{W}\right) = \frac{3\left(a/W\right)^{1/2}\left[1.99 - (a/W)\left(1 - a/W\right)\left(2.15 - 3.93\left(a/W\right) - 2.7\left(a^2/W^2\right)\right)\right]}{2\left(1 + 2a/W\right)\left(1 - a/W\right)^{3/2}}, \tag{7}$$

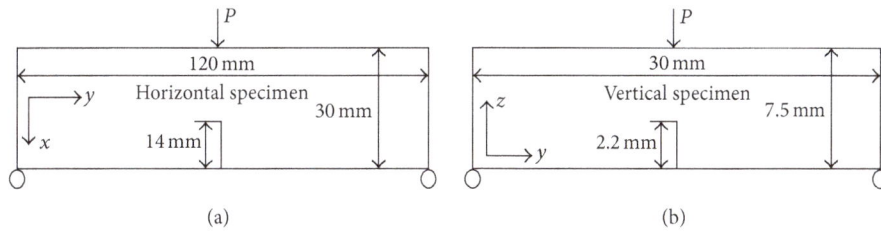

FIGURE 2: Dimensions of the three-point bending beams: (a) horizontal specimen; (b) vertical specimen.

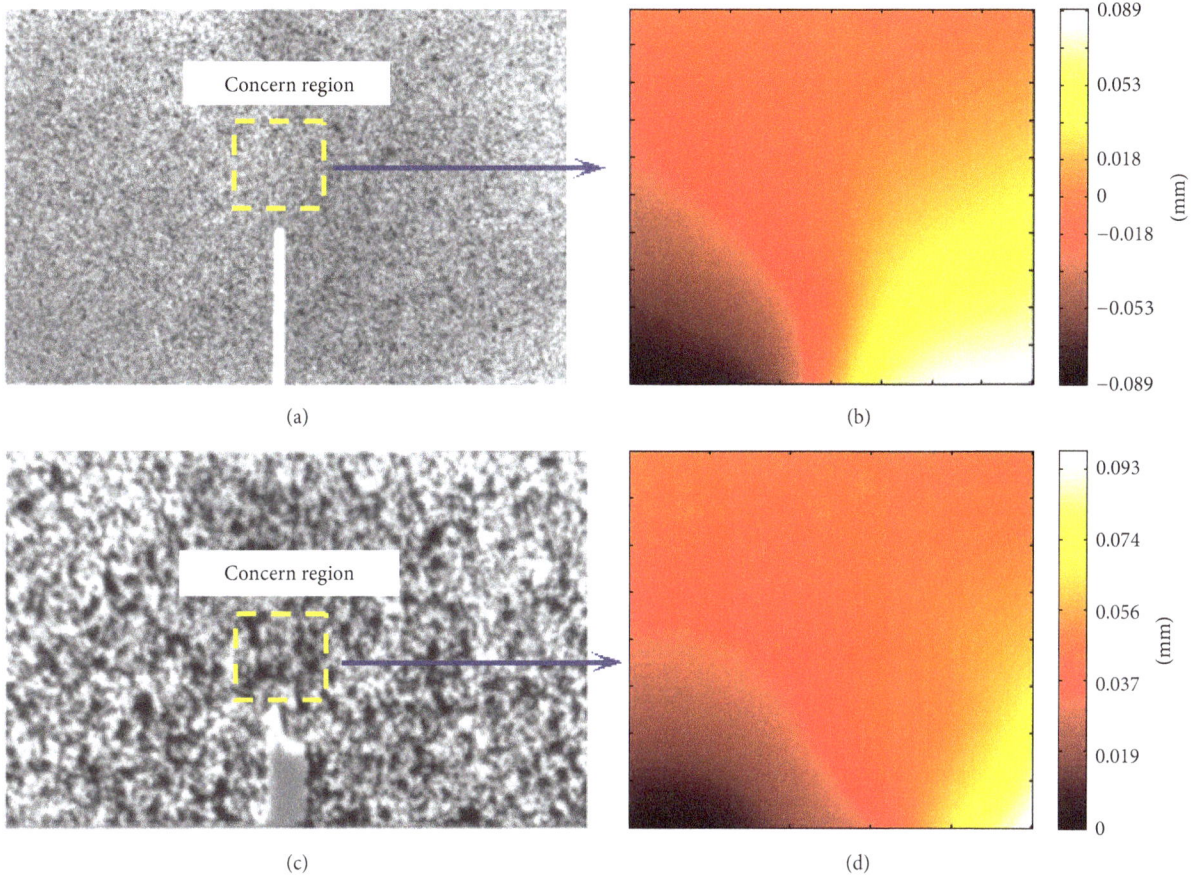

FIGURE 3: Experimental results: (a) gray image of the horizontal specimen; (b) y direction field of the horizontal specimen; (c) gray image of vertical specimen; (d) y direction displacement field of the vertical specimen.

where B is the thickness, $W = 2B$ is the width, $S = 4W$ is the length, a is the length of crack, and P is the load.

The length of the horizontal specimen is 120 mm; the width is 30 mm; the crack length is 14 mm. The crack direction is in the oriented direction. The length of the vertical specimen is 30 mm; the width is 7.5 mm; the crack length is 2.2 mm.

The goal of fracture toughness testing is to find the critical load for crack propagation using P-V curve. Load P was tested with a load sensor, and the crack opening displacement mouth V was tested using the extensometer normally. The specimen is very small, so there is not enough space to attach the extensometer. Hence, digital image correlation, which

replaces the extensometer, is applied to the test crack opening mouth displacement V. The tester and digital CCD camera are synchronous, so CCD could collect speckle patterns of different loads, which were used to draw P-V curve shown in Figure 4.

The critical load of the horizontal specimen was 817 N. The fracture toughness value was 194.87 N/mm$^{3/2}$ for the horizontal specimen which was obtained using the critical load and specimen size using (6). The critical load of the vertical specimen was 208 N. Similarly, the fracture toughness value 91.32 N/mm$^{3/2}$ of the vertical specimen was obtained. Due to the restrictions of the plate thickness, the vertical specimen is small and the sizes of horizontal and vertical

(a)

(b)

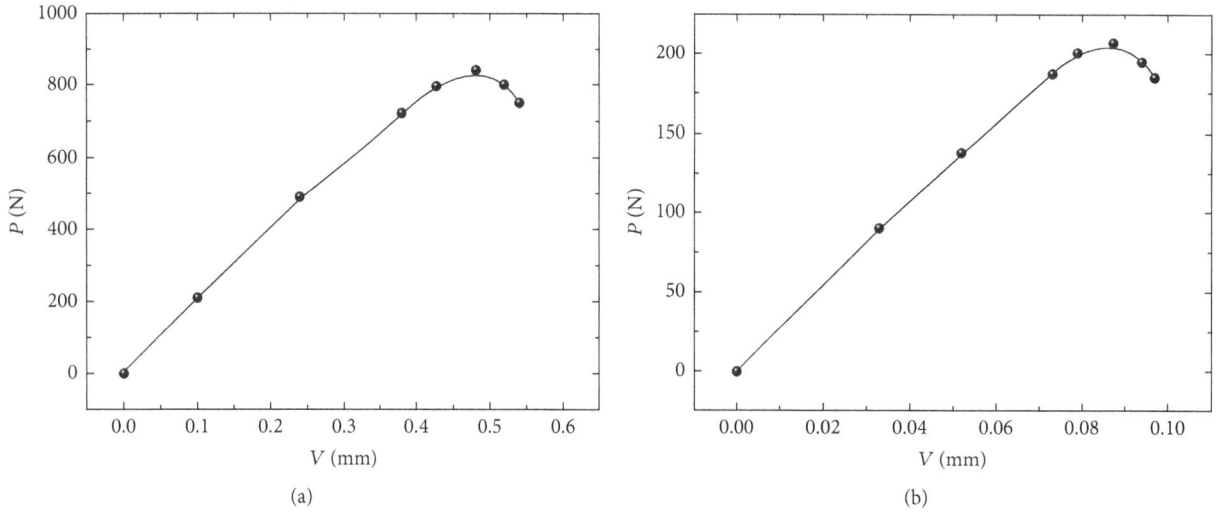

FIGURE 4: P-V curves: (a) horizontal specimen; (b) vertical specimen.

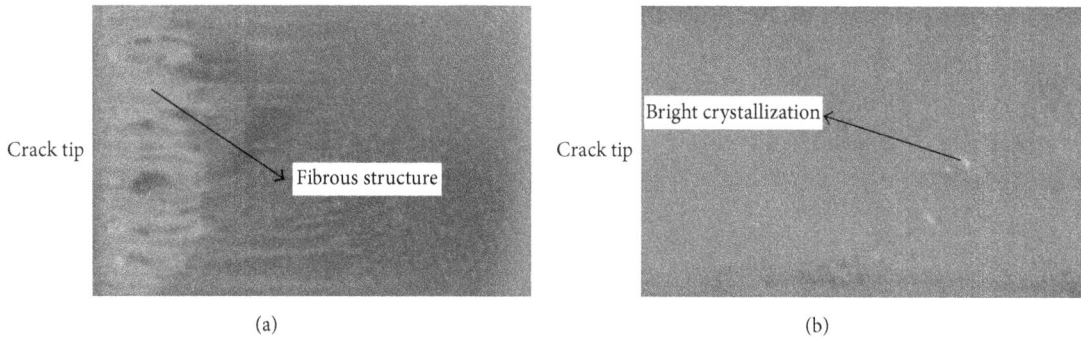

(a)

(b)

FIGURE 5: Macroscopic fracture morphology: (a) fracture morphology of the horizontal specimen; (b) fracture morphology of the vertical specimen.

TABLE 1: Critical loads for different specimens.

Specimens	Horizontal specimen	Vertical specimen
Critical load (N)	817	383

specimens are different. In this paper, the fracture toughness is used to calculate the critical loads of the same size, which can reflect the fracture mechanical properties in different directions clearly. Considering (6), if the vertical specimen is as big as the horizontal specimen, the critical load is 383 N shown in Table 1. The experimental results show that the critical load of the horizontal specimen is larger than that of the vertical one. Hence, the crack perpendicular to the directional tension grows more easily.

3.3. Fracture Morphology. The fracture morphology is a record of the material fracture. The analysis of the fracture morphology is used to analyze the fracture process according to the observation and analysis of the macroscopic and microscopic characteristics. The mechanism of the fracture can be found via analysis of the fracture morphology. The fracture morphology of both the horizontal and the vertical

specimen is shown in Figures 5(a) and 5(b). The fracture surface of the horizontal specimen has a fibrous structure, suggesting a ductile fracture. The fracture surface of the horizontal specimen shows clear plastic deformation, while the fracture surface of the vertical specimen shows no plastic deformation but reveals bright crystallization indicating a brittle fracture. Based on the analysis of the fracture morphology we conclude that the horizontal specimen has superior fracture mechanical properties compared to the vertical specimen because of the oriented molecular chains.

4. Molecular Chain Model

The crack propagation direction of the horizontal and vertical specimen is shown in Figures 6(a) and 6(c). Because the molecular chains in x and y directions are oriented, the mechanical properties are enhanced. The molecular chains in x and y directions are represented by thick solid lines in Figures 6(b) and 6(d). Because the molecular chains in z direction are not oriented, the mechanical properties are weak compared with x and y directions. The molecular chains in z direction are represented using thin dotted lines in Figure 6(d).

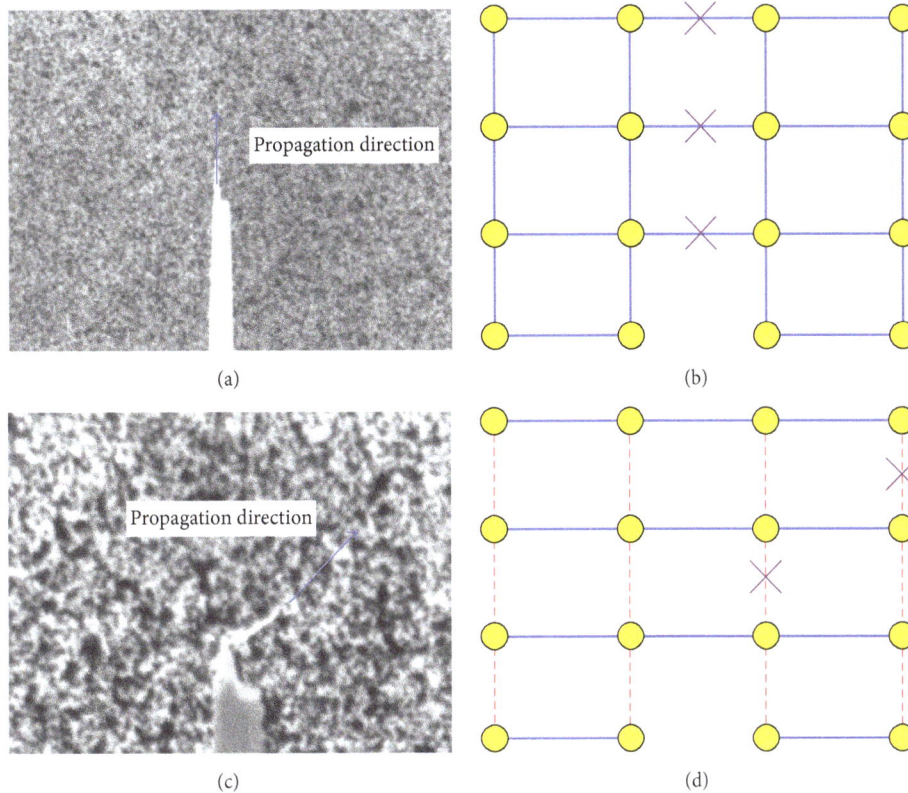

FIGURE 6: Crack propagation direction: (a) propagation direction of the horizontal specimen; (b) propagation direction schematic of the horizontal specimen; (c) propagation direction of the vertical specimen; (d) propagation direction schematic of the vertical specimen.

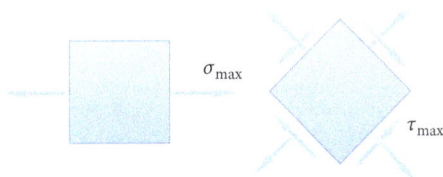

FIGURE 7: Stress state of the tip of the crack.

For the horizontal specimen, when the maximum tensile stress is equal to the value of the tensile strength shown in Figure 7, the crack begins to grow. Because of the large tensile stress, the molecular chains of the tip of the crack are damaged. They are represented by × symbol in Figure 6(b). The crack of the horizontal specimen grows in crack direction, while the nonoriented molecular chains of the vertical specimen are weak. When the maximum shear stress of the tip of the crack equals the shear strength shown in Figure 7, the molecular chains in vertical direction represent the shear failure indicated by × symbol in Figure 6(d). The direction of the maximum shear stress occurs at a 45-degree angle shown in Figure 7. Hence, the crack of the vertical specimen grows at 45 degrees. The conclusions based on the molecular chain model are consistent with the experimental results.

5. Conclusion

Extended digital image correlation was used to measure the displacement field near the tip of the crack when the cracks grow in different directions in oriented PMMA. The crack of the horizontal specimen grows in the crack direction, while the crack of the vertical specimen grows in an oblique direction. Because the molecular chains perpendicular to the oriented direction are not oriented, the mechanical properties are weaker. The nonoriented molecular chains of the vertical specimen are damaged first, and the crack grows in an oblique direction.

The critical load of different specimens was measured. The experiment shows that the critical load of the horizontal specimen is larger than that of the vertical specimens. Therefore, the crack perpendicular to the directional tension grows more easily.

Based on the analysis of the fracture morphology we conclude that the horizontal specimen has better fracture mechanical properties than the vertical specimen because of the oriented molecular chains.

Based on our experiments, the molecular chain model of oriented PMMA could be established. The analysis results using the molecular chain model are consistent with the results of the experiment, which confirms the reliability of the molecular chain model. Using the molecular chain model, the surface crack of the real component should be paid more attention to because the crack in the thickness direction grows more easily.

Competing Interests

The authors declare that there are no competing interests regarding the publication of this article and regarding the funding that they have received.

Acknowledgments

The authors acknowledge the financial support of the National Natural Science Foundation of China (Grants nos. 11402166, 51301057, 51208336, and 51408400) and the General Projects in Tianjin Research Program of Application Foundation and Advanced Technology (Grants nos. 13JCYBJC39100 and 14JCYBJC21900).

References

[1] J. P. Berry, "Fracture processes in polymeric materials. I. The surface energy of poly(methyl methacrylate)," *Journal of Polymer Science*, vol. 50, no. 153, pp. 107–115, 1961.

[2] B. Mukherjee and D. J. Burns, "Fatigue-crack growth in polymethylmethacrylate," *Experimental Mechanics*, vol. 11, no. 10, pp. 433–439, 1971.

[3] C. W. Woo and C. L. Chow, "Fatigue crack propagation in aluminium and PMMA," *International Journal of Fracture*, vol. 26, no. 2, pp. R37–R42, 1984.

[4] W. M. Cheng, G. A. Miller, J. A. Manson, R. W. Hertzberg, and L. H. Sperling, "Mechanical behaviour of poly(methyl methacrylate)—part 1 Tensile strength and fracture toughness," *Journal of Materials Science*, vol. 25, no. 4, pp. 1917–1923, 1990.

[5] W. M. Cheng, G. A. Miller, J. A. Manson, R. W. Hertzberg, and L. H. Sperling, "Mechanical behaviour of poly (methyl methacrylate). Part 2 the temperature and frequency effects on the fatigue crack propagation behaviour," *Journal of Materials Science*, vol. 25, no. 4, pp. 1924–1930, 1990.

[6] H.-S. Kim and Y.-W. Mai, "Effect of temperature on fatigue crack growth in unplasticized polyvinyl chloride," *Journal of Materials Science*, vol. 28, no. 20, pp. 5479–5485, 1993.

[7] H. S. Kim and X. M. Wang, "Temperature and frequency effects on fatigue crack growth of UPVC," *Journal of Materials Science*, vol. 29, no. 12, pp. 3209–3214, 1994.

[8] F. Ramsteiner and T. Armbrust, "Fatigue crack growth in polymers," *Polymer Testing*, vol. 20, no. 3, pp. 321–327, 2001.

[9] X. F. Yao, G. C. Jin, K. Arakawa, and K. Takahashi, "Experimental studies on dynamic fracture behavior of thin plates with parallel single edge cracks," *Polymer Testing*, vol. 21, no. 8, pp. 933–940, 2002.

[10] X. F. Yao, W. Xu, M. Q. Xu, K. Arakawa, T. Mada, and K. Takahashi, "Experimental study of dynamic fracture behavior of PMMA with overlapping offset-parallel cracks," *Polymer Testing*, vol. 22, no. 6, pp. 663–670, 2003.

[11] W. Xu, X. F. Yao, M. Q. Xu, G. C. Jin, and H. Y. Yeh, "Fracture characterizations of V-notch tip in PMMA polymer material," *Polymer Testing*, vol. 23, no. 5, pp. 509–515, 2004.

[12] S. Sahraoui, A. El Mahi, and B. Castagnède, "Measurement of the dynamic fracture toughness with notched PMMA specimen under impact loading," *Polymer Testing*, vol. 28, no. 7, pp. 780–783, 2009.

[13] R. Zhang, R. Guo, and S. Wang, "Mixed mode fracture study of PMMA using digital gradient sensing method," *Engineering Fracture Mechanics*, vol. 119, pp. 164–172, 2014.

[14] M. R. Ayatollahi, A. R. Torabi, and M. Firoozabadi, "Theoretical and experimental investigation of brittle fracture in V-notched PMMA specimens under compressive loading," *Engineering Fracture Mechanics*, vol. 135, pp. 187–205, 2015.

[15] F. Berto, D. A. Cendon, P. Lazzarin, and M. Elices, "Fracture behaviour of notched round bars made of PMMA subjected to torsion at −60°C," *Engineering Fracture Mechanics*, vol. 102, pp. 271–287, 2013.

[16] F. Berto, M. Elices, P. Lazzarin, and M. Zappalorto, "Fracture behaviour of notched round bars made of PMMA subjected to torsion at room temperature," *Engineering Fracture Mechanics*, vol. 90, pp. 143–160, 2012.

[17] W. Shang, X. J. Yang, and L. N. Zhang, "Study of anisotropic mechanical properties for aeronautical PMMA," *Latin American Journal of Solids and Structures*, vol. 11, no. 10, pp. 1777–1790, 2014.

[18] W. H. Peters and W. F. Ranson, "Digital imaging techniques in experimental stress analysis," *Optical Engineering*, vol. 21, no. 3, pp. 427–431, 1982.

[19] L. C. S. Nunes and J. M. L. Reis, "Experimental investigation of mixed-mode-I/II fracture in polymer mortars using digital image correlation method," *Latin American Journal of Solids and Structures*, vol. 11, no. 2, pp. 330–343, 2014.

[20] B. Pan, L. Yu, and D. Wu, "Accurate ex situ deformation measurement using an ultra-stable two-dimensional digital image correlation system," *Applied Optics*, vol. 53, no. 19, pp. 4216–4227, 2014.

[21] G. Besnard, F. Hild, and S. Roux, "'Finite-element' displacement fields analysis from digital images: application to Portevin–Le Châtelier bands," *Experimental Mechanics*, vol. 46, no. 6, pp. 789–803, 2006.

[22] J. Réthoré, F. Hild, and S. Roux, "Extended digital image correlation with crack shape optimization," *International Journal for Numerical Methods in Engineering*, vol. 73, no. 2, pp. 248–272, 2008.

[23] S. Roux and F. Hild, "Digital image mechanical identification (DIMI)," *Experimental Mechanics*, vol. 48, no. 4, pp. 495–508, 2008.

[24] X. C. Zhang, G. Yang, N. Zhan, and H. Ji, "Gray change detection method for damage monitoring in materials," *Applied Optics*, vol. 54, no. 4, pp. 934–939, 2015.

[25] J. L. Chen, N. Zhan, X. C. Zhang, and J. X. Wang, "Improved extended digital image correlation for crack tip deformation measurement," *Optics and Lasers in Engineering*, vol. 65, pp. 103–109, 2015.

[26] N. Moës, J. Dolbow, and T. Belytschko, "A finite element method for crack growth without remeshing," *International Journal for Numerical Methods in Engineering*, vol. 46, no. 1, pp. 131–150, 1999.

[27] Z. He, A. Kotousov, A. Fanciulli, F. Berto, and G. Nguyen, "On the evaluation of stress intensity factor from displacement field

affected by 3D corner singularity," *International Journal of Solids and Structures*, vol. 78-79, pp. 131–137, 2016.

[28] A. Kotousov, Z. He, and A. Fanciulli, "Application of digital image correlation technique for investigation of the displacement and strain fields within a sharp notch," *Theoretical and Applied Fracture Mechanics*, vol. 79, pp. 51–57, 2015.

[29] L. P. Pook, "A 50-year retrospective review of three-dimensional effects at cracks and sharp notches," *Fatigue & Fracture of Engineering Materials & Structures*, vol. 36, no. 8, pp. 699–723, 2013.

Polymer-Cement Mortar with Quarry Waste as Sand Replacement

D. N. Gómez-Balbuena,[1,2] T. López-Lara (iD),[3] J. B. Hernandez-Zaragoza (iD),[3]
R. G. Ortiz-Mena,[2] M. G. Navarro-Rojero,[1] J. Horta-Rangel (iD),[3] R. Salgado-Delgado,[4]
V. M. Castano (iD),[5] and E. Rojas-Gonzalez[3]

[1]CIATEQ A.C., Av. Del Retablo No. 150, Col. Constituyentes Fovissste, 76150 Santiago de Querétaro, QRO, Mexico
[2]Instituto Tecnológico Superior de Huichapan, Domicilio conocido sn Col. El Saucillo, 42411 Huichapan, HGO, Mexico
[3]División de Estudios de Posgrado, Facultad de Ingeniería, Universidad Autónoma de Querétaro, Cerro de las Campanas S/N, Col. Niños Héroes, 76010 Santiago de Querétaro, QRO, Mexico
[4]División de Estudios de Posgrado e Investigación, Instituto Tecnológico de Zacatepec, Calzada Tecnológico No. 27, Col. Centro, 62780 Zacatepec, MOR, Mexico
[5]Centro de Física Aplicada y Tecnología Avanzada, Universidad Nacional Autónoma de México, Boulevard Juriquilla 3001, 76230 Santiago de Querétaro, QRO, Mexico

Correspondence should be addressed to T. López-Lara; lolte@uaq.mx

Academic Editor: Antonio Gilson Barbosa de Lima

The activities of carved Quarry extraction generate problems of landscape pollution such is the case of solid waste discharged into open land dumps in central Mexico. This article presents the technological application of this solid waste in a new polymeric material with properties similar to those of a traditional mortar. It is concluded that the polymeric material uses low amounts of cement with respect to the traditional mortar, and it is elaborated with the recycled quarry as they are presented in its granulometry. The polymer used favored a low water/cement ratio (0.3) which did not allow to decrease resistance due to the fine nature of the materials (residues and cement) in addition to maintaining the workability of the material. The quarry residue was classified as silt with low plasticity and was characterized by X-ray diffraction and Fluorescence to identify 76% of SiO_2, which is why it was used as a stone aggregate even though the fines content was approximately 93%. The maximum compression resistance obtained at 28 days were 8 Mpa with the polymer/solid ratios of 0.10, water/solids of 0.30, and quarry/solids of 0.67. Linear equations were analyzed for more representative values with R squared adjustment.

1. Introduction

The extraction activities of quarries (volcanic tuffs of the Riolitica type), as shown in Figure 1, generate a huge amount of solid waste that pollutes the environment and generate a lot of dust in the environment. The economic-mining activity in the region of Huichapan, Hidalgo, Mexico, represented by material banks for different industries ranks second in importance in the nonmetallic mining district of Hidalgo State. Companies extract carved rock that is marketed in the national and international market. The final products are tile for floors and facades, columns, blocks, and handcrafted pieces [1]. The volume of quarry mining production in the state of Hidalgo, Mexico, in the last 5 years averages 58×106 kg with an annual value over $ 214,000 USD [2]. It is estimated that about 40% of the production volume is wasted [3], which represents an annual volume of 23.2×103 kg of waste. The current waste management strategy consists of unloading them in landfills in the open, regardless of the potential use that these by-products may present to other industries. Such waste is classified into two types: solid waste resulting from quarry sites or processing units and sludge originated in the processes of cutting and detailed by the water used to cool and lubricate the machines

Figure 1: Accumulation of sludge (own source).

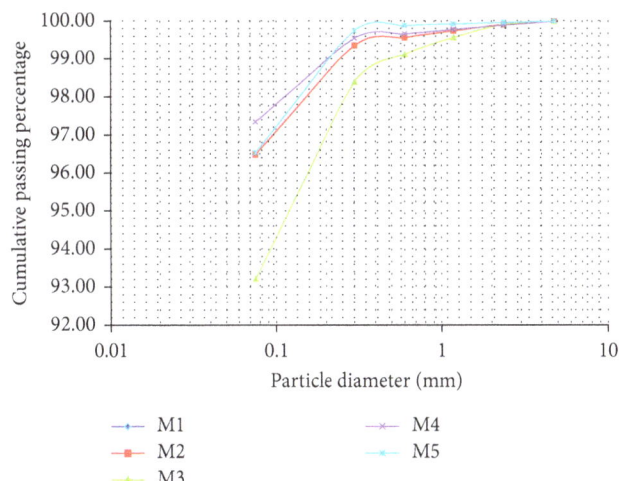

Figure 2: Quarry sieve analysis (own source).

used in said processes. These sludges accumulate gradually, reducing productive space within the company, or are thrown on the sides of the roads, accumulated in unused lands, which over time be leached or dragged, and obstruct the flow of aquifers or drainages. The large quantities already accumulated of the waste demand a prompt solution, which can be sustainable and economically beneficial for the quarry industry: as indicated by Galetakis and Soultana [4], the key to the successful use of quarry dust is its adequate characterization and the development of a simple and economically viable process to convert this waste material into marketable products.

The production and dumping of solid waste has exacerbated carbon emissions and increased pollution in metropolitan cities around the world. Waste management remains a global challenge for both developed and developing countries [5]. A significant number of researchers have studied the use of quarry waste in construction offering viable solutions to this problem. The predominant proposed applications are concrete production (42%), self-compacting concrete production (26%), and block production (18%) [4]. Almeida et al. [3] produced high-performance concrete using recycled stone mud and substituted 5% sand with quarry dust improving strength and durability values in all mixtures containing less than 20% dust. Balamurugan and Perumal [6] used quarry dust in the Tamil Nadu region, India as sand replacement material for concrete production, with a maximum increase in compressive strength (19.18%), tensile strength (21.43%), and resistance to bending (17.8%) with 50% replacement of sand by quarry dust. Sureshchandra et al. [7] replaced sand with quarry dust for the production of hollow concrete blocks. The blocks with replacement of 50% of sand by quarry dust had better performance than those with complete replacement of sand. Arunachalam et al. [8] used quarry powder as a light aggregate and aluminum powder as an air entraining agent for the production of lightweight concrete obtaining resistance of 3–7 MPa for mixtures with quarry dust. Adajar et al. [9] investigated the structural performance of concrete with quarry waste as a substitute for fine aggregates in a concrete mix. They formulated a model to predict the compressive strength of the mixtures made. Lohani et al. [10] replaced partially sand in concrete production. The dust content up to 30% increases the compressive strength of concrete. If the dust content exceeds 30%, the resistance

gradually decreases. Safiuddin et al. [11] concluded that the addition of fine quarry waste can be used as a good substitute for sand in the production of concrete. Galetakis et al. [12] developed a laboratory method for the production of recycled quarry construction elements. Venkatakrishnaiah and Rajkumar [13] reinforced concrete with plastic waste fibers replacing natural sand with quarry dust from the Tamil Nadu region, India. The maximum resistance and best workability were with 30% of sand replacement.

Cement mortar and concrete have disadvantages such as delayed hardening, low tensile strength, shrinkage by drying, and low chemical resistance. To reduce these disadvantages, the use of polymers to modify the properties of mortar and cement has been dominant materials in the construction industry since the 1980s, which are now popularly used in advanced countries [14, 15]. Polymer-modified cement mortars are used in civil infrastructures, bridges, insulation for walls, self-leveling mortars, and concrete for fracture repair due to their excellent resistance, environmental protection and workability [2]. There is a wide variety of commercial latex polymers mostly based on elastomeric and thermoplastic polymers that form continuous films of polymer when dehydrated [16, 17]. The latex polymers include butylbenzene latex, neoprene emulsion, polyvinyl chloride-vinylidene chloride emulsion, styrene-acrylic emulsion, styrene-butadiene carboxy latex, polyacrylate latex, and so on [18]. The dust contents that are normally handled are less than 30% in order not to affect the workability and compressive strength [19].

Polymers such as latex, redispersible polymer powders, water-soluble polymers, liquid resins, and monomers are used for the modification of mortar or cement. Latex is the most widely used additive [20]. In general, latex-type polymers are copolymer systems consisting of two or more monomers, and their total solid content corresponds to 40% or 50% of their weight [21]. The hydration of the cement precedes the process of forming thin films of polymer which leads to the monolithic comatrix phase in which the organic polymer matrix and the cement gel matrix are

FIGURE 3: Sample of recycled Quarry (own source).

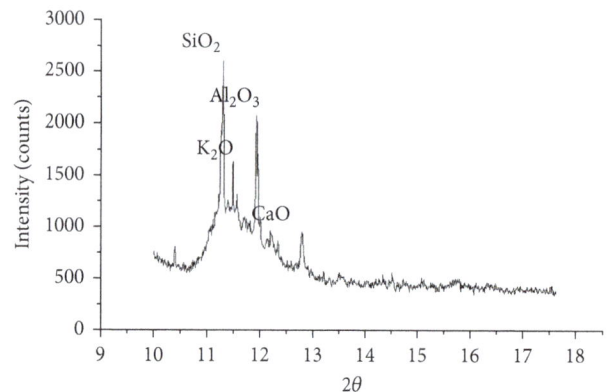

FIGURE 4: X-ray diffraction spectrum of the Quarry waste used (own source).

homogenized [22, 23]. Usually, a polymer/cement ratio from 5% to 15% and a water/cement ratio from 30% to 50% of the latex modified concrete depend on the workability [24].

The growth of the construction industry has led to overexploitation of natural resources such as gravel and river sand by production of concrete. So, the global trend is to use alternative materials (recycled materials) in the construction industry to make rational and sustainable use of natural materials and therefore reduce costs of construction [9].

2. Materials and Methods

2.1. Materials. The following materials were used:

(a) Cement CPC 30R (Ordinary Portland Cement) which complies with the characteristics established in the Mexican standard NMX-C-414-ONNCCE.

(b) Quarry waste (rhyolitic volcanic tuff) extracted from the solid waste of stone of "Jaramillo" quarries in Maney town, Huichapan Hidalgo, Mexico. It was used Quarry as a shorthand notation of Quarry waste or Quarry residue on the text.

(c) The polymer used was a synthetic latex emulsion and acrylic resins that had the specified requirements in the ASTM-1059-99 Type I Standard.

(d) Water used for the mixing and curing of the material with a pH value of 7 (determined by a test strip).

2.2. Methods. The methods used for the experimentation were the following:

(a) Geotechnical characterization of the waste. The field identification [25] of the material was done as well as the grain-size distribution [26], plasticity properties [27], and soil classification [28].

(b) Physical-chemical characterization of the waste. The mineralogical characterization for the determination of primary mineral species (deposit mineral associations) was done by X-ray diffraction (XRD) with Bruker D8-Advance equipment using Göebel mirror (nonflat samples), high-temperature chamber (up to 900°C), X-ray generator KRISTALLOFLEX K 760-80F (power: 3000 W, voltage: 20–60 kV, and current: 5–80 mA), and a Seifert model JSO-DEBYEFLEX 2002 fitted with a copper cathode and a nickel filter.

(c) Analysis and comparison of grain-size distributions [26] of different types of sands as well as their mineralogical composition determined by X-ray diffraction and fluorescence.

(d) Compression test according to [29]. The compression strength of the mixtures was determined according the ASTM C39/C39M-2016b Standard Test Method for Compressive Strength of cylindrical concrete specimens at 3, 7, 14, and 28 days. For the compression test [29], a hydraulic press of 20 tons was used, with pressure sensor WIKA model A10, from 0 to 200 bar and analog output from 0 to 10 Vdc, Fluke Brand Model 115 multimeter.

3. Results and Discussion

3.1. Geotechnical Characterization of the Waste. According to the field identification, grain-size distribution analysis [26], determination of the plasticity limits (liquid and plastic) [27], and classification of soils [28], the following results were obtained:

From the field identification, the recycled quarry residues were materials with low tenacity and slow dilatation and had very low resistance in the dry state. No odor was perceived. The material color was brown to white in light tones. From the classification of soils, the material was a rock powder with little content of slightly plastic inorganic clay located below the "A" line in the plasticity chart. Figure 2 shows the grain-size distribution analysis curves of five waste samples [26]; the data showed that more than 90% of the material passed the 200 mesh. The liquid limit was 24.98%, and plastic limit was 21.25%. Plastic index was 4% on average. So, the classification of soil [28] was ML (inorganic low-compressibility lime, material whose particles have certain cohesion between them in the presence of water). Based on [24], the waste Quarry has the following important engineering properties when the material is compacted and saturated: permeability from semipermeable to impermeable, acceptable shear strength, medium compressibility, and acceptable workability as a construction material. Figure 3 shows the waste Quarry used.

3.2. Geological and Physical-Chemical Characterization of the Quarry Waste. The geological analysis of the waste [30]

TABLE 1: % of chemical composition of recycled Quarry.

Chemical composition of Quarry	SiO$_2$	Al$_2$O$_3$	K$_2$O	CaO	Na$_2$O	TiO$_2$	SO$_3$	MgO	Cl	P$_2$O$_3$
Normalized means	75.958	10.796	8.5099	2.4279	0.9631	0.4791	0.323	0.1486	0.1144	0.0567

TABLE 2: % of chemical composition of recycled Quarry.

Chemical composition of Quarry	BaO	ZrO$_2$	Rb$_2$O	ZnO	SrO	Y$_2$O$_3$	CeO$_2$	Ga$_2$O$_3$	Nb$_2$O$_3$
Normalized means	0.0541	0.0536	0.0307	0.0256	0.0209	0.0172	0.0169	0.0023	0.0021

indicated that the geology of the Huichapan caldera corresponds to an upper ignimbrite with columnar fracture and partially welded. Ignimbrite contains lithic fragments of andesite, quartz, and feldspar in a vitreous matrix (lightly crushed vitreous fragments).

The qualitative analysis by means of X-ray diffraction of Quarry dust is observed in Figure 4. The major component in the Quarry waste was silicon dioxide.

Tables 1 and 2 show the results of the X-ray fluorescence analysis of the recycled Quarry (results expressed as a percentage of the compounds present in the waste). Tables 1 and 2 show the following: (1) Silicon dioxide is the major component and is present in 76%. According to [31], the silicates are the most important component of the hydrated cement and the cause of their resistance. Silicon dioxide compound presents a significant difference between the Quarry waste and the cement, since the latter requires only 25% of content. Based on [32], this excess of silicon dioxide will favor the reduction of the porosity of the mixture to improve the interface of the Portland cement paste adhered to the aggregate. Therefore, the strength and compactness of the final product are increased. (2) CaO is the compound that provides the greatest resistance to cement [31]: in this study, the Quarry waste has much lower amounts than cement, 2.4% and 67%, respectively. It would be expected that mixtures containing high percentages of Quarry waste present low resistances. (3) Alkaline compounds (such as Na$_2$O) cause disintegration of concrete and affect the rate of increase of cement strength [31]. Na$_2$O compound (0.96%) in the Quarry waste is within of the allowable range of 0.2 to 1.3% of the cement. (4) Magnesium oxide (MgO) [31] is a substance that often accompanies calcium oxide. MgO is not combined during the cooking process of Portland cement and therefore it does not form hydraulic components but remains as free MgO. MgO is similar to lime. So water hydrates and increases the volume of MgO. A high percentage of MgO implies the risk of expansion [33]. Expansion by MgO is more dangerous because it appears very slowly over the years. For this reason, the cement standard stipulates a maximum limit of 5% for the MgO content. In this study, the MgO content was favorable with only 0.15%. The gray-green color of Portland cement is due to MgO [31].

3.3. Analysis and Comparison of Grain-Size Distributions and Mineralogical Composition of Sands.
Figure 5 shows a comparative analysis of the chemical compositions obtained by X-ray diffraction of the Quarry, river sands of Boye-HUI-53

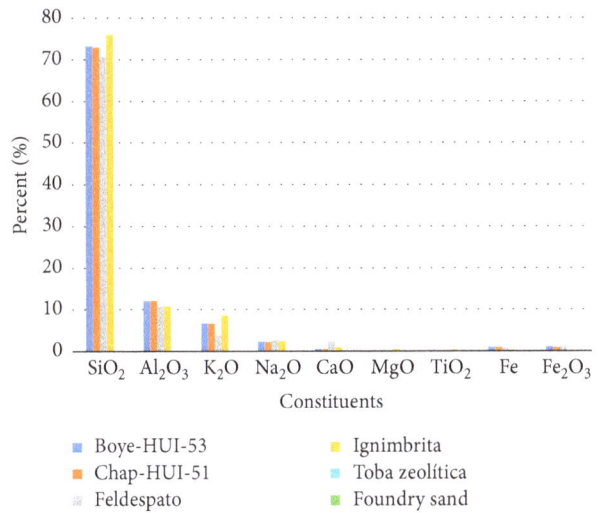

FIGURE 5: Chemical composition for different sands and ignimbrita (own source).

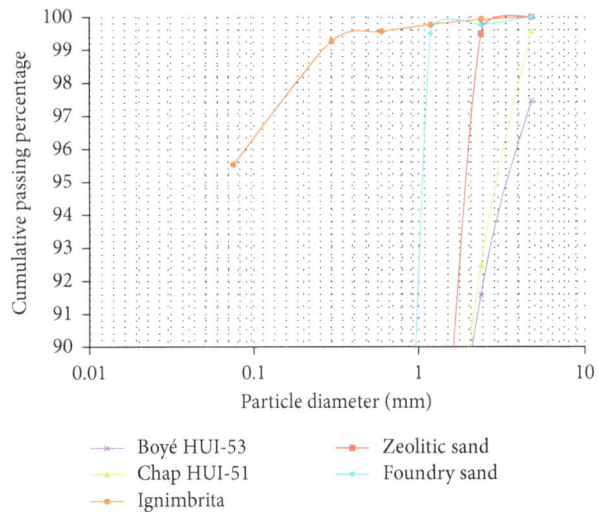

FIGURE 6: Sieve analysis for different sands [24, 29] (own source).

and Chap-HUI-51 [2], feldspathic sand, zeolitic tuff sand, and silica sand foundry. The river sands are from regions near the place where the ignimbrite (Quarry waste) was extracted. Feldspathic sand is used in the ceramics industry [34] and zeolitic tuff sand is used for lining mortar [35]. High-quality silica sand foundry is a by-product generated by the ferrous and nonferrous metal foundry [36]. The

TABLE 3: Quantities in grams for the different concrete mixtures (own source).

MIX	Cement (g)	Quarry (g)	Water (g)	Polymer (g)	Quarry/solids	Water/solids	Polymer/solids
M1				30	0.67	0.30	0.10
M2			90	45	0.67	0.30	0.15
M3		200		60	0.67	0.30	0.20
M4				30	0.67	0.40	0.10
M5			120	45	0.67	0.40	0.15
M6	100			60	0.67	0.40	0.20
M7				40	0.75	0.30	0.10
M8			120	60	0.75	0.30	0.15
M9		300		80	0.75	0.30	0.20
M10				40	0.75	0.40	0.10
M11			160	60	0.75	0.40	0.15
M12				80	0.75	0.40	0.20

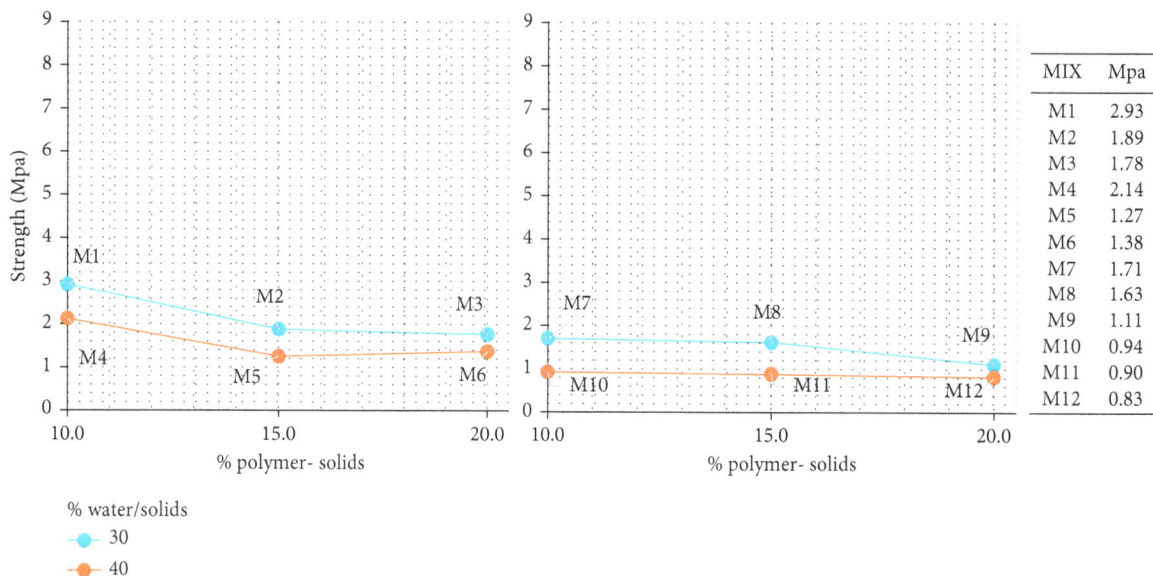

MIX	Mpa
M1	2.93
M2	1.89
M3	1.78
M4	2.14
M5	1.27
M6	1.38
M7	1.71
M8	1.63
M9	1.11
M10	0.94
M11	0.90
M12	0.83

% water/solids
— 30
— 40

FIGURE 7: Compression resistance at 3 days (own source).

results showed that the mineralogical compositions of all the sands and quarries are very similar.

A comparative grain-size analysis distribution [26] corresponding to zeolitic sand [35], foundry sand [36], ignimbrite rhyolitic, and 2 types of sands for construction in the regions of Chapantongo and Boyé in Hidalgo México was reported in Figure 6.

From the comparison of the sands (Figure 6), it was observed that the sands of Chapantongo and Boyé had similar grain-size distributions with average grain size of 2.36 mm passing both 80% content by mesh number 16. Chapantongo sand was slightly thinner since 39% of the material analyzed passed the 50 mesh (grain size of 0.3 mm) compared to 26% of the Boyé sand that passed the same mesh. The zeolitic sand and foundry sand had a finer granulometry whose aggregates in both cases passed in 60% the 30 mesh (0.60 mm). The ignimbrite curve showed a very smooth slope which indicates that their grain size differs significantly from the other sands. More than 95% of the ignimbrite passed the 200 mesh (0.075 mm).

Based on the comparative analysis of the physico-chemical characterization of the Quarry waste and various sands, it was observed that it could be viable to replace the sand by a fine aggregate by the similarity of chemical composition. From the grain-size distributions curves, it was observed that the sands had more uniform dimensions (sandy aggregates with very few fines). On the other hand, the ignimbrite was a fine soil, and therefore the mechanical behavior can be less favorable. However, derived from the similarity in chemical composition and granulometry between the Quarry and various sands studied, the feasibility of replacing 100% of sand as a fine aggregate by the ignimbrite can be used in the production of mortars and concretes. Based on [37], this substitution

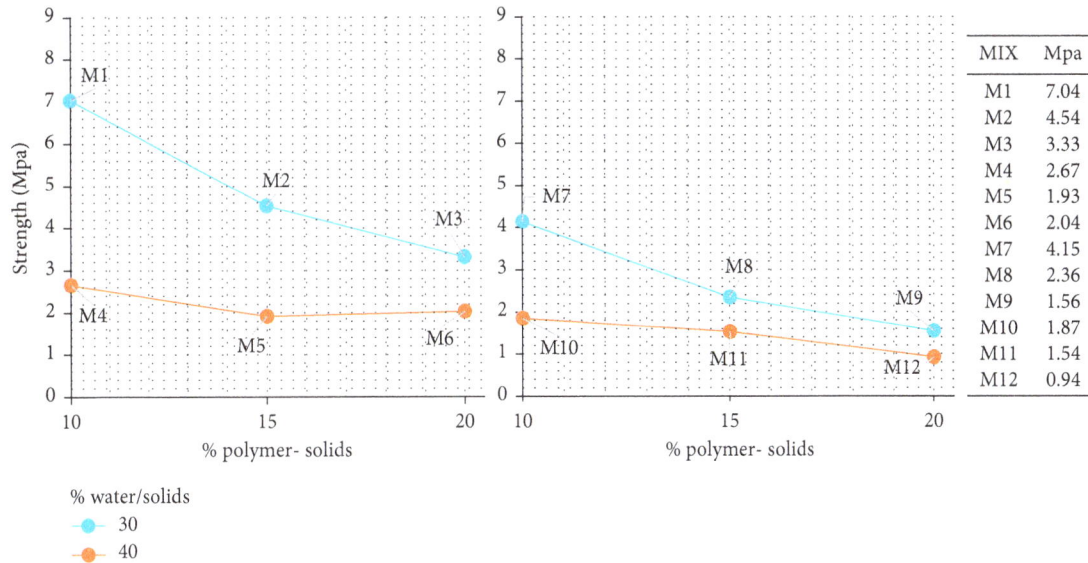

MIX	Mpa
M1	7.04
M2	4.54
M3	3.33
M4	2.67
M5	1.93
M6	2.04
M7	4.15
M8	2.36
M9	1.56
M10	1.87
M11	1.54
M12	0.94

FIGURE 8: Compression resistance at 7 days (own source).

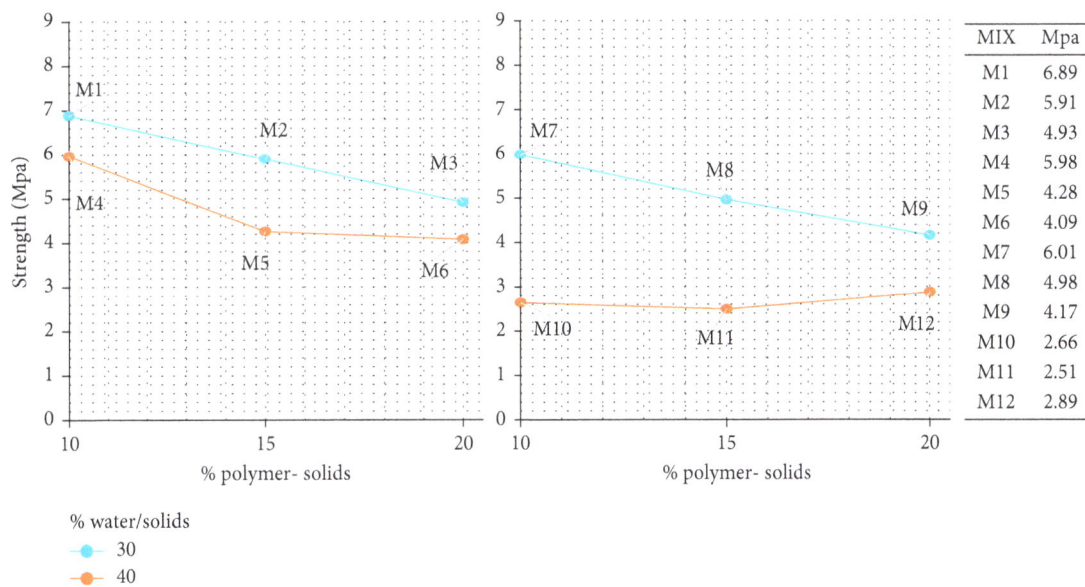

MIX	Mpa
M1	6.89
M2	5.91
M3	4.93
M4	5.98
M5	4.28
M6	4.09
M7	6.01
M8	4.98
M9	4.17
M10	2.66
M11	2.51
M12	2.89

FIGURE 9: Compression resistance at 14 days (own source).

generates in the products increased tenacity and impact resistance and reduced shrinkage by drying and cracking in its hardened state.

3.4. Compression Test Analysis. Compressive strength testing was undertaken at 3, 7, 14, and 28 days of age upon 0.051 m diameter and 0.102 m length cylindrical specimens, maintaining a length to diameter ratio of 2, according to [38]. 12 mixtures were elaborated of different proportions based on the following premises: (1) use the largest amount of quarry dust; (2) use the least amount of water without affecting aspects such as the

workability of the sample and without using additives such as superplasticizers; and (3) use the least amount of polymer. Table 3 shows the 12 proportions for the different samples considered for the present study. The samples were numbered according to the MIX column, being M1 the sample number one and so on. The columns cement, quarry, water, and polymer show the quantities used in grams for each mixture. The amount of 100 g of cement was constant in all the mixtures, adding double amount of the Quarry (200 g) for the mixtures M1 to M6 and the triple amount of Quarry (300 g) for the mixtures M7 to M12. In the Quarry/solids column, the high content of ignimbrite residue is highlighted with respect to the quantity of

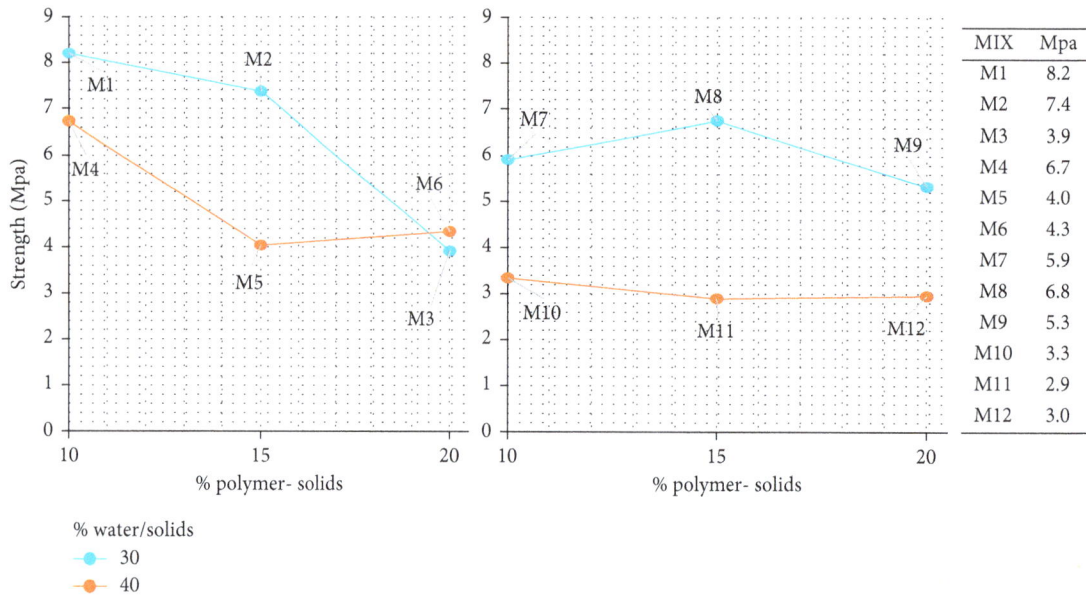

FIGURE 10: Compression resistance at 28 days (own source).

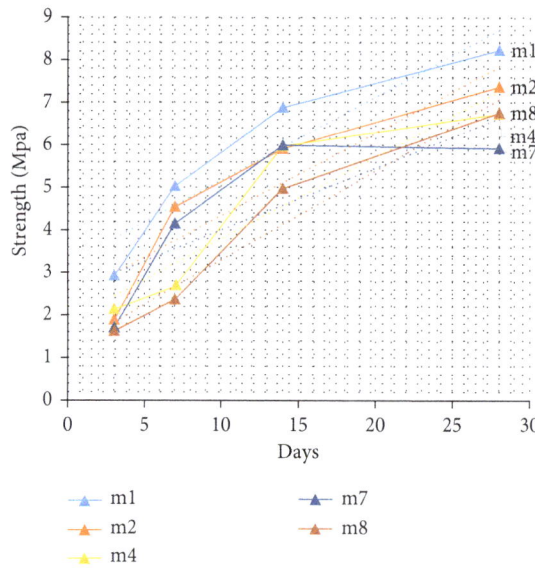

FIGURE 11: Most representative values and trend lines (own source).

solids handling values of 0.67 and 0.75 (ratios calculated by (2)). The amounts of 90, 120, and 160 g of water used in the mixtures correspond to ratios of 0.3 and 0.4 of water with respect to the amount of solids (indicated in the water/solids column and calculated by (3)). The polymer was mixed in ratios of 0.10, 0.15, and 0.20 with respect to the amount of solids (indicated in the column polymer/solids and calculated by (4)). The above-mentioned solids are indicated as the sum of cement and Quarry in (1).

$$\text{Solids} = \text{grams quarry} + \text{grams cement}, \tag{1}$$

$$\frac{\text{Quarry ratio}}{\text{Solids ratio}} = \frac{(\text{quarry grams})}{(\text{quarry grams} + \text{cement grams})}, \tag{2}$$

$$\text{Water ratio solids} = \frac{(\text{grams of water})}{(\text{grams of quarry} + \text{grams of cement})}, \tag{3}$$

$$\frac{\text{Polymer ratio}}{\text{Solids ratio}} = \frac{(\text{polymer grams})}{(\text{grams of quarry} + \text{grams cement})}. \tag{4}$$

TABLE 4: Linear equations and R-square value for most representative mixtures (own source).

Linear equations	R square value
$y1 = 0.1957x + 3.2262;$	$R^2 = 0.8707$
$y2 = 0.1943x + 2.4037;$	$R^2 = 0.8368$
$y4 = 0.1923x + 1.8775;$	$R^2 = 0.8339$
$y7 = 0.1458x + 2.5507;$	$R^2 = 0.6328$
$y8 = 0.2096x + 1.2071;$	$R^2 = 0.9445$

For compression tests according to [29] at 3, 7, 14, and 28 days, the reported results are the average of three compression tests of each of the 12 mixtures performed. The load was applied axially and continuously until failure of the specimen recording the maximum load applied and the type of fracture according to [29]. At three days, the highest resistance of 3 MPa occurred in the M1 mixture with quarry/solids ratio of 0.67, water/solids of 0.30, and polymer/solids of 0.10. The results are shown in Figure 7.

Figure 8 shows results obtained at 7 days highlighting M1 and M7 with respect to the resistance measured at 3 days with increases of 130% and 142% reaching 7 and 4 Mpa, respectively. The results at 14 days are observed in Figure 9 highlighting M4 that increased the resistance obtained to 7 days from 2.6 Mpa up to 6 MPa in 14 days with a water/solids ratio of 0.4. Another significant resistance increase occurred in M7 that reached 4 Mpa with low water/solids ratios of 0.3 and polymer/solids of 0.1 but significant Quarry/solids ratio of 0.75.

Figure 10 shows that M1 reached the highest resistance at 28 days in the order of 8 Mpa and M2 shows resistance close to M1 with a polymer/solids ratio of 0.15. M8 also stands out with resistance close to 7 MPa and the highest Quarry/solids ratio of 0.75.

Figure 11 shows the increase in resistance as the age of the specimens increases, being in all cases M1 the best behavior reaching a resistance of 8 MPa at the age of 28 days. A rapid increase in resistance is noted between the ages of 7 and 14 days. After 14 days, there is a gradual and slow increase of resistance until 28 days. Trend lines were annexed to Figure 11. The corresponding linear equations and the value of R square of trend lines are shown in Table 4. The value of R square shows that the values are attached to a straight line of trend with a deviation margin of less than 20% in most cases.

Compressive load [29] was applied at 14 days of age. Figure 12 includes only the samples of highest recorded load (M1, M3, M7, M8, and M11). The aforementioned specimens show cone-type fracture patterns, that is, cone well defined only at one end and vertical fractures through the cylindrical column (also called type two fractures). This fracture pattern is common and representative of cementing material. Only in M9, vertical fractures were formed across the ends, without well-defined cone formation or fracture type three. Figure 13 shows fractures at the age of 28 days of the specimens M1, M3, M5, M7, and M8, observing in all the cases conical fractures type two.

4. Conclusions

The waste derived from the Quarry stone was classified as inorganic low compressibility lime according the plasticity properties and grain-size distribution.

FIGURE 12: Types of fractures in samples at 14 days. Mixtures (a) M1, (b) M3, (c) M7, (d) M8, and (e) M11 with higher performance, respectively, from the left side (own source).

The geological analysis of the ignimbrite (Quarry stone) indicated that it contains lithic fragments of andesite, quartz, and feldspar in a vitreous matrix (lightly crushed vitreous fragments).

Based on the comparative analysis of the physico-chemical characterization of the Quarry waste and various sands, it was observed that it could be viable to replace the sand by a fine aggregate by the similarity of chemical composition. The major component in the waste is silicon dioxide (76%), and so it was considered pertinent to replace the content of sands that generally constitutes a mortar.

It is concluded that the polymeric material uses low amounts of cement with respect to the traditional mortar, and it is elaborated with the recycled quarry as they are presented in its granulometry, which saves the process of size selection.

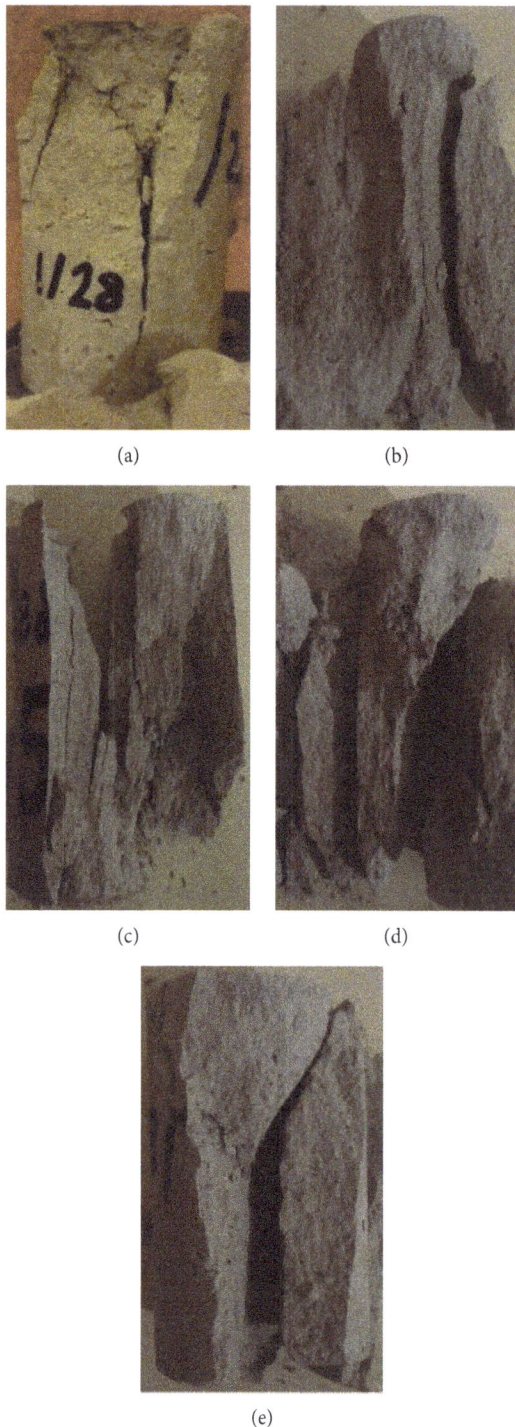

(a)   (b)

(c)   (d)

(e)

FIGURE 13: Types of fractures in 28-day samples. Mixtures (a) M1, (b) M3, (c) M5, (d) M7, and (e) M8 with higher performance, respectively, from the left side (own source).

According to the greater resistance obtained in the compression tests to specimens with age of 28 days, it is concluded that it is possible to replace the use of sand as fine aggregate by Quarry dust without significant reduction of the compressive strength of mortars.

The utilization of ignimbrite residues (Quarry waste) in the manufacture of mortars presents the best resistance to compression at 28 days of age with a resistance of 8 MPa with the following optimum proportions: polymer/solids of 0.1, water/solids of 0.3, and Quarry/solids of 0.67. Quantities greater than 0.15 of the polymer/solids ratio significantly decrease compressive strength [20]. Likewise, the compressive strength was considerably affected when quarry/solids ratios greater than 0.67 were used.

The polymer used in the mortar allowed a low water/solids ratio (0.3) and acceptable workability.

This strategy of reusing of ignimbrite waste has the advantage of using a large amount of these residues with respect to the amount of cement without using additional processes in the waste. The Quarry waste was used as it was collected directly from the deposits. This effective and sustainable solution is considered for the solid waste management of the Quarry industry and presents an alternative of raw material for the production of mortars.

Conflicts of Interest

The authors declare that they have no conflicts of interest.

References

[1] Servicio Geológico Mexicano, "gob.mx," August 2017, http://mapserver.sgm.gob.mx/Cartas_Online/geologia/1653_F14-C78_GM.pdf.

[2] Servicio Geológico Mexicano, "SGM," August 2017, http://www.sgm.gob.mx/pdfs/HIDALGO.pdf.

[3] N. Almeida, F. Branco, J. de Brito, and J. R. Santos, "High-performance concrete with recycled stone slurry," *Cement and Concrete Research*, vol. 37, no. 2, pp. 210–220, 2006.

[4] M. Galetakis and A. Soultana, "A review on the utilisation of quarry and ornamental stone industry fine by-products in the construction sector," *Construction and Building Materials*, vol. 105, pp. 769–781, 2015.

[5] A. K. Choudhary, J. N. Jha, K. S. Gill, and S. K. Shukla, "Utilization of fly ash and waste recycled product reinforced with plastic wastes as construction materials in flexible pavement," in *Proceedings of the Geo Congress 2014*, pp. 3890–3902, Atlanta, GA, USA, February 2014.

[6] G. Balamurugan and P. Perumal, "Behaviour of concrete on the use of quarry dust to replace sand–an experimental study," *Engineering Science and Technology*, vol. 3, no. 6, pp. 776–781, 2013.

[7] H. S. Sureshchandra, G. Sarangapani, and B. G. Naresh Kumar, "Experimental investigation on the effect of replacement of sand by quarry dust in hollow concrete block for different mix proportions," *International Journal of Environmental Science and Development*, vol. 5, no. 1, pp. 15–19, 2014.

[8] N. Arunachalam, V. Mahesh, P. Dileepkumar, and V. Sounder, "Development of innovative building blocks," *Journal of Mechanical and Civil Engineering*, vol. 1, no. 1, pp. 1–7, 2014.

[9] M. A. Q. Adajar, E. de Guzman, R. Ho, C. Palma Jr. III, and D. Sindico, "Utilization of aggregate quarry waste in construction industry," *International Journal of GEOMATE*, vol. 12, no. 31, pp. 16–22, 2017.

[10] T. K. Lohani, M. Padhi, K. P. Dash, and S. Jena, "Optimum utilization of Quarry dust as partial replacement of sand in concrete," *International Journal of Applied Sciences and Engineering Research*, vol. 1, no. 2, pp. 391–404, 2012.

[11] M. Safiuddin, S. Raman, and M. Zain, "Utilization of quarry waste fine aggregate in concrete mixtures," *Journal of Applied Sciences Research*, vol. 3, no. 3, pp. 202–208, 2003.

[12] M. Galetakis, G. Alevizos, and K. Leventakis, "Evaluation of fine limestone quarry by-products, for the production of building elements," *Construction and Building Materials*, vol. 26, no. 1, pp. 122–130, 2012.

[13] R. Venkatakrishnaiah and P. Rajkumar, "Effect of quarry dust on waste plastic fiber reinforced concrete – an experimental study," *Research in Civil and Environmental Engineering*, vol. 1, no. 4, pp. 234–238, 2013.

[14] Y. Ohama, "Recent progress in concrete-polymer composites," *Advanced Cement Based Materials*, vol. 5, no. 2, pp. 31–40, 1997.

[15] L. Czarnecki, "Concrete-polymer composites: trends shaping the future," *Journal of the Society of Materials Engineering for Resources of Japan*, vol. 15, no. 1, pp. 1–5, 2007.

[16] M. Frigione, "Concrete with polymers," *Eco-Efficient Concrete*, pp. 386–436, Woodhead Publishing limited, Cambridge, UK, 2003.

[17] Y. Ohama, "Polymer based admixtures," *Cement and Concrete Composites*, vol. 20, no. 2-3, pp. 189–212, 1998.

[18] W. Meng, "Basic mechanical properties research of SAE latex-modified mortar," *Advanced Materials Research*, vol. 1088, pp. 621–625, 2015.

[19] S. R. Bothra and Y. M. Ghugal, "Polymer-modified concrete: review," *International Journal of Research in Engineering and Technology*, vol. 4, no. 4, pp. 845–848, 2015.

[20] Y. Ohama, *Handbook of Polymer-Modified Concrete and Mortars*, William Andrew Inc., Norwich, NY, USA, 1995.

[21] Y. Ohama, "Principle of latex modification and some typical properties of latex modified mortars and concretes," *ACI Materials Journal*, vol. 84, no. 6, pp. 511–518, 1987.

[22] H. B. Wagner, "Polymer-modified hydraulic cements," *Industrial and Engineering Chemistry, Product Research and Development*, vol. 4, no. 3, pp. 191–196, 1965.

[23] L. K. Aggarwal, P. C. Thapliyal, and S. R. Karade, "Properties of polymer–modified mortars using epoxy and acrylic emulsions," *Construction and Building Materials*, vol. 21, no. 2, pp. 379–383, 2007.

[24] T. Fukuchi and Y. Ohama, "Process technology and properties of 2500 kg/cm2 strenght polymer impregnated concrete," in *Proceedings of the Second International Congress on Polymers in Concrete*, D. W. Fowler and D. R. Paul, Eds., Polymers in Concrete, Austin, TX, USA, 45–56, 1978.

[25] P. M. Cabrera Castro and E. Beira Fontaine, "Geotechnical characterization of the experimental field at the civil engineering from the University of Oriente," *Ingeniería*, vol. 11, no. 2, pp. 57–66, 2007.

[26] ASTM, ASTM D422-63, *Standard Test Method for Particle-Size Analysis of Soils*, ASTM International, West Conshohocken, PA, USA, 2007.

[27] ASTM, ASTM Standard D 4318-10, *Standard Test Methods for Liquid Limit, Plastic Limit, and Plasticity Index of Soils*, Annual Book of ASTM Standards, West Conshohocken, PA, USA, 2010.

[28] ASTM Standard D 2487-93, *Standard Test Method for Classification of Soils for Engineering Purposes*, Annual Book of ASTM Standards, West Conshohocken, PA, USA, 1993.

[29] ASTM International, "Standard test method for compressive strength of cylindrical concrete specimens," March 2017, http://www.astm.org.

[30] M. Milán, C. Yañez, L. Navarro, S. P. Verma, and G. Carrasco Nuñez, "Geología y qeoquímica de elementos mayores de la Caldera de Huichapan, Hidalgo, México," *Geofísica Internacional*, vol. 32, no. 2, pp. 261–276, 1993.

[31] A. Neville and J. J. Brooks, *Tecnología del Concreto*, Trillas, Mexico City, México, 2010.

[32] S. Valdés Constantino, "La Adición de Humo de Sílice para Concretos de Alta Resistencia," 2015.

[33] E. F. Irassar, V. L. Bonavetti, and G. Menéndez, "Cementos con material calcáreo: formación de thaumasita por ataque de sulfatos," *Revista de la Construcción*, vol. 9, no. 1, 2010.

[34] D. Laverde, J. Pedraza, S. Ospina et al., "El beneficio de arenas feldespáticas: una solución para la industria cerámica colombiana," *Dyna*, vol. 71, no. 143, pp. 45–54, 2004.

[35] G. del Carmen Tiburcio Munive, M. Sandez Aguilar, S. Aguayo Salinas, and D. BUrgos Flores, "Morteros para revestimiento de muros, utilizando arena zeolítica," *Épsilon*, vol. 1, no. 10, pp. 57–65, 2008.

[36] B. Bhardwaj and P. Kumar, "Waste foundry sand in concrete: a review," *Construction and Building Materials*, vol. 156, pp. 661–674, 2017.

[37] PINACAL, *Planes de Competitividad Innovación y Entorno*, Unión Europea, Valladolid, Spain, 2007.

[38] ASTM, ASTM C31/C31M-17, *Standard Practice for Making and Curing Concrete Test Specimens in the Field*, ASTM International, West Conshohocken, PA, USA, 2017.

13

Needleless Electrospinning and Electrospraying of Mixture of Polymer and Aerogel Particles on Textile

Abu Shaid ⓘ, Lijing Wang ⓘ, Rajiv Padhye ⓘ, and Amit Jadhav

School of Fashion and Textiles, RMIT University, Melbourne, VIC, Australia

Correspondence should be addressed to Abu Shaid; s3304096@student.rmit.edu.au

Academic Editor: Frederic Dumur

Needleless electrospinning and electrospraying of aerogel particles in comparatively lower electric voltage (9 kV) have been demonstrated in the paper. Aerogel particles were dispersed in polymer solution and then needlelessly electrospun/sprayed by creating high electric charge at the syringe tip using a curved wire. FTIR spectra and SEM images proved that aerogel particles were deposited onto the base textile. In case of electrospinning, a nanofibre web holding the aerogel particles covered the fabric surface, whereas in case of electrospraying, aerogel particles deposited with the microdroplets of the polymer. The electrospraying process showed great potential for fabric surface functionalization due to the high amount of particle deposition on fabric. The new approach can be applicable for transferring other particulate materials on fabric surface through needleless electrospinning and electrospraying processes.

1. Introduction

Electrospinning is a method where high electric force is used as a mean to pull polymer solution and to produce very fine fibres with the diameter in nanometre range. On the contrary, if the polymer molecules do not have sufficient molecular cohesion to withstand the electric pull to form nanofibre, then the ejected stream breaks and creates drops of polymer and the process is referred as electrospraying. Though the method of electrospinning was invented in 1934 [1], it took more than 40 years to come to the needleless method of electrospinning. The first needleless electrospinning was based on a ring spinneret [2], and gradually various other shapes of spinneret were evolved. In the most basic electrospinning setup, a high voltage is applied to a metallic needle and polymer solution is extruded through the hole of the needle. The voltage is continuously increased to the point when the repulsive force between same charge molecules exceeds the surface tension and a jet of polymer solution erupts towards the grounded collector. The ejected polymer jets experience instability in their path to the collector which causes bending and stretching of the jet and results in the production of very fine fibres. In needleless

electrospinning method, the needle is replaced with roller, ball, disc, cylinder, beaded wire, and so on. All these needleless electrospinning methods can be divided primarily into two groups: rotary and stationary spinneret needleless electrospinning [3]. Rotary needleless electrospinning includes rotary cylinder, disc, ball, spiral wire coil, splashing spinneret, cone, edge of metal plate, rotating roller, bowel edge, rotary beaded chain [4–9], and so on. In stationary spinneret needleless electrospinning, the use of vertical rod [10], air bubble [11], conical wire coil [12], and metallic tube electrode [13] has been reported. In current approach, we used a curved wire inside the syringe and syringe tip to condense high electric charge inside the tip where the mixture of polymer and aerogel particles gets charged and extruded to the collector.

Aerogel is a nanoporous material with extremely light weight, high heat insulation, and impressive absorption capability which can be utilized in numerous fields of application like heat protection, oil absorption, and acoustic insulation. Application of aerogel on textiles is a growing interest, and gradually the field of application is expanding. There are few possible ways how the aerogel can be combined with textiles. Textile fibre, yarn, or fabric can be added into the aerogel precursor solution during its synthesis, or already synthesized aerogel can be applied on fibre, yarn, or fabric surface. There are plenty of literatures available where the fibres are immersed into the silica solution to prepare fibrous aerogel composite [14, 15]. In case of the use of synthesized aerogel on textiles, the aerogel particles are mostly used to maintain the flexibility of textile materials. Several reports are available where aerogel particles are either mixed with binder and knife coated on textiles [16, 17], or padded with textile material [18], or thermally bonded in nonwoven textile [19].

A very limited number of reports are available where the electrospinning technique was used to combine both aerogel and textile [20, 21]. Such approach is mainly to add electrospun nanofibre into the aerogel precursor before aerogel synthesis to produce fibrous aerogel. To the best of our understanding, there was no literature available regarding the use of synthesized aerogel particles in polymer solution to electrospin aerogel containing nanofibre until 2015. The application of aerogel particles in polymer solution to produce nanofibre has recently been reported by Mazrouei-Sebdani et al. [22]. Here, aerogel particles were mixed with polymer solution and extruded through the hole of the electrospinning needle by using high voltage. For the first time, the current study aims to apply aerogel particle on textile surface by using an innovative needleless electrospinning and electrospraying method. Without limiting the polymer type, here one water-based polymer and one solvent-based polymer solution were used to demonstrate the concept of such electrospinning technique. In traditional needled electrospinning of particles, clogging of the needle hole is a major concern, whereas the needleless method requires very high voltage as high as 120 kV [23]. In this respect, the current method has a significant advantage that it is a needleless approach and requires comparatively lower power supply (9 kV).

2. Experimental

2.1. Chemicals.
Laboratory-grade dimethylformamide (DMF) was purchased from Merck KGaA (Germany), PVA (polyvinyl alcohol) was obtained from Chem-Supply (Australia), thermoplastic polyurethane (TPU) chip was purchased from Pacific Urethanes (Australia), and Enova aerogel particles were sourced from Cabot Corporation (USA). 100% Nomex woven fabric was used as base textile. Cationic surfactant Parudul BS was used to disperse aerogel particles. Aerogel particles were grinded to smaller size by using the ball milling method prior to the addition in the polymer solution. 9% PVA polymer solution was prepared in deionised water, and then, the small amount of binder material (Tubicoat WP1665 HT) was added in the solution. 5% TPU solution was prepared in DMF at 40°C and continued to be stirred for 8 hours.

2.2. Instruments.
A KDS 200 digital syringe pump (from KD Scientific Inc., Holliston, MA, USA) was used to maintain a constant flow of solution. High voltage power supply, ES30P-5W, with a maximum output voltage of 30 kV was used to generate high voltage. A 10 ml plastic syringe was used. A solid metallic wire of 0.8 mm diameter was inserted into the syringe or the tip of the syringe in a formation of loop while the tail of the loop was extended into the solution. The morphology of the fabric after electrospinning and spraying was observed using a FEI Quanta 200 scanning electron microscope (SEM). A PerkinElmer 400 FTIR spectrometer was used for the characterisation of the sample within the range of 650 cm^{-1} and 4000 cm^{-1}.

2.3. Needleless Electrospinning Method.
A mixture of 1.5% aerogel and polymer solution was prepared by very slow addition of aerogel particles in the polymer solution with continuous stirring. Then, few drops of ionic surfactant and binder were added, and the mixture was continued to be stirred for at least 4 hours before electrospinning. As shown in Figure 1, a metallic wire was inserted into the tip of a plastic syringe where the tail of the wire was extended to the polymer mix and a curved loop was formed inside the tip. The wire was connected with the high voltage source. The highly charged wire inside the polymer mix charged the molecules before reaching to the tip while the curved wire loop in the tip condensed high electric charge. A rotating drum collector, covered with Nomex fabric as base textile, was attached with a grounded wire. The collector was placed at 8 cm working distance from the syringe tip. Aerogel and polymer mixture was supplied through the plastic syringe, and the syringe piston was attached with the movable jaw of the metering pump to maintain a constant rate of liquor flow. When the high voltage source was turned on, high electric charges were condensed on the curved wire inside the syringe tip and charged the polymer-aerogel mixture. As soon as the repulsive force of similar charges exceeded the surface tension (at 9 kV), thin polymer jets with dispersed aerogel particles were ejected towards the collector as shown in Figure 2. The whipping instability thinned the polymer jet and produced nanofibres. As the aerogel particles were also ejected with the polymer jet, the produced nanofibre webs deposited on the particles and bonded them on base textile.

2.4. Needleless Electrospraying.
The same apparatus was used to electrospray aerogel particle on fabric surface. To electrospray rather than electrospinning, a lower concentration of polymer solution was used. In this case, 1% aerogel particle was dispersed in 5% TPU solution in DMF. The aerogel-polymer mixture jet is ejected by the high electric pull but breaks into numerous droplets while travelling towards the collector due to the lack of sufficient molecular cohesion. Aerogel-mixed polymer droplets deposited on the base fabric surface and fixed therein by the physical bonding of polymer and fabric.

2.5. Measurement of Fibre Diameter and Particle Size.
The particle size of transferred aerogel particle and the diameter of formed nanofibre were measured by using ImageJ®, a recognized software for porosity, particle sizing, and measuring in an image.

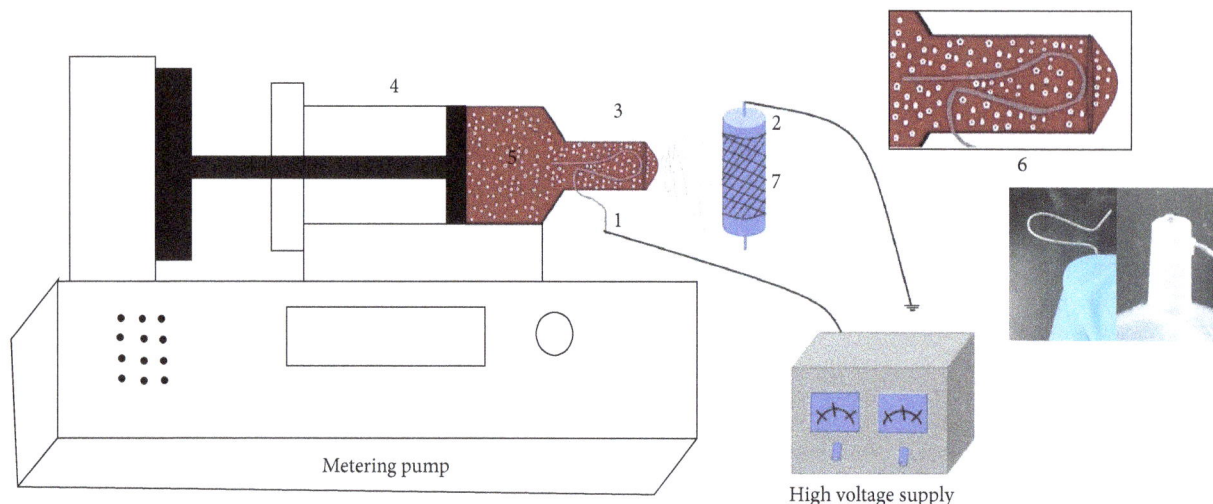

FIGURE 1: Schematic diagram of the process setup. A curved wire (1) was inserted inside the syringe tip (3). The high voltage source was connected to this wire. The textile fabric (7) was wrapped around the rotating drum collector (2) to act as base textile. The syringe (4) was placed on the metering pump. When the high voltage was turned on, aerogel containing polymer mix (5) became charged and attracted towards the collector. The syringe tip with wire is shown insight (6).

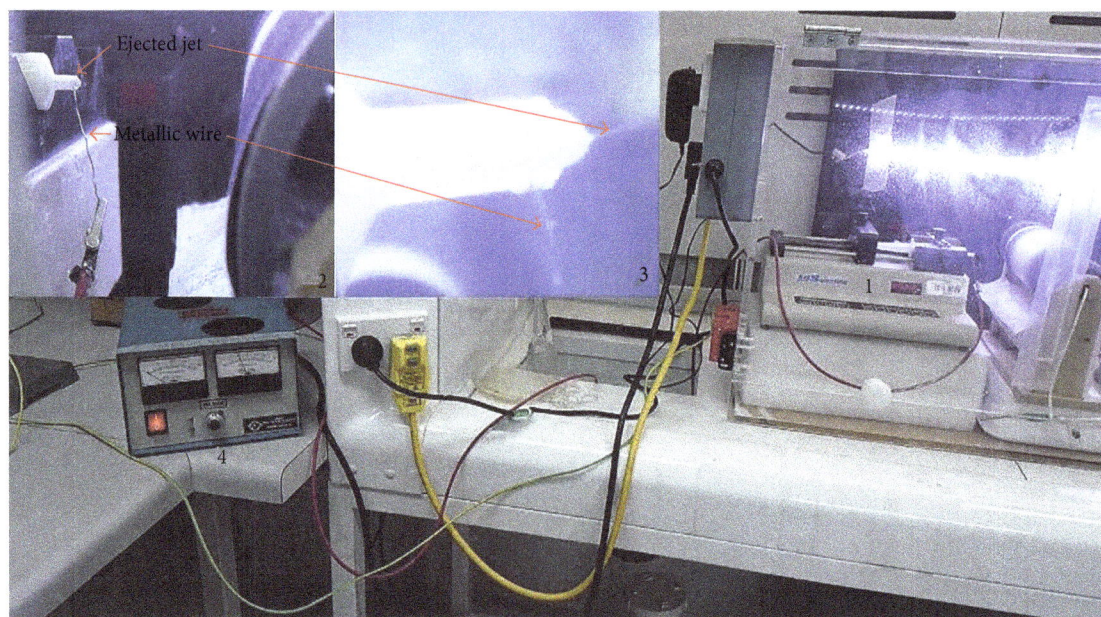

FIGURE 2: Electrospinning/spraying setup. The plastic syringe was placed attached with the arm of the metering pump (1) and connected to the high voltage source (4). A rotating drum collector (2) was connected to the grounded wire. When the voltage was turned on, the polymer jet was ejected (3) from the syringe tip.

3. Results and Discussion

3.1. Transfer of Aerogel Particles. The fabric sample after electrospinning showed a white layer of nonwoven web on base textile. When this sample was observed under the SEM, electrospun nanofibres were clearly visible with randomly deposited aerogel particles.

It was observed that aerogel particles have been transferred through electrical force along with nanofibres, deposited onto the base textile, and bonded therein with the nanofibre webs as hundreds of nanofibres covered the particles to hold them on the fabric surface (Figure 3). Thus, the aerogel

particles were trapped between woven base fabric and nonwoven electrospun nanofibre web. Like any other usual electrospun web, here aerogel nonwoven was formed as layers of electrospun web on base fabric. In general, the fastness of such type of deposition will depend on the features of the polymer (chemical composition, molecular weight, solid content, etc.), the surface characteristics of the base textile (fibre type, fabric construction, surface finish, etc.), and electrospinning parameters. Thus, the bonding strength will totally depend on the selection of material and process parameters chosen by any individual. Moreover, in many applications (such as filtration), the

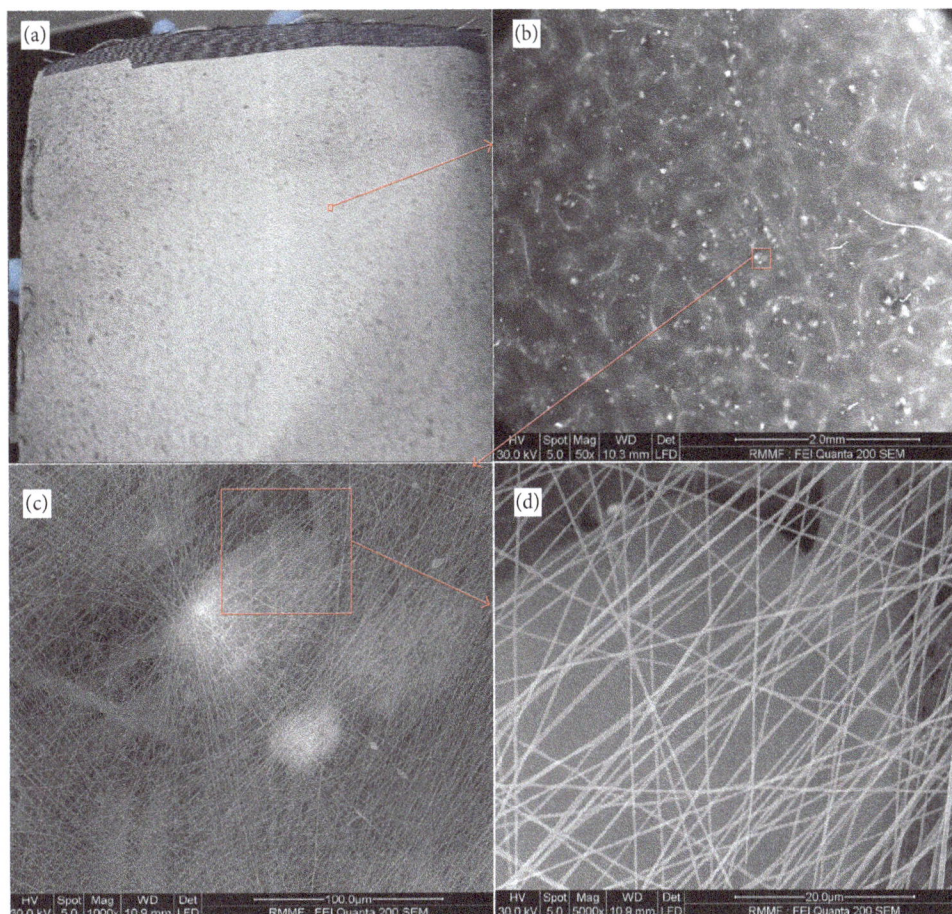

FIGURE 3: Electrospun layer on base textile in normal view (a), under SEM (b), and at higher magnifications (c, d).

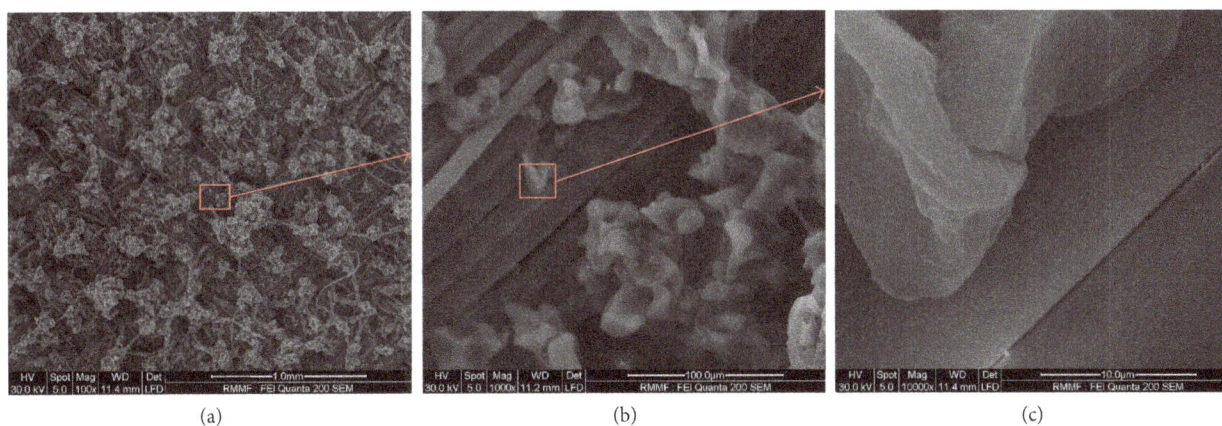

FIGURE 4: Electrosprayed aerogel particles on fabric surface. Higher magnification shows the sprayed drops of polymer containing numbers of aerogel particles.

firm bonding of deposited nonwoven layers is not a major concern as the nonwoven stays between other layers of fabric or material in sandwich construction. In any case, it is obvious from the above discussion that the developed method is an efficient way to needlelessly electrospin aerogel particles and with comparatively lower voltage.

In case of electrospraying, aerogel particles were transferred through the microdrops of the polymer. In Figure 4, the SEM micrograph clearly shows numerous such drops on fabric surface covering almost entire base textile. To ensure that the transferred drops are not just the polymer clots, the FTIR analysis had been carried out. Figure 5 shows the FTIR graphs of aerogel particles, fabric electrosprayed

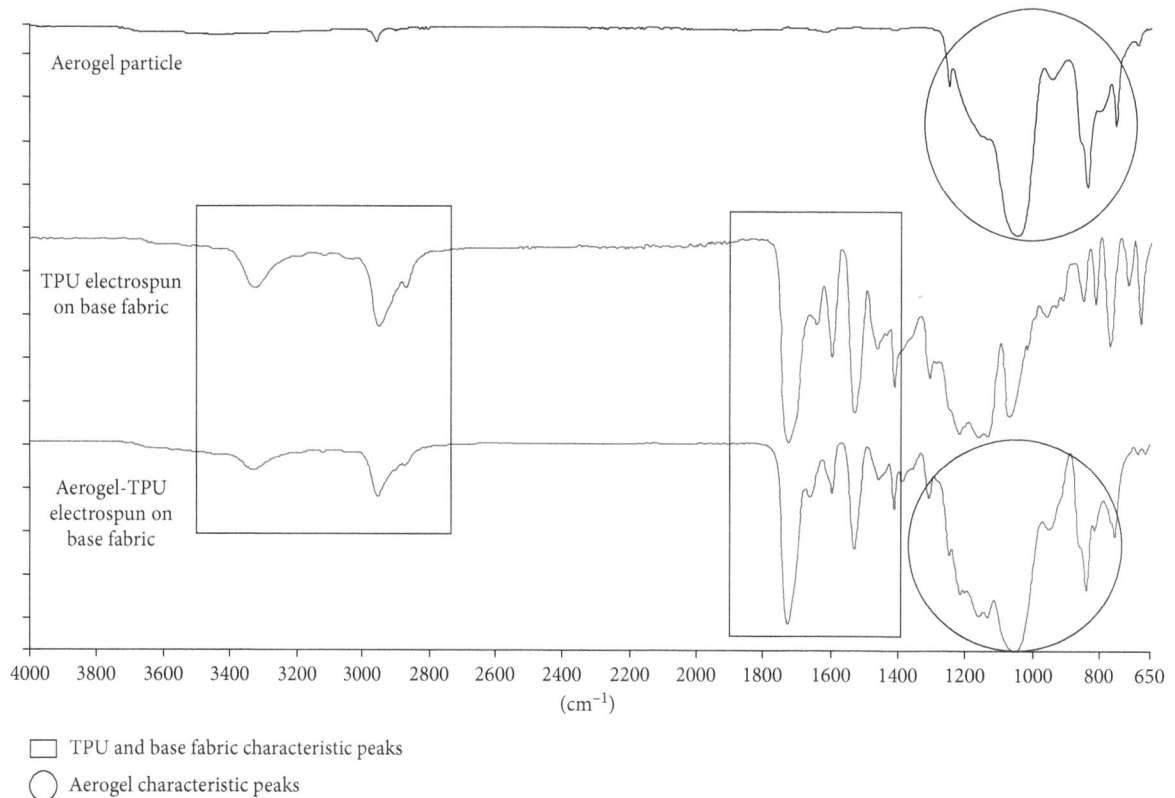

FIGURE 5: FTIR spectra of aerogel particle, polymer spun, and polymer-aerogel-mixed electrosprayed fabric sample.

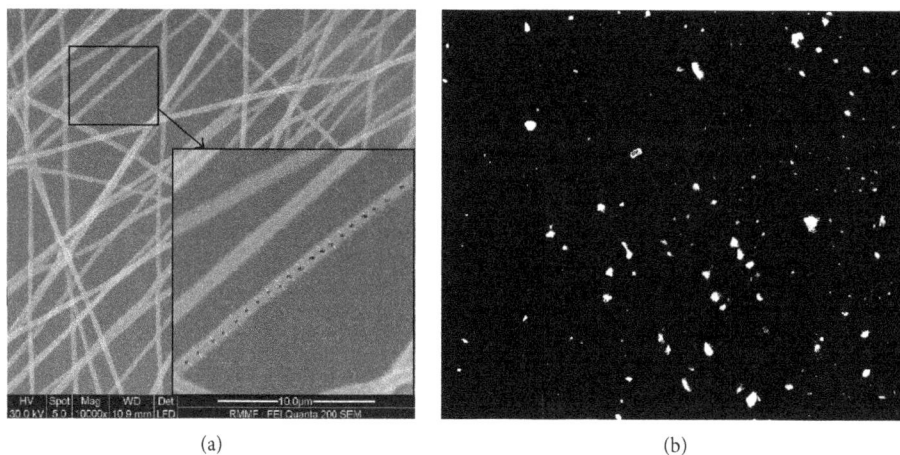

FIGURE 6: Measurement of fibre diameter (a) and particle size (b) by using ImageJ.

only with the TPU polymer, and abovementioned electro-sprayed sample which contains both aerogel particles and TPU polymer.

From the FTIR spectra, it can be clearly seen that peaks from both aerogel particle and TPU polymer-base fabric are present in the electrosprayed sample. The FTIR spectra of aerogel show an intense peak around $1089 \, cm^{-1}$ which is the typical bending vibration of Si-O [23]. The peak at $843 \, cm^{-1}$ is due to the bending vibration of Si-O [23, 24], and the peak at $948 \, cm^{-1}$ is for the stretching vibration of Si-OH [25]. All these peaks can also be seen at the electrospun sample which

evidences that aerogel particles were electrosprayed on base fabric by the TPU polymer droplets. The amount of particles transferred by the spraying method was higher than the electrospinning method.

3.2. Fibre Diameter and Particle Size. The SEM images were converted to 8 bit bmp images to analyze the particle size and fibre diameter. The scale bar shown in each SEM image was measured and taken to set scale in ImageJ. Fibre diameter was measured in 20 different positions as shown in Figure 6(a).

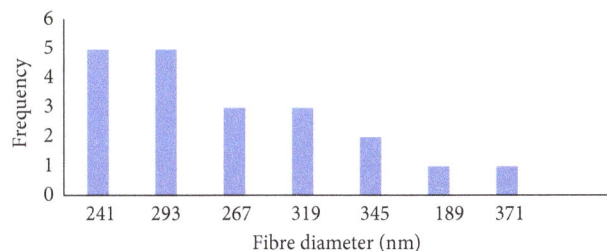

FIGURE 7: Diameter of the produced nanofibre on top of aerogel particles.

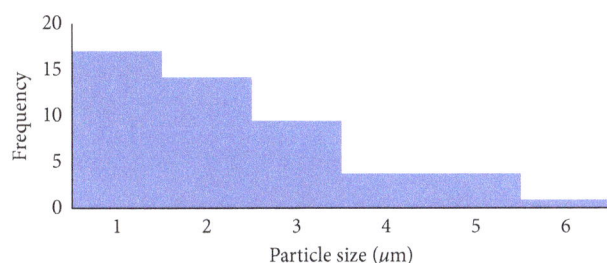

FIGURE 8: Particle size of transferred aerogel particles.

For particle size measurement, the threshold of the SEM image was first adjusted to nullify the noises other than particles themselves as shown in Figure 6(b). The diameter of the electrospun nanofibre had seen uniform and mostly regular without beads formation. The histogram in Figure 7 shows that the fibre diameter mostly laid between 241 and 293 nm. The diameter ranged from 189 nm to 368 nm with an average value of 270 nm.

In the case of the particle size of transferred aerogel particles, it was found that mostly the smaller sized particles were transferred to the base textiles. The average particle size of most of the particles ranges from 1 to 3 μm as seen in Figure 8.

4. Conclusion

This paper demonstrates an innovative technique to apply particulate material, such as aerogel particles, on textile surface by using the needleless electrospinning/ electrospraying method. A curved wire was used inside the syringe tip as an electrode to charge the particles and polymer mix to generate electric force for discharging of the particles. The SEM micrograph of resultant electrospun web revealed that aerogel particles were transferred to the fabric surface and bonded by the nanofibre web. The electrosprayed surface showed no such fibre deposition, but the higher amount of particle transfer was evident. The findings have numerous opportunities to be utilized for the application of aerogel and other particulate materials on textile or other surfaces using the needleless electrospinning/spraying technique.

Conflicts of Interest

The authors declare that they have no conflicts of interest.

Acknowledgments

The authors would like to thank Mister Martin Gregory for his support regarding instruments and testings. The authors also thank Dr. Xin Wang for his valuable advice.

References

[1] F. Anton, "Process and apparatus for preparing artificial threads," Google Patents, 1934.

[2] W. Simm, C. Gosling, R. Bonart, and B. V. Falkai, "Fibre fleece of electrostatically spun fibres and methods of making same," Google Patents, 1979.

[3] M. Yu, R. H. Dong, X. Yan et al., "Recent advances in needleless electrospinning of ultrathin fibers: from academia to industrial production," *Macromolecular Materials and Engineering*, vol. 302, no. 7, pp. 1–19, 2017.

[4] D. Wu, X. Huang, X. Lai, D. Sun, and L. Lin, "High throughput tip-less electrospinning via a circular cylindrical electrode," *Journal of Nanoscience and Nanotechnology*, vol. 10, no. 7, pp. 4221–4226, 2010.

[5] N. Thoppey, J. Bochinski, L. Clarke, and R. Gorga, "Edge electrospinning for high throughput production of quality nanofibres," *Nanotechnology*, vol. 22, no. 34, p. 345301, 2011.

[6] S. Tang, Y. Zeng, and X. Wang, "Splashing needleless electrospinning of nanofibres," *Polymer Engineering & Science*, vol. 50, no. 11, pp. 2252–2257, 2010.

[7] O. Jirsak, F. Sanetrnik, D. Lukas, V. Kotek, L. Martinova, and J. Chaloupek, "Method of nanofibres production from a polymer solution using electrostatic spinning and a device for carrying out the method," Google Patents, 2009.

[8] B. Lu, Y. Wang, Y. Liu et al., "Superhigh-throughput needleless electrospinning using a rotary cone as spinneret," *Small*, vol. 6, no. 15, pp. 1612–1616, 2010.

[9] N. M. Thoppey, J. R. Bochinski, L. I. Clarke, and R. E. Gorga, "Unconfined fluid electrospun into high quality nanofibres from a plate edge," *Polymer*, vol. 51, no. 21, pp. 4928–4936, 2010.

[10] H. U. Shin, Y. Li, A. Paynter, K. Nartetamrongsutt, and G. G. Chase, "Vertical rod method for electrospinning polymer fibers," *Polymer*, vol. 65, pp. 26–33, 2015.

[11] Y. Liu and J.-H. He, "Bubble electrospinning for mass production of nanofibres," *International Journal of Nonlinear Sciences and Numerical Simulation*, vol. 8, no. 3, pp. 393–396, 2007.

[12] X. Wang, H. Niu, T. Lin, and X. Wang, "Needleless electrospinning of nanofibres with a conical wire coil," *Polymer Engineering & Science*, vol. 49, no. 8, pp. 1582–1586, 2009.

[13] P. Pokorny, E. Kostakova, F. Sanetrnik et al., "Effective AC needleless and collectorless electrospinning for yarn production," *Physical Chemistry Chemical Physics*, vol. 16, no. 48, pp. 26816–26822, 2014.

[14] J. Wang, J. Kuhn, and X. Lu, "Monolithic silica aerogel insulation doped with TiO$_2$ powder and ceramic fibers," *Journal of Non-Crystalline Solids*, vol. 186, pp. 296–300, 1995.

[15] X. Yang, Y. Sun, D. Shi, and J. Liu, "Experimental investigation on mechanical properties of a fiber-reinforced silica

aerogel composite," *Materials Science and Engineering: A*, vol. 528, no. 13-14, pp. 4830–4836, 2011.

[16] A. Shaid, L. Wang, and R. Padhye, "The thermal protection and comfort properties of aerogel and PCM-coated fabric for firefighter garment," *Journal of Industrial Textiles*, vol. 45, no. 4, pp. 611–625, 2015.

[17] A. Shaid, M. Furgusson, and L. Wang, "Thermophysiological comfort analysis of aerogel nanoparticle incorporated fabric for fire fighter's protective clothing," *Chemical and Materials Engineering*, vol. 2, no. 2, pp. 37–43, 2014.

[18] L. Jin, K. Honga, and K. Yoona, "Effect of aerogel on thermal protective performance of firefighter clothing," *Journal of Fiber Bioengineering and Informatics*, vol. 6, no. 3, pp. 315–324, 2013.

[19] M. Venkataraman, R. Mishra, D. Jasikova, T. M. Kotresh, and J. Militky, "Thermodynamics of aerogel-treated nonwoven fabrics at subzero temperatures," *Journal of Industrial Textiles*, vol. 45, no. 3, pp. 387–404, 2014.

[20] L. Li, B. Yalcin, B. N. Nguyen, M. A. B. Meador, and M. Cakmak, "Flexible nanofibre-reinforced aerogel (xerogel) synthesis, manufacture, and characterization," *ACS Applied Materials & Interfaces*, vol. 1, no. 11, pp. 2491–2501, 2009.

[21] H. Wu, Y. Chen, Q. Chen, Y. Ding, X. Zhou, and H. Gao, "Synthesis of flexible aerogel composites reinforced with electrospun nanofibres and microparticles for thermal insulation," *Journal of Nanomaterials*, vol. 10, pp. 1–8, 2013.

[22] Z. Mazrouei-Sebdani, A. Khoddami, H. Hadadzadeh, and M. Zarrebini, "Synthesis and performance evaluation of the aerogel-filled PET nanofibre assemblies prepared by electro-spinning," *RSC Advances*, vol. 5, no. 17, pp. 12830–12842, 2015.

[23] X. Zhou, H. Xiao, J. Feng, C. Zhang, and Y. Jiang, "Preparation and thermal properties of paraffin/porous silica ceramic composite," *Composites Science and Technology*, vol. 69, no. 7, pp. 1246–1249, 2009.

[24] Z. Xiangfa, X. Hanning, F. Jian, Z. Changrui, and J. Yonggang, "Preparation, properties and thermal control applications of silica aerogel infiltrated with solid–liquid phase change materials," *Journal of Experimental Nanoscience*, vol. 7, no. 1, pp. 17–26, 2012.

[25] B. Shokri, M. A. Firouzjah, and S. Hosseini, "FTIR analysis of silicon dioxide thin film deposited by metal organic-based PECVD," in *Proceedings of the 19th international symposium on plasma chemistry society*, Bochum, Germany, July 2009.

14

Deproteinization of Nonammonia and Ammonia Natural Rubber Latices by Ethylenediaminetetraacetic Acid

N. Moonprasith,[1] A. Poonsrisawat,[2] V. Champreda,[2]
C. Kongkaew,[1] S. Loykulnant,[1] and K. Suchiva[1]

[1]Natural Rubber Focus Unit, National Metal and Materials Technology Center, 114 Thailand Science Park, Phahonyothin Road, Khlong Nueng, Khlong Luang, Pathum Thani 12120, Thailand
[2]Enzyme Technology Laboratory, National Center for Genetic Engineering and Biotechnology, 113 Thailand Science Park, Phahonyothin Road, Khlong Nueng, Khlong Luang, Pathum Thani 12120, Thailand

Correspondence should be addressed to S. Loykulnant; surapicl@mtec.or.th

Academic Editor: Angela De Bonis

Deproteinized natural rubber (DPNR) was made from sodium hydroxymethylglycinate latex (SH-latex) and ammonia latex (NH_3-latex) by mixing with different forms of ethylenediaminetetraacetic acid (EDTA) in the presence of sodium dodecyl sulfate (SDS). The mixtures were stirred at room temperature followed by centrifugation to separate the denatured proteins. The optimized reaction contained 0.01 wt% ethylenediaminetetraacetic acid tetrasodium (EDTA-4Na) with 1% SDS. The nitrogen contents of the DPNR from SH-latex and NH_3-latex were reduced to 0.005 wt% and 0.008 wt%, respectively, compared to 0.551 wt% in the starting SH-latex and 0.587 wt% in the NH_3-latex. SDS-PAGE analysis and FT-IR spectroscopy showed decomposition of latex proteins to peptides of smaller molecular weight. Physical properties of the DPNR rubber were studied. The novel EDTA-4Na treatment is considered an effective deproteinization method with potential application on both ammonia and nonammonia preservative systems.

1. Introduction

Natural rubber (NR) field latex (*Hevea brasiliensis*) contains 30–40% rubber by weight dispersed as rubber latex particles in water with some minor nonrubber constituents such as proteins, lipids, carbohydrates, sugars, and metal ions. The rubber particles of fresh rubber latex contain *cis*-1,4-polyisoprene surrounded by proteins and lipids on the outer surface. The protein component of the natural rubber latex particles is believed to be exclusively associated with the particle surface as an adsorbed layer [1]. These proteins present on the surface of NR particles in the latex state may sometimes bring about latex allergy mediated by type I immunoglobulin E [2, 3]. The potential protein allergy risk in NR products for sensitized users has resulted in an increasing demand on deproteinized NR latex, particularly for medical devices and hypoallergic products.

Different chemical and enzymatic methods have been reported for deproteinization of NR. NR was effectively deproteinized in the latex stage by enzymatic degradation which removed proteins present on the surface of the rubber particle [4, 5]. This resulted in marked decrease in the nitrogen content of NR to less than 0.02%, which was approximately 5% of the starting material [6–8]. However, the enzymatic process requires long incubation time (usually more than 24 h) while the remaining proteins, peptides, or amino acids may result in intraoperative anaphylactic reactions of hypersensitive patients of allergy [3, 9]. NR treatment with urea and saponification led to reduction in the total nitrogen content to lower than 0.02 wt% within 1 hour under suitable conditions [6].

The conventional ammonia system for latex preservation using chemicals such as TMTD (Tetra methyl thiuram disulfide) and ZnO as secondary preservatives that are highly toxic can potentially cause allergy. This leads to an increasing trend on using nonammonia preservatives for fresh latex. Development of an efficient method for deproteinization of NR preserved by nonammonia system is thus of great interest.

SH preservative has no strong smell, is free of nitrosamine, and is nontoxic to cells. The pH of SH-latex can be adjustable to 8–11 which is considered advantageous for subsequent application.

In this present work, the use of ethylenediaminetetraacetic acid tetrasodium (EDTA-4Na) in conjunction with surfactant on deproteinization of nonammonia and ammonia latices was studied. This can potentially cause chelation of the metal ions coordinate with NR proteins, resulting in denaturation and removal of the proteins from NR surface. The deproteinized rubber was characterized for physical properties. This work demonstrates the novel EDTA treatment as an effective alternative for NR deproteinization for preparation of deproteinized latex for allergenic free applications.

2. Experimental

2.1. Materials. Fresh natural rubber (FNR) latex was collected by tapping from mature *Hevea* rubber trees of Thai Rubber Latex Co., Ltd., plantation in Chonburi province, Thailand. Sodium hydroxymethylglycinate (SH) was provided from ISP Chemical (Columbus, Ohio, USA). Ethylenediaminetetraacetic acid, Coomassie Brilliant Blue R-250, sodium dodecyl sulfate, lauric acid, potassium hydroxide, and ammonium hydroxide were analytical grade and supplied by Sigma-Aldrich. Ammonium persulfate, N,N,N′,N′-tetramethylethylene-diamine (TEMED), and acrylamide were procured from Biorad Laboratories, Inc. (Hercules, CA).

2.2. Methods

2.2.1. Preservation of NR Latex. The fresh natural rubber (FNR) latex was divided into two fractions. The first fraction was suddenly preserved by the addition of 0.4 wt% cosmetic grade preservative, sodium hydroxymethylglycinate (SH) [10], and adjusted the pH to 10 using potassium hydroxide (KOH) and used as the nonammonia latex sample. The second fraction was added with 0.4 wt% ammonium hydroxide and used as the ammonia latex sample.

2.2.2. NR Deproteinization. The DPNR latex was prepared by treating FNR latex with ethylenediaminetetraacetic acid sodium (EDTA-Na), ethylenediaminetetraacetic acid disodium (EDTA-2Na), ethylenediaminetetraacetic acid trisodium (EDTA-3Na), and ethylenediaminetetraacetic acid tetrasodium (EDTA-4Na) at various concentrations (0.05, 0.10, 0.20 wt%) in the presence of 1 wt% sodium dodecyl sulfate (SDS) at room temperature for 1 h with continuous stirring. Then, the treated latex was centrifuged at 10,000 rpm for 50 min. The cream fraction was redispersed in 1 wt% SDS solution to make 30% dry rubber content (DRC) latex and washed twice by centrifugation. The obtained cream fraction was redispersed in distilled water to make DPNR of 30% dry rubber content (DRC) latex. The samples of untreated and deproteinized latices were air-dried. The DPNR latex was coagulated with methanol for characterization of physical properties. The effects of different EDTA salt complexes were studied at 0.10 wt% of the chelating agent using the

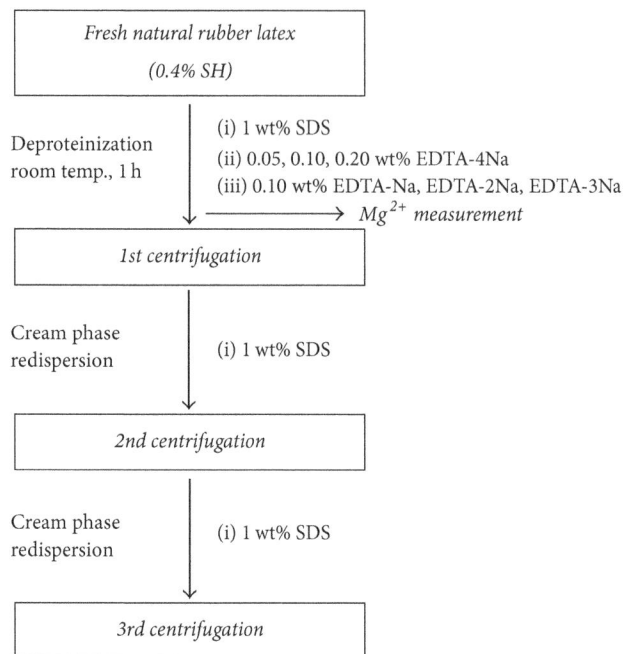

FIGURE 1: Schematic representation of deproteinization procedure.

method described above. The deproteinization procedure is presented in Figure 1.

2.2.3. Analytical Methods. Protein profiles of DPNR were determined by sodium dodecyl sulfate polyacrylamide gel electrophoresis (SDS-PAGE) according to the standard procedure [11] with 4% stacking gels and 15% separating gels at a constant current of 15 mA for 1 h. The molecular weight reference marker was run simultaneously in all experiments. The protein bands were visualized by staining Coomassie Blue R-250.

The total nitrogen content of the NR samples was measured by using the Kjeldahl method according to the Rubber Research Institute of Malaysia (RRIM) test method [12]. Functional groups of NR samples were characterized by Fourier transform infrared (FT-IR) spectroscopy using a Perkin Elmer Spectrum Spotlight 300 spectrometer (Perkin Elmer, USA). Physical properties of DPNR rubber were determined in accordance with SMR Bulletin No. 7-1992 for dirt content, volatile matter content, ash content, initial Wallace plasticity (P_0), and plasticity retention index (PRI) and Mooney viscosity.

3. Results and Discussion

3.1. Effects of EDTA-4Na Concentration. In order to determine the optimal dosage of EDTA-4Na on deproteinization, the nonammonia and ammonia latices were incubated with EDTA-4Na at various concentrations (0.05, 0.10, and 0.20 wt%) in the presence of 1 wt% SDS at room temperature for 1 h. Separation and washing of the cream fraction in the presence of 1 wt% SDS were later carried out. The final cream

TABLE 1: Nitrogen content of FNR and deproteinized FNR (DPNR) by using EDTA-4Na at various concentrations.

Samples	Preservative	pH	EDTA-4Na (wt%)	Nitrogen content (wt%)
FNR	—	6.5	—	0.638
FR-NH$_3$	Ammonia	10	—	0.587
FR-SH	SH	10	—	0.551
FR-NH$_3$	Ammonia	10	0.05	0.043
FR-NH$_3$	Ammonia	10	0.10	0.009
FR-NH$_3$	Ammonia	10	0.20	0.008
FR-SH	SH	10	0.05	0.012
FR-SH	SH	10	0.10	0.005
FR-SH	SH	10	0.20	0.005

TABLE 2: Magnesium ion content of FNRL and FNRL deproteinized by using EDTA-4Na.

Samples	Preservatives	EDTA-4Na (wt%)	Mg^{2+} content (ppm)
FNRL	—	—	657
FR-NH$_3$	Ammonia	—	294
FR-SH	SH	—	635
FR-NH$_3$ DPNR	Ammonia	0.05	152
FR-NH$_3$ DPNR	Ammonia	0.10	123
FR-NH$_3$ DPNR	Ammonia	0.20	89
FR-SH DPNR	SH	0.05	417
FR-SH DPNR	SH	0.10	341
FR-SH DPNR	SH	0.20	289

fraction was redispersed in distilled water to make DPNR latex with 30% dry rubber content (DRC).

The total nitrogen contents of the fresh natural rubber (FNR) and deproteinized natural rubber (DPNR) is shown in Table 1. The nitrogen content of the air-dried FNR film was 0.638%. Increasing dosage of EDTA-4Na led to a decreasing trend of nitrogen content in the deproteinized NR preserved by both methods. Treatment of FR-NH$_3$ with 0.20 wt% EDTA-4Na led to a remarkable decrease of the nitrogen content from 0.587 wt% to 0.008 wt%. Moreover, the total nitrogen content of FR-SH DPNR was found to be the lowest when 0.10 wt% EDTA-4Na was used which resulted in the final nitrogen content of 0.005 wt% compared with 0.551 wt% of the untreated sample, FR-SH.

According to the results in Table 2, the Mg^{2+} content of FNRL was decreased from 657 to 294 ppm after the fresh latex was preserved with NH$_4$OH. This was due to the complex formation of magnesium ammonium phosphate (MgNH$_4$PO$_4$) when magnesium and phosphate ions of the fresh latex reacted with ammonia ion. The Mg^{2+} content of FR-NH$_3$ DNPR samples tended to reduce with increasing EDTA-4Na dosage, resulting in the lowest Mg^{2+} content (89 ppm) at 0.20 wt% EDTA-4Na. The Mg^{2+} content of FR-SH sample was slightly reduced compared to that of the starting material (FNRL). Again, the Mg^{2+} content of the nonammonia latex tended to decrease with increasing EDTA-4Na content; that is, the lowest Mg^{2+} content (289 ppm) was found in the latex treated with 0.20 wt% EDTA-4Na. The

results demonstrated that the magnesium ions present in the fresh NR latex could be scrapped by EDTA-4Na chelating agent under the same conditions used for deproteinization.

Effect of different EDTA salt complexes on deproteinization of nonammonia and ammonia NR latices was studied. As shown in Table 3, the highest deproteinization efficiency was observed with EDTA-4Na and tended to be lower with decreasing Na number in the complex. The lowest nitrogen contents of 0.008 wt% and 0.005 wt% were obtained for FR-NH$_3$ DPNR and FR-SH-DPNR treated with 0.10 wt% EDTA-4Na, respectively. EDTA-1Na, EDTA-2Na, and EDTA-3Na could also remove proteins by denaturation. The highest efficiency of EDTA-4Na on deproteinization could be due to its higher solubility in water compared to the other EDTA complexes. Removal of proteins in NR by EDTA suggested that most proteins present in fresh NR are attached to the rubber via weak attractive forces, which can be disturbed by chelation. Huang and coworkers reported that the coordination of metal ion with protein was destroyed by scrapping of metal by EDTA resulting in the denaturation of NR proteins [13].

Removal of divalent metal ions can result in denaturation of the proteins associated with the latex particles due to disturbance of the protein's structure [14]. The highest efficiency of EDTA-4Na on deproteinization of NR followed by EDTA-3Na, EDTA-2Na, and EDTA-1Na, respectively, can be explained by the chemical characteristics of the different EDTA complexes. EDTA in different forms have different capability on chelating divalent metal ions. According to

TABLE 3: Nitrogen content of FNR and deproteinized FNR prepared using various EDTA complexes at 0.10 wt%.

Samples	Preservatives	pH	0.1 wt% EDTA complexes	Nitrogen content (wt%)
FNR	—	6.5	—	0.638
FR-NH$_3$	Ammonia	10	—	0.587
FR-SH	SH	10	—	0.551
FR-NH$_3$	Ammonia	10	EDTA-1Na	0.029
FR-NH$_3$	Ammonia	10	EDTA-2Na	0.022
FR-NH$_3$	Ammonia	10	EDTA-3Na	0.013
FR-NH$_3$	Ammonia	10	EDTA-4Na	0.008
FR-SH	SH	10	EDTA-1Na	0.022
FR-SH	SH	10	EDTA-2Na	0.012
FR-SH	SH	10	EDTA-3Na	0.009
FR-SH	SH	10	EDTA-4Na	0.005

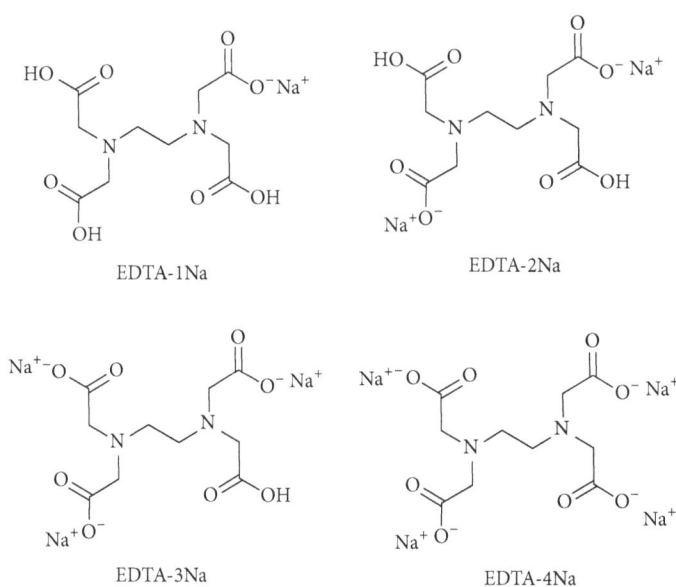

FIGURE 2: Chemical structures of EDTA.

their chemical structures (see Figures 2 and 3), EDTA-4Na can chelate two divalent metal ions on stoichiometry basis. Lower chelating efficiency is expected for EDTA-3Na, EDTA-2Na, and EDTA-1Na, respectively. Solubility of the EDTA complexes in water also follows this order where EDTA-4Na showed the highest solubility. In addition, solubilization of EDTA-4Na led to pH between 10 and 11, which was very close to the pH of the NR samples while solubilization of EDTA-3Na, EDTA-2Na, and EDTA-1Na resulted in solution with less alkaline pH, respectively. Altogether, this resulted in the highest deproteinization efficiency of EDTA-4Na as observed in this study.

3.2. Determination of Proteins Pattern of DPNR-EDTA.
Protein patterns in the cream fractions of NR samples deproteinized by different forms of EDTA were determined using SDS-PAGE.

Figures 4 and 5 show the protein patterns in the FNRL and deproteinized ammonia and nonammonia NR samples separated according to their molecular weight. Overall treatment with EDTA under the experimental conditions led to decomposition of proteins to fragments with lower molecular weights. The protein patterns of the controls (FR-SH and FR-NH$_3$ FNRL) are shown in lane 1. The major bands at 14.4 and 25 kDa are known to be derived from proteins covering the surface of the rubber latex particle [15]. Slight degradation of proteins was observed in FNRL treated with EDTA (in lane 3). Distinct decomposition of proteins was found in DPNR with EDTA-4Na, EDTA-3Na, EDTA-2Na, and EDTA-1Na which resulted in protein bands with lower molecular weight and smear. No significant difference was observed between samples treated with different forms of EDTA complexes.

3.3. FT-IR Spectra of DPNR-EDTA-4Na.
The residual proteins of ammonia and nonammonia fresh latices treated with 0.10 wt% EDTA-4Na were analyzed by FT-IR (Figure 6). For fresh NR latex, the clear peak at 3280 cm^{-1} assigned to NH stretching was found indicating the existence of proteins or

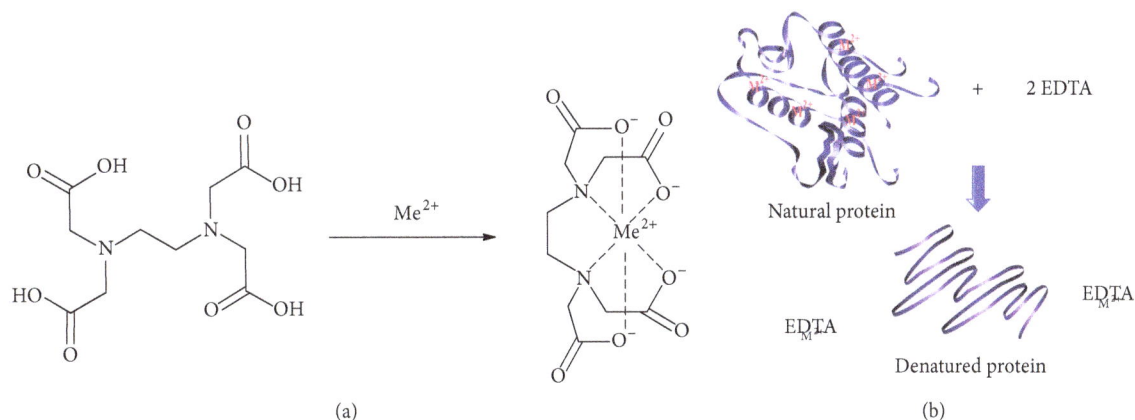

FIGURE 3: Mechanism of chelation reaction with metal ions and metal ions on protein: (a) mechanism of chelation reaction; (b) metal ions in the protein layer was scraped by EDTA.

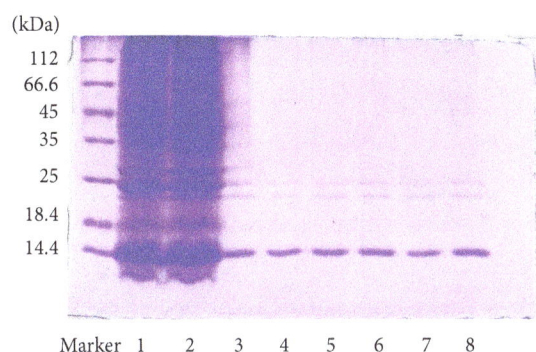

FIGURE 4: SDS-PAGE of redispersed cream fraction from deproteinized nonammonia FNRL with EDTA complexes. FN-SH samples were incubated with 0.1 wt% EDTA in different forms in the presence of 1 wt% SDS for 1 h. Lane 1: FR-SH, lane 2: mixture of FR-SH and EDTA-4Na without centrifugation, lane 3: washed FR-SH with SDS followed by centrifugation, lane 4: FR-SHDPNR-EDTA, lane 5: FR-SHDPNR-EDTA-1Na, lane 6: FR-SHDPNR-EDTA-2Na, lane 7: FR-SHDPNR-EDTA-3Na, and lane 8: FR-SHDPNR-EDTA-4Na.

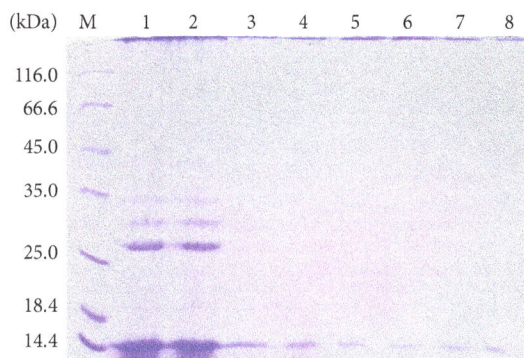

FIGURE 5: SDS-PAGE of redispersed cream fraction from deproteinized ammonia FNRL with EDTA complexes. FN-NH$_3$ was incubated with 0.10 wt% of EDTA in different forms for 1 hour at room temperature in the presence of 1 wt% SDS. Lane 1: FR-NH$_3$, lane 2: mixture of FR-NH$_3$ and EDTA-4Na without centrifugation, lane 3: washed FR-NH$_3$ with SDS followed by centrifugation, lane 4: FR-NH$_3$DPNR-EDTA, lane 5: FR-NH$_3$DPNR-EDTA-1Na, lane 6: FR-NH$_3$DPNR-EDTA-2Na, lane 7: FR-NH$_3$DPNR-EDTA-3Na, and lane 8: FR-NH$_3$DPNR-EDTA-4Na.

peptides. After deproteinization with 0.10 wt% EDTA-4Na, both FR-SH DPNR-EDTA-4Na and FR-NH$_3$ DPNR-EDTA-4Na show their peak at 3320 cm^{-1} instead of 3280 cm^{-1}. The newly appeared band around 3320 cm^{-1} in the purified NR was assigned to the NH stretching of oligopeptides such as dipeptide [5, 16, 17]. The absence of NH stretching band in FR-SH DPNR-EDTA-4Na and FR-NH$_3$ DPNR-EDTA-4Na indicated that the amount of residual proteins in these samples was less than the detection limit of FT-IR analysis.

3.4. Physical Properties of DPNR.

Physical properties of the DPNR prepared using the two different preservative systems with 0.10 wt% EDTA-4Na are compared with relevant international specifications according to the Standard Thai Rubber STR5L in Table 4.

The dirt content, volatile matter content, ash content, nitrogen content, and Mooney viscosity of both deproteinized NR samples were conformed to the specifications.

Exception was found for the PRI; that is, the PRI values for FR-SH DPNR-EDTA-4Na and FR-NH$_3$ DPNR-EDTA-4Na were 40.5 and 37.6, respectively. The results imply that the two DPNR samples possessed relatively low thermal oxidative aging resistance. This could be explained by the low protein content in the samples because it has been reported that natural proteins or amino acids in NR could act as natural antioxidant [18]. The presence of remnant oligopeptides or small protein fragments in DPNR and the nearly complete removal of proteins by surfactant washing could be assessed indirectly by measuring the resistance to thermal oxidative aging represented in terms of plasticity retention index (PRI).

4. Conclusions

The deproteinized natural rubber (PDNR) was successfully prepared in the latex stage for both preservative systems by

(a)

(b)

(c)

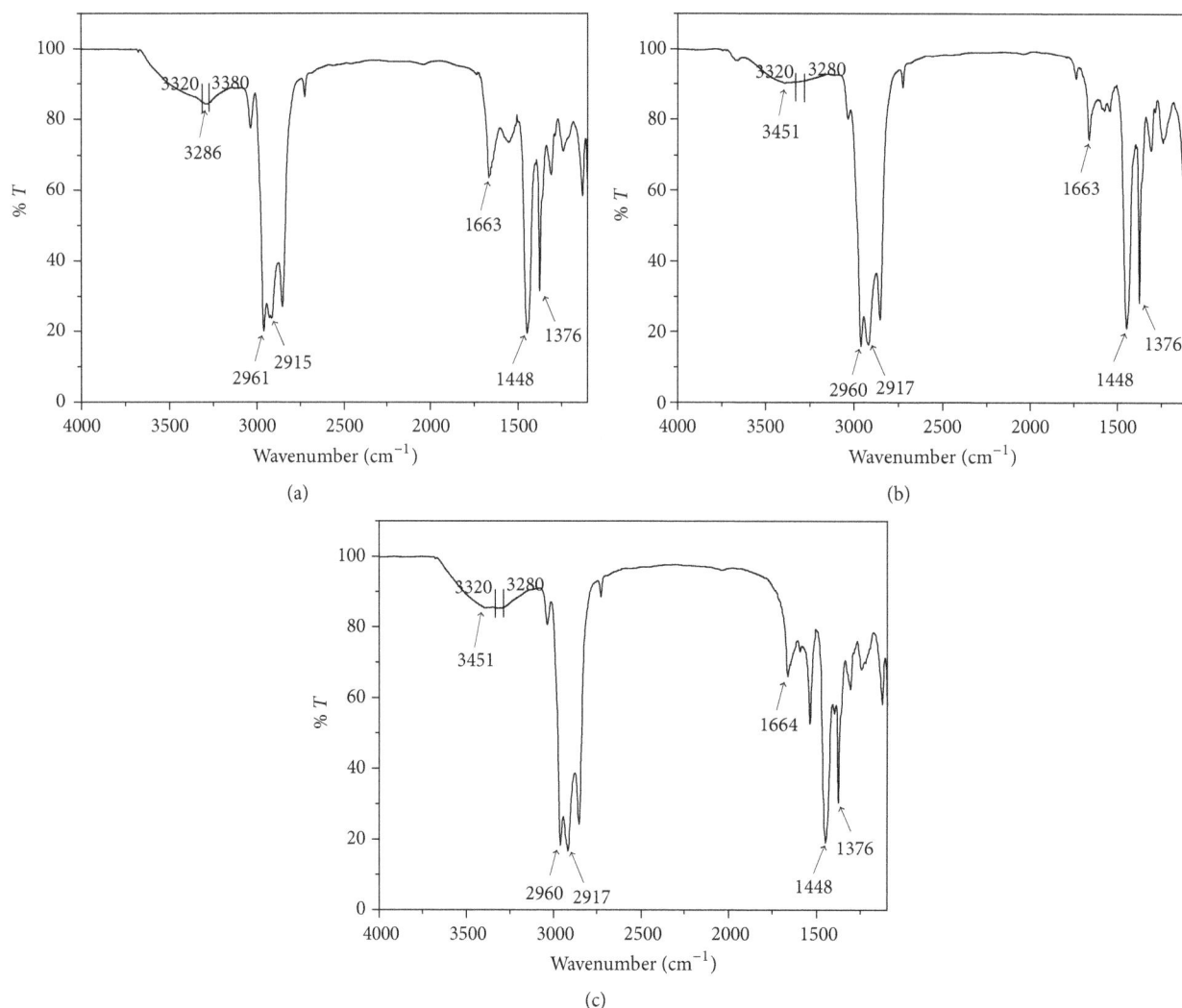

FIGURE 6: FT-IR spectra of DPNR-EDTA-4Na. (a) FNR, (b) FR-SH DPNR-EDTA-4Na, and (c) FR-NH$_3$ DPNR-EDTA-4Na.

TABLE 4: Physical properties of DPNR.

Properties	Samples		
	Standard of STR 5L	FR-SH DPNR-EDTA-4Na	FR-NH$_3$ DPNR-EDTA-4Na
Dirt content (wt%)	0.04 max	0.004	0.003
Volatile matter content (wt%)	0.8 max	0.03	0.03
Ash content (wt%)	0.4 max	0.11	0.13
Nitrogen content (wt%)	0.6 max	0.005	0.008
Initial Wallace plasticity (P_0)	35 min	55.5	50.0
Plasticity retention index (PRI)	60 min	40.5	37.6
Mooney viscosity, ML(1 + 4) 100°C	—	67.0	66.3
Color (Lovibond)	6.0 max	1.0	1.5

using EDTA-4Na in the presence of surfactant. The optimum concentration of EDTA-4Na was 0.1 wt% while EDTA-1Na, EDTA-2Na, and EDTA-3Na could also remove protein but with lower efficiency. The magnesium ion content of the FNRL was reduced with increasing EDTA-4Na concentration. The nonammonia preservative has no strong smell, is free of nitrosamine, and is nontoxic to cell with adjustable pH

(8–11). These properties are considered useful for application in producing healthcare products.

Conflicts of Interest

The authors declare that there are no conflicts of interest regarding the publication of this paper.

References

[1] Y. Tanaka, "Structural characterization of natural polyisoprenes: solve the mystery of natural rubber based on structural study," *Rubber Chemistry and Technology*, vol. 74, no. 3, pp. 355–375, 2001.

[2] P. W. Allen and G. F. Bloomfield, *Chemistry and Physics of Rubber Like Substances*, L. Bateman, Ed., vol. 51, Maclaren and Sons, London, UK, 1963.

[3] J. W. Yunginger, R. T. Jones, A. F. Fransway, J. M. Kelso, M. A. Warner, and L. W. Hunt, "Extractable latex allergens and proteins in disposable medical gloves and other rubber products," *The Journal of Allergy and Clinical Immunology*, vol. 93, no. 5, pp. 836–842, 1994.

[4] P. A. J. Yapa and W. S. E. Fernando, "Enhancement of deproteinization of Hevea rubber by maturation of papain treated Hevea latex," *Vidyodaya Journal of Science*, vol. 7, pp. 93–100, 1997.

[5] A. H. Eng, Y. Tanaka, and S. N. Gan, "Short communication FTIR-studies on amino group in purified Havea rubber," *Journal of Natural Rubber Research*, vol. 7, p. 152, 1992.

[6] J. Yunyongwattanakorn, Y. Tanaka, J. Sakdapipanich, and V. Wongsasutthikul, "Highly-purified natural rubber by saponification of latex: Analysis of residual proteins in saponified natural rubber," *Rubber Chemistry and Technology*, vol. 81, no. 1, pp. 121–137, 2008.

[7] N. Ichikawa, A. H. Eng, and Y. Tanaka, 1993, Proc. Inter. Rubber Technol. Conf., Kuala Lumpur, Malaysia, 101.

[8] S. Nakade, G. Kuga, M. Hayashi, and Y. Tanaka, "Highly purified natural rubber IV. Preparation and characteristics of gloves and condoms," *Journal of Natural Rubber Research*, vol. 12, no. 1, pp. 33–42, 1997.

[9] P. K. Viswanath, J. K. Kevin, R. Abe, K. B. Naveen, and N. F. Jordan, "Characterization of latex antigen and demonstration of latex-specific antibodies by enzyme-linked immunosorbent assay in patients with latex hypersensitivity," *Allergy and Asthma Proceedings*, vol. 13, no. 6, pp. 329–334, 1992.

[10] S. loykulnant, C. Kongkaew, and O. Chaikumpollert, *A method for preservation of natural rubber latex using methylol compounds*, National Science and Technology Development Agency, 2010, Malaysia patent MY142541 (A).

[11] U. K. Laemmli, "Cleavage of structural proteins during the assembly of the head of bacteriophage T4," *Nature*, vol. 227, no. 5259, pp. 680–685, 1970.

[12] 1973, Rubber Res Inst Malaysia SMR Bull, 17.

[13] Z. Huang, H. Zhang, W. Yao, Y. Liu, J. Zhou, and L. Tang, "Raman spectroscopy study of zinc finger ZNF191(243-368)," *Chinese Science Bulletin*, vol. 48, no. 16, pp. 1722–1727, 2003.

[14] M. M. Rippel, C. A. P. Leite, L.-T. Lee, and F. Galembeck, "Formation of calcium crystallites in dry natural rubber particles," *Journal of Colloid and Interface Science*, vol. 288, no. 2, pp. 449–456, 2005.

[15] G. M. Bristow, "The huggin's k' parameter for polyisoprenes," *Journal of Polymer Science*, vol. 62, no. 174, 168 pages, 1962.

[16] A. H. Eng, S. Kawahara, and Y. J. Tanaka, "Structural characteristic of natural rubber-roles of ester groups," *Journal of Applied Polymer Science. Applied polymer symposium*, vol. 53, no. 5, 1994.

[17] Y. Tanaka, S. Tangpakdee, and J. Kautsch, GummiKunstst, 1997, 69, 6.

[18] R. F. A. Altman, "Natural antioxidants in hevea latex," *Rubber Chemistry and Technology*, vol. 21, no. 3, pp. 752–764, 1948.

Experimental and Numerical Study of Texture Evolution and Anisotropic Plastic Deformation of Pure Magnesium under Various Strain Paths

Hamad F. Alharbi ⓘ, Monis Luqman, Ehab El-Danaf ⓘ, and Nabeel H. Alharthi

Mechanical Engineering Department, College of Engineering, King Saud University, P.O. Box 800, 11421 Riyadh, Saudi Arabia

Correspondence should be addressed to Hamad F. Alharbi; harbihf@ksu.edu.sa

Academic Editor: Ivan Gutierrez-Urrutia

The deformation behavior and texture evolution of pure magnesium were investigated during plane strain compression, simple compression, and uniaxial tension at room temperature. The distinctive stages in the measured anisotropic stress-strain responses and numerically computed strain-hardening rates were correlated with texture and deformation mechanisms. More specifically, in plane strain compression and simple compression, the onset of tensile twins and the accompanying texture-hardening effect were associated with the initial high strain-hardening rates observed in specimens loaded in directions perpendicular to the crystallographic c-axis in most of the grains. The subsequent drop in strain-hardening rates in these samples was correlated with the exhaustion of tensile twins and the activation of pyramidal <c+a> slip systems. The falling strain-hardening rates were observed in simple compression and plane strain compression with loading directions parallel to the c-axis where the second pyramidal <c+a> slip systems were the only slip families that can accommodate deformation. For uniaxial tension with the basal plane parallel to the tensile axis, the prismatic <a> and second pyramidal <c+a> slips are the main deformation mechanisms. The predicted relative slip and twin activities from the crystal plasticity simulations clearly showed the effect of texture on the type of activated deformation mechanisms.

1. Introduction

Magnesium and its alloys have potential applications in automobile and aerospace industries due to their low density, high specific strength, and good castability [1, 2]. However, magnesium materials have low ductility at room temperature compared to other ductile metals [3, 4]. The low ductility in magnesium can be generally attributed to the lack of independent easy-activated slip systems that can accommodate large plastic strains.

Magnesium has a close-packed (hcp) crystal structure and exhibits a high degree of plastic anisotropy due to the possible activation of various slip and twin families and the resulting pronounced texture. The slip systems include the basal <a>, prism <a>, pyramidal <a>, and second pyramidal <c+a> slip systems [5–7]. The basal <a> family is the easiest slip system to activate in magnesium with lower slip resistance compared to the other systems. The second pyramidal <c+a> slip system is the only slip family that can accommodate plastic strain in the basal normal direction (all the other slip systems operate in the basal <a> direction only), but it is also very hard to activate. Twinning is an alternative deformation mechanism in magnesium that can also accommodate plastic strains in the nonbasal directions. The $\{10\bar{1}2\}<\bar{1}011>$ tensile twins and the $\{10\bar{1}1\}<10\bar{1}2>$ contraction twins are the most common twin systems observed in magnesium and its alloys [8–11]. The extension twins give extension along the crystallographic c-axis in magnesium and reorient the crystal by 86° around $<1\bar{2}10>$ directions, while the contraction twins supply contraction along the crystallographic c-axis and reorient the crystal by 56° around $<1\bar{2}10>$ directions [8, 9].

The extension twins are readily observed in magnesium and its alloys during simple compression tests when the loading direction is perpendicular to the crystallographic *c*-axis of the crystal. Such deformation modes are commonly triggered in textured (polycrystalline) magnesium when compressed parallel to the extrusion direction of an extruded bar or compressed perpendicular to the normal direction of a rolled sheet [10–12]. The extension twins can also be activated in magnesium during a uniaxial tension if the tensile loading direction is parallel to the crystallographic *c*-axis of the lattice crystals [13, 14]. The activation of extension twins in magnesium has important implications on the texture evolution and strain-hardening rates. The initial high strain-hardening rates observed in magnesium are commonly linked to the activation of extension twins in the early stage of deformation. The widely accepted hypothesis for the physical mechanism by which deformation twinning contributes to the unusually high strain-hardening rates is the texture hardening, where tensile twins usually reorient the grains into hard (unfavored slip) orientations. An alternative hypothesis for explaining the high strain-hardening rates associated with the onset of deformation twinning is the Hall–Petch-like effect, where twinning divides the grains thereby reducing the effective slip distance [15–18]. However, the latter explanation is not likely to be effective due to the fast growth (thickening) of tensile twins in magnesium until it consumes the entire grain.

The current study presents detailed investigations of the mechanisms that govern the anisotropic plastic response of pure magnesium under various strain paths at room temperature. We have measured the anisotropic stress-strain response in different mechanical tests including plane strain compression, simple compression, and uniaxial tensile tests in different sample directions that were initially processed by extrusion and rolling. We have also measured the texture of the initial and deformed samples using electron back-scattered diffraction (EBSD) technique. The differences in the strain-hardening responses and texture evolution between uniaxial loading and plane strain compression tests under certain loading directions were discussed. The highly observed plastic anisotropy in pure Mg was related to the type of predominant slip and/or twin deformation modes in each sample orientation and supported by predictions from crystal plasticity simulation. Such information is important for developing more accurate crystal plasticity models for Mg alloys.

2. Experimental Procedure

The material used in this study was an as-cast billet of high-purity magnesium (99.9%). A cylindrical bar with 35 mm (diameter) × 50 mm (long) and a plate of 70 mm × 58 mm × 19 mm were cut from the billet for extrusion and rolling processing, respectively. The cylindrical bar and plate were first annealed at 250°C in a vacuum tube furnace for 4 hrs. The cylindrical bar was then hot extruded at 250°C from 35 mm to 15 mm (~5.4 extrusion ratio), and the plate was hot rolled at about 200°C from 19 mm to 8.6 mm (~55% reduction in thickness) in multiple passes. The extruded bar and rolled plate were subsequently used to fabricate five sets of samples with different orientations to investigate the effect of strain path change and the associated texture on the anisotropic

mechanical response of pure magnesium under various mechanical tests.

Five sets of samples with different orientations were machined from the cylindrical extruded bar and rolled plate for subsequent mechanical tests as depicted in Figure 1. The orientations of the five set of samples were designed such that the loading directions in the mechanical tests align either perpendicular or parallel to the typical crystallographic orientations of the *c*-axis resulting from extrusion and rolling processes. These samples were subjected to plane strain compression (PSC) ($ND90ED_0$ and $ND0ED_0$ samples), simple compression ($CA0ED_0$ and $CA0ND_0$ samples), and uniaxial tensile ($TA90ND_0$ sample) tests at room temperature. The orientations of these samples can be distinguished by their labels, where the number in the label illustrates the angle between the loading direction in the mechanical test and the original processing direction (extrusion direction ED_0 of the extruded bar or normal direction ND_0 of the rolled plate). As an example, the label of $ND90ED_0$ sample indicates that the loading (normal) direction ND in the PSC test makes an angle of 90° with respect to the original extrusion direction ED_0. The labels also show the rolling (RD), transverse (TD), and normal (ND) directions in plane strain compression, compression axis (CA) in simple compression, and tensile axis (TA) in uniaxial tensile tests. Two specimens for each set of orientations were mechanically tested, and the variance in the measured stress-strain responses between the two samples tested for each group was found to be very small.

The plane strain compression (PSC) tests were carried out in a channel die that allows an elongation of the specimens in RD direction only, whereas deformation is suppressed in TD direction. Teflon sheet and high-pressure grease were used between the sample and the die surfaces to reduce frictional effect. The dimensions of the specimens for the PSC tests were 11 mm (RD) × 8 mm (TD) × 7 mm (ND). As the labels of the PSC specimens indicate, the loading direction ND is perpendicular to the original extrusion direction ED_0 in $ND90ED_0$ samples, whereas it is parallel to ED_0 in $ND0ED_0$ samples. For simple compression test, cylindrical samples were cut with dimensions of 13 mm (diameter) × 20 mm (height) and 5 mm (diameter) × 8 mm (height) for $CA0ED_0$ and $CA0ND_0$, respectively. The $CA0ED_0$ samples were machined from the cylindrical extruded bar with the compression axis CA parallel to ED_0, while $CA0ND_0$ samples were cut from the rolled plate with CA parallel to the plate's normal direction ND_0. The tensile specimens $TA90ND_0$ were machined from the rolled plate in the TD_0-RD_0 plane with the tensile axis TA parallel to the plate's rolling direction RD_0. The tensile samples were machined as per ASTM E8 with a gauge length of 25 mm. All mechanical tests were performed at room temperature with a constant initial strain rate of about 10^{-3} s^{-1}, using a screw-driven INSTRON machine. The raw data of the PSC and simple compression tests were corrected for machine compliance. The change of gauge length in the uniaxial tensile tests was measured using an axial extensometer (Epsilon 3542-0125M).

The texture and microstructure evolutions of all samples were characterized using a field-emission scanning electron microscope (JEOL JSM-7600F) equipped with an electron

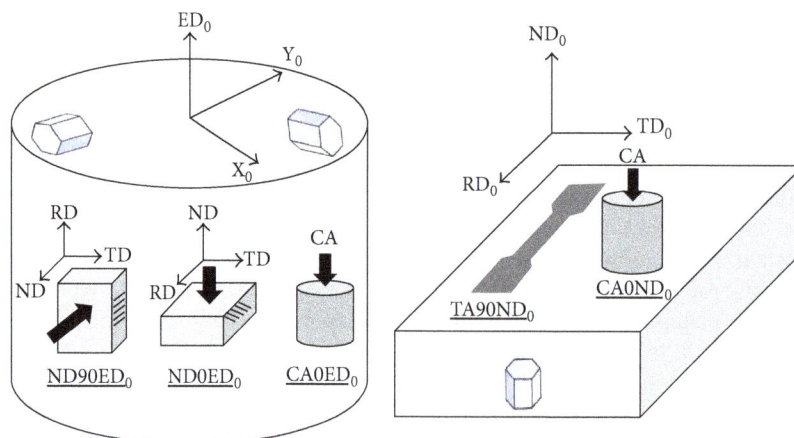

FIGURE 1: Schematic illustration of the orientations of the samples machined from the cylindrical extruded bar and rolled plate and used in plane strain compression (ND90ED$_0$ and ND0ED$_0$), simple compression (CA0ED$_0$ and CA0ND$_0$), and uniaxial tensile (TA90ND$_0$) tests. The typical crystallographic orientations of the c-axis resulting from extrusion and rolling processes are also shown.

backscattered diffraction (EBSD) camera supplied by Oxford Instruments. The specimens were prepared for EBSD scanning by mechanical polishing using subsequent oil-based diamond suspensions of 9, 3, and 1 μm, respectively. This was followed by 4–6 s chemical etching using a solution of 60 ml ethanol, 20 ml distilled water, 15 ml acetic acid, and 5 ml nitric acid. The EBSD scan areas for all samples were taken as large as possible ($>700\,\mu$m $\times 700\,\mu$m) to provide a more representative texture information, but with a slightly large step size of 1 μm to be able to complete the scanning in a reasonable time.

3. Experimental Results

3.1. True Stress-Strain Curves and Strain-Hardening Rates. The true stress-strain curves measured on different pure magnesium samples under various mechanical tests including PSC, simple compression, and uniaxial tensile are shown in Figure 2. In order to easily compare the mechanical responses between the different tests, the flow curves are plotted based on the von Mises definition of the equivalent stress and equivalent strain [19]. The strong anisotropic plastic deformation responses shown in Figure 2 have been reported in prior literature for magnesium and its alloys (e.g. [11, 13, 14, 20, 21]). Figure 3 shows the corresponding strain-hardening rates computed numerically from the stress-strain curves. The strain-hardening rates were normalized by the shear modulus of magnesium. This figure clearly indicates the major differences in the strain-hardening behavior between the different specimens tested in this study.

3.2. Microstructure and Texture Evolution. Microstructure and texture evolutions of all specimens were investigated by EBSD scanning which reveals both the structure and crystallographic orientations of the grains. Figure 4 shows the EBSD inverse pole figure (IPF) maps and pole figures of the cylindrical extruded bar and rolled plate that were used to fabricate all five sets of samples shown in Figure 1. The

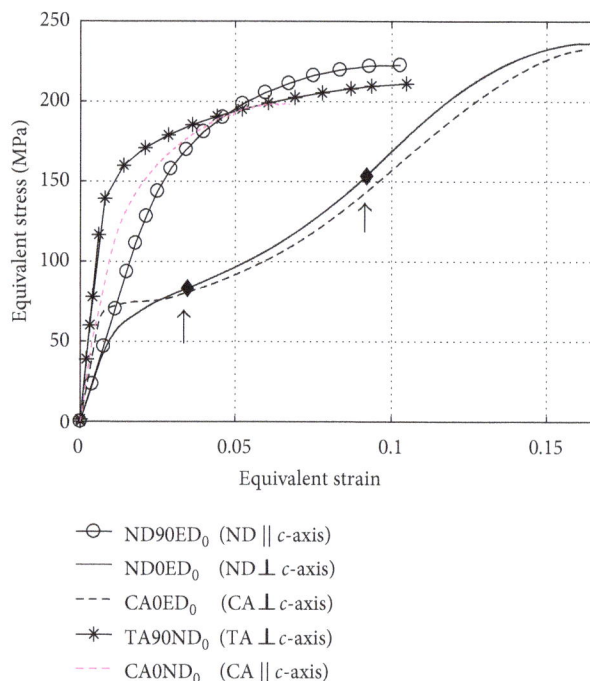

FIGURE 2: Flow curves of pure magnesium at room temperature under different mechanical tests including PSC (ND90ED$_0$ and ND0ED$_0$), simple compression (CA0ED$_0$ and CA0ND$_0$), and uniaxial tensile (TA90ND$_0$). The loading directions with respect to the typical crystallographic c-axis in most grains for each specimen are shown between parentheses.

IPF maps are collected on planes perpendicular to the extrusion direction (ED$_0$) and normal direction (ND$_0$) of the cylindrical extruded bar and rolled plate, respectively. Therefore, the IPF maps reflect the orientations of the extruded bar's extrusion direction (ED$_0$) and rolled plate's normal direction (ND$_0$) with respect to the local crystal frames. The measured pole figures clearly indicate that the extruded bar exhibited a typical prismatic $10\bar{1}0$ fiber texture, where basal planes oriented parallel to the extrusion

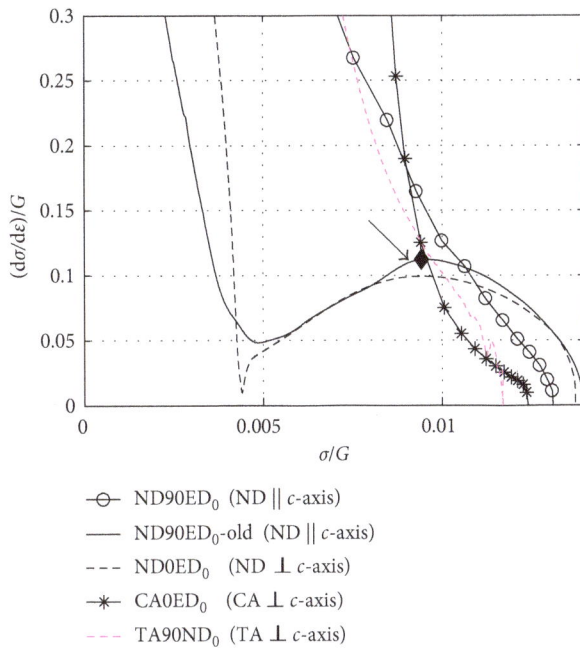

FIGURE 3: Normalized strain-hardening responses of pure magnesium at room temperature under different mechanical tests including PSC ($ND90ED_0$ and $ND0ED_0$), simple compression ($CA0ED_0$ and $CA0ND_0$), and uniaxial tensile ($TA90ND_0$).

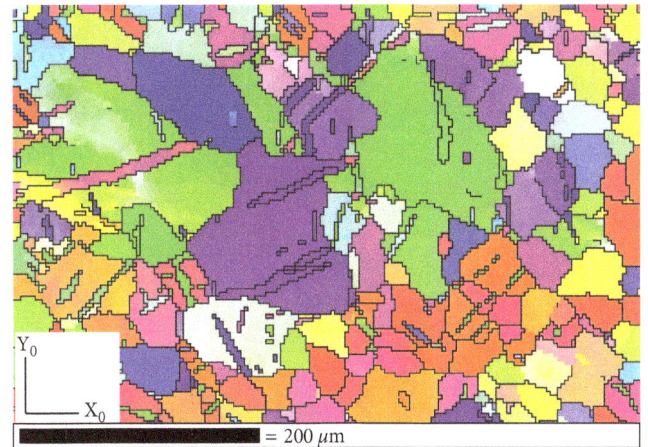

(a)

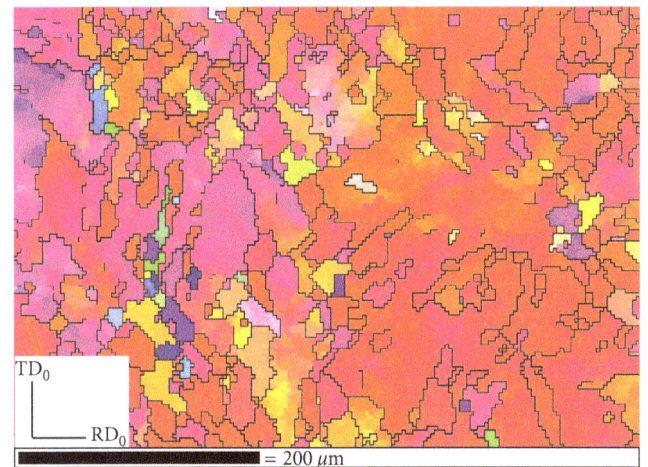

(b)

FIGURE 4: EBSD inverse pole figure (IPF) maps (top) and pole figures (bottom) showing microstructure and texture of the (a) cylindrical extruded bar and (b) rolled plate of the pure magnesium. The IPF maps reflect the orientations of the bar's extrusion direction (ED_0) and rolled plate's normal direction (ND_0) with respect to the local crystal frames. The measured pole figures show the initial texture of the processed (extruded and rolled) material. The reference frames (X_0, Y_0, ED_0) and (RD_0, TD_0, ND_0) are the same as in Figure 1.

direction ED_0 (or the crystallographic c-axis in most grains oriented in a radial direction that is perpendicular to ED_0) (e.g. [22–27]). The rolled plate also showed a typical fiber texture with the crystallographic c-axis in most grains aligned parallel to plate's normal direction ND_0 (e.g. [21, 28, 29]). Based on this texture, the orientations of the five sets of samples were designed such that the loading directions in the mechanical tests align either perpendicular or parallel to the typical crystallographic c-axis in most of the grains. More specifically, the loading direction was selected to be (1) parallel to the crystallographic c-axis in most grains in the samples $ND90ED_0$ and $CA0ND_0$ and (2) perpendicular to the crystallographic c-axis in most grains in the samples $ND0ED_0$, $CA0ED_0$, and $TA90ND_0$. In order to make this clear, we show in Figure 5 the $\{0001\}$ and $\{10\bar{1}0\}$ pole figures of the initial texture for all samples using the new reference frame for each mechanical test which were obtained by appropriate rotations of the pole figures in Figure 4.

The texture and microstructure of all specimens after each mechanical test are collected by EBSD scanning. For the purpose of simplicity and clarity, the texture will be described in terms of the orientation of the crystallographic c-axis in the crystal with respect to the loading direction in the mechanical test. The texture of $ND90ED_0$ and $ND0ED_0$ samples after PSC tests at room temperature is shown in Figure 6. The EBSD scans were acquired in the RD-TD plane in the midsection of the sample. It is clear that the crystallographic c-axis in most grains aligns parallel to ND direction in both $ND90ED_0$ and $ND0ED_0$ samples after PSC tests (the poles reappear at the center of the pole figures for both samples). It should be noted that even though the

starting texture for $ND90ED_0$ and $ND0ED_0$ samples was different (Figures 5(a) and 5(b)), the final texture after PSC tests for the two set of samples was found to be similar. The

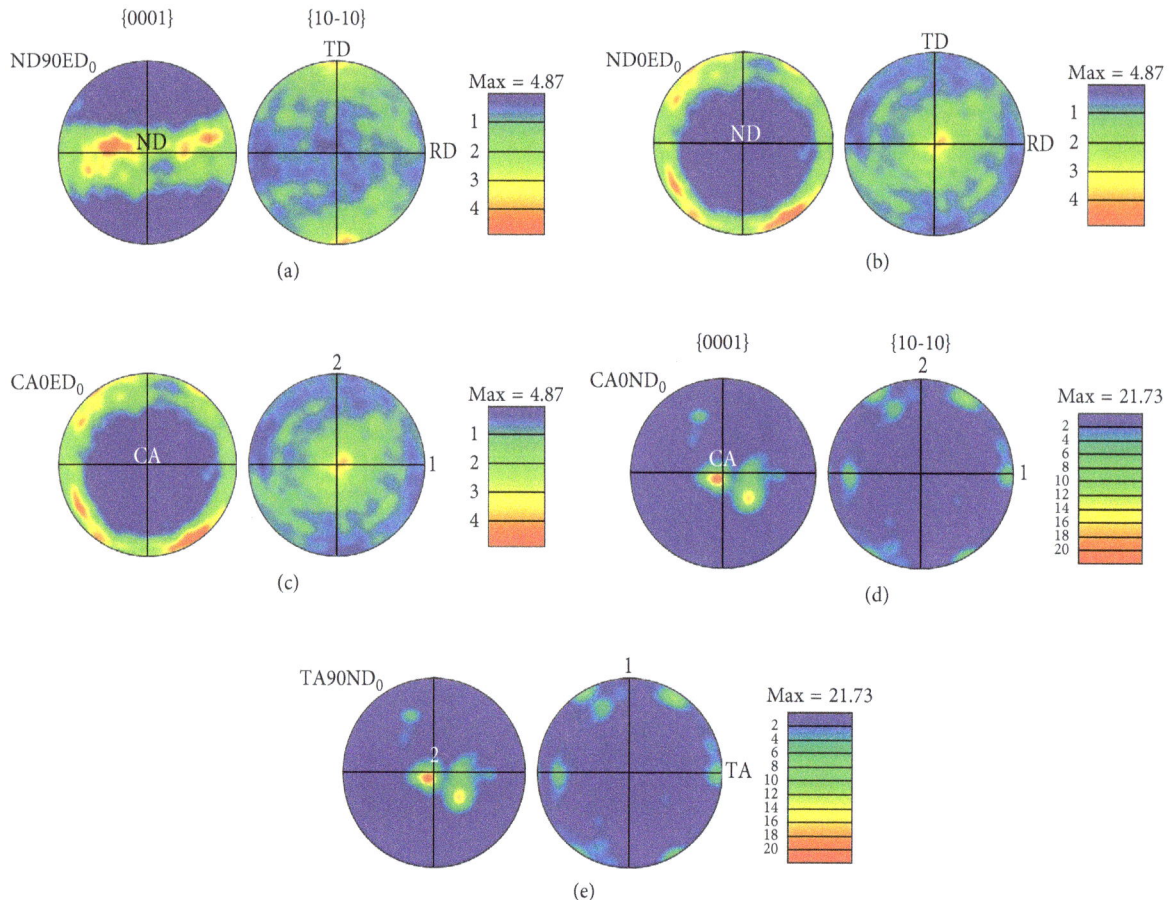

FIGURE 5: {0001} and {10$\bar{1}$0} pole figures of the initial texture for the five set of samples shown in Figure 1 using the reference frame for each mechanical test: (RD, TD, ND) for PSC, (1, 2, CA) for simple compression and (1, 2, TA) for uniaxial tension, where 1 and 2 refer to the lateral directions of the compression or tension test. The pole figures were obtained by appropriate rotations of the pole figures shown in Figure 4 with respect to the new reference frame. (a) ND90ED$_0$. (b) ND0ED$_0$. (c) CA0ED$_0$. (d) CA0ND$_0$. (e) TA90ND$_0$.

reason for such texture change is related to the type of deformation mechanisms, which will be explained in more detail in Discussion. Figures 7(a) and 7(b) present the texture after simple compression test at room temperature for CA0ED$_0$ and CA0ND$_0$ samples, respectively. The pole figures are collected in a plane normal to the compression axis CA in the midsection of the sample, with CA lying at the center of the pole figures. It is seen once again that even though the starting texture for CA0ED$_0$ and CA0ND$_0$ samples is different (Figures 5(c) and 5(d)), the final texture after simple compression test is similar to the crystallographic c-axis in most grains that align parallel to the loading direction CA in both CA0ED$_0$ and CA0ND$_0$ samples. Lastly, the texture after uniaxial tensile test for the TA90ND$_0$ sample is illustrated in Figure 7(c) which is collected in the tensile plane about 3 mm apart from the fracture surface in the gauge length region. It is noted that the texture of TA90ND$_0$ sample after the tensile test has little change compared to the initial texture.

4. Discussions

4.1. Deformation Mechanisms and Texture Evolution. The deformation behavior of magnesium and its alloys can

be very different depending on the initial texture and the type of plastic deformation mechanisms. The mechanical deformation behavior of the pure magnesium tested in this study as described by the flow curves shown in Figure 2 can be divided into two groups each having a special characteristic in common. The first group (ND0ED$_0$ and CA0ED$_0$ samples) has a concave stress-strain curve consisting of an initial drop in hardening rate followed by a marked increase up to a true strain of 0.1 where another drop is noticed. On the other hand, the second group (ND90ED$_0$, CA0ND$_0$, and TA90ND$_0$ samples) has the typical falling strain-hardening rates similar to the saturation-type hardening observed in high and medium staking fault energy face-centered cubic metals. The offset yield point for the samples in the first group is about 75 MPa, whereas it is about 130 MPa for the samples in the second group. All flow curves in both groups have a sharp loss of strength at the end of the mechanical tests accompanied by the occurrence of inhomogeneous deformation in terms of shear cracks. It should be clear that this classification does not assume the same deformation mechanisms for all samples within the same group. The deformation mechanisms and texture evolution for the samples in each group are discussed next.

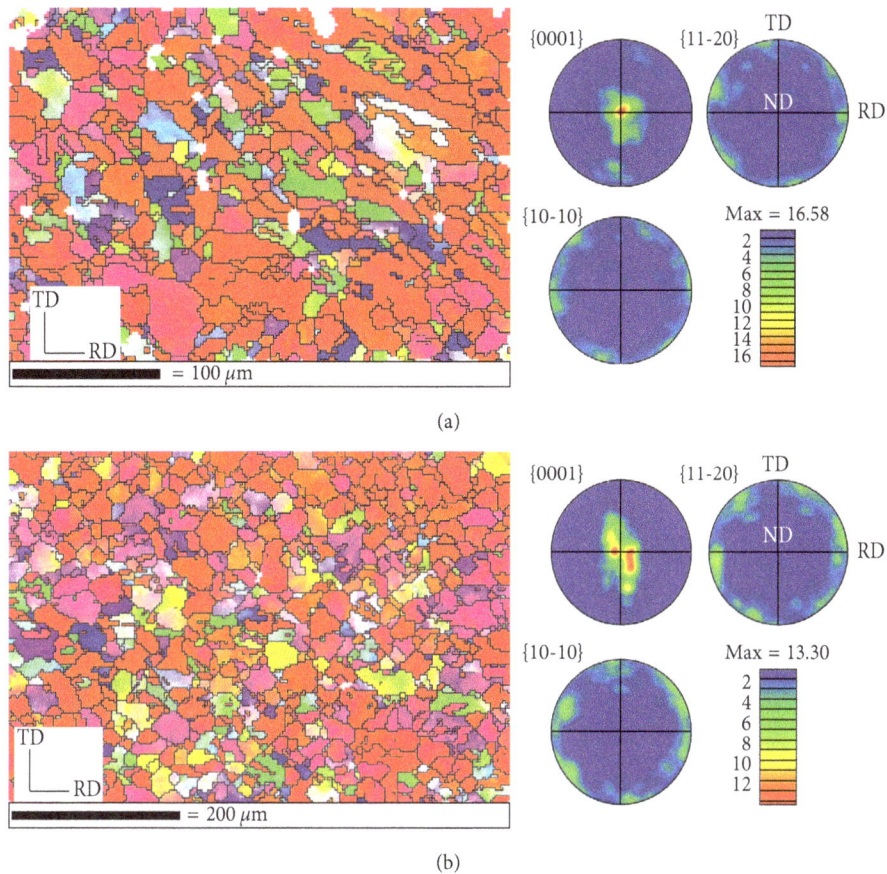

FIGURE 6: EBSD inverse pole figure (IPF) maps (left) and pole figures (right) showing microstructure and texture in RD-TD plane of the (a) ND0ED$_0$ and (b) ND90ED$_0$ pure magnesium samples after plane strain compression. The color legend is the same as in Figure 4.

For the first group consisting of ND0ED$_0$ and CA0ED$_0$ samples, the loading (compression) direction in the mechanical tests (PSC for ND0ED$_0$ and simple compression for CA0ED$_0$) aligns nearly perpendicular to the crystallographic c-axis in most of the grains as the texture indicated in Figures 5(b) and 5(c). This type of deformation mode is known to activate the $\{10\bar{1}2\}<\bar{1}011>$ tensile twins extensively at the early stage of deformation [21, 30]. This twin mode reorients the crystal by 86° around $<1\bar{2}10>$ directions bringing the basal plane to be nearly perpendicular to the loading direction. The formation of this basal texture ($\{0001\}$-fiber texture) leads to texture hardening where most grains become hard to deform with the c-axis nearly parallel to the loading axis. The $\{10\bar{1}2\}$ tensile twinning in our samples appears to start at an axial true strain of about 0.03 in ND0ED$_0$ and CA0ED$_0$ samples. This strain value corresponds to the points at which the strain-hardening rates start to increase as illustrated in Figures 2 and 3 (marked using arrows, note that Figure 2 shows the equivalent strain). This observation is in agreement with previously reported experiments in Mg alloys that detected the initial appearance of tensile twins at strains in the range of 0.02 to 0.04 [21, 31]. It should be clear that, in this study, we have not tried to detect the initial appearance of tensile twins experimentally at this small-strain level since this would require interrupting the mechanical tests frequently. The

production of extension twins and the accompanied texture hardening lead to the sharp increase in flow stresses only after a small amount of true strain, where the flow stresses increased from about 80 MPa to about 160 MPa by a true strain of about 0.07 in both ND0ED$_0$ and CA0ED$_0$ samples. Therefore, the very high strain-hardening rates observed in samples ND0ED$_0$ and CA0ED$_0$ can be attributed to the activation of extension twins and the associated texture-hardening effect. Furthermore, the exhaustion of extension twins in these samples is estimated to occur at about 0.08 strain. This strain value corresponds to the peak strain-hardening rates shown in Figure 3. It should be noted that this value is very close to the maximum shear strain that can be produced by extension twins, estimated to be 0.1289 shear strain (or about 0.065 axial strain). In other words, most of the grains in ND0ED$_0$ and CA0ED$_0$ samples are expected to form a basal texture due to extension twins at about 0.08 strain. The plastic strain after the exhaustion of extension twins beyond an equivalent strain of about 0.08 is most likely accommodated by the second pyramidal $<c+a>$ slip. Once slip and/or twinning cannot accommodate the shape change, a sudden loss of strength occurred due to the development of an inhomogeneous deformation in terms of shear bands that leads to softening and eventually cracking.

It is interesting to note that the flow responses and strain-hardening rates in ND0ED$_0$ and CA0ED$_0$ samples are quite

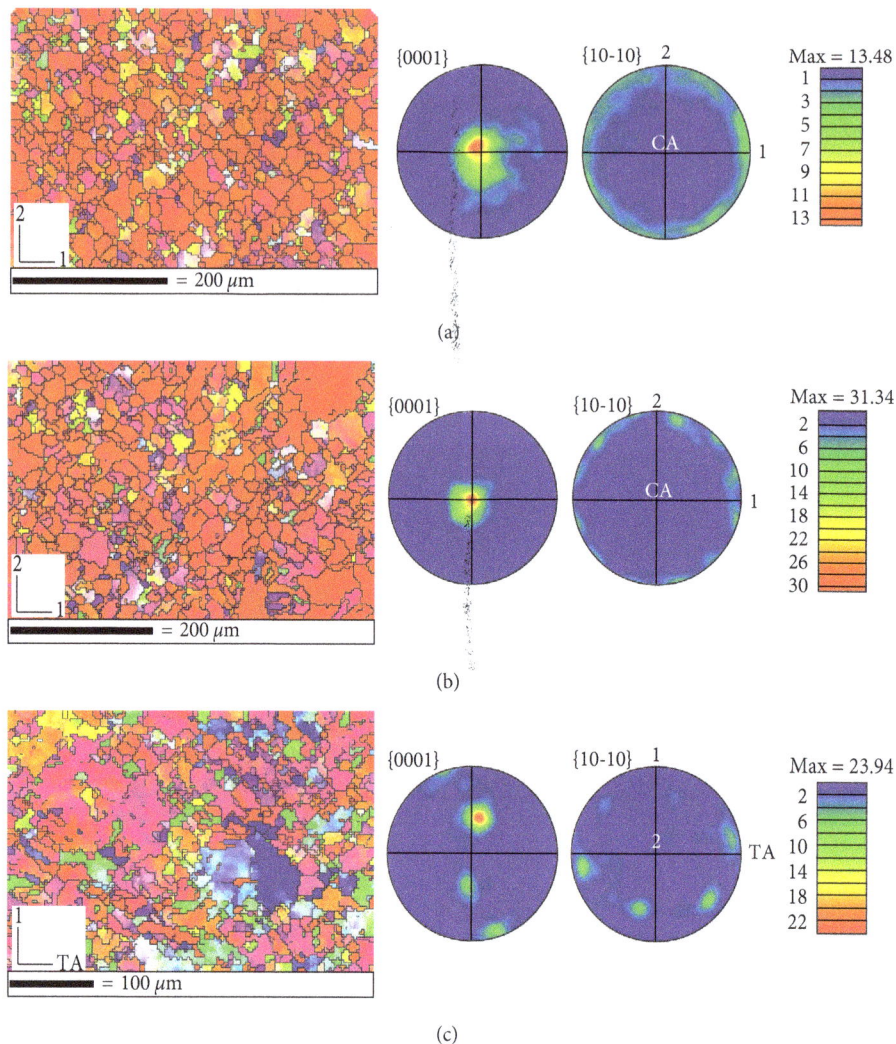

FIGURE 7: EBSD inverse pole figure (IPF) maps (left) and pole figures (right) showing texture of (a) CA0ED$_0$ and (b) CA0ND$_0$ pure magnesium samples after simple compression collected in a plane normal to the compression axis CA. The IPF and pole figures in (c) illustrate the texture of r sample after uniaxial tensile test collected in the tensile plane. The labels 1 and 2 in (c) represent the lateral directions of the tension test. The color legend is the same as in Figure 4.

similar in spite of the difference in the applied boundary conditions for the two mechanical tests. To explain this, let us first consider the geometrical constraints in the two mechanical tests. In simple compression with CA normal to the c-axis, the sample has no lateral constraint that would suppress the extension of crystals along the c-axis as a result of extension twins. However, in plane strain compression with ND normal to the c-axis, the crystals with c-axis parallel to TD are not expected to twin because twinning would cause broadening in the constrained direction, TD. Thus, only the crystals with c-axis not perfectly aligned with TD can twin (especially those with c-axis parallel to the free direction, RD). In other words, the amount of twin activities is expected to be higher in simple compression compared to plane strain compression when the loading direction aligns perpendicular to the c-axis. This observation has been numerically investigated using crystal plasticity models for ideal orientations in prior literature (e.g. [20, 30, 32]). However, in the current study, it is noted that the flow responses and

strain-hardening rates of the plane strain compressed sample (ND0ED$_0$) are quite similar to the simple compressed sample (CA0ED$_0$). This can be attributed to the existence of many grains in ND0ED$_0$ sample with enough scatter of the c-axis about TD and therefore the easy activation of extension twins compared to other slip systems. This is clearly evidenced by the fact that the final texture of ND0ED$_0$ sample at the end of the mechanical test had a strong basal texture and was also supported by the rapid activation of extension twins as predicted by crystal plasticity simulations. It is worth mentioning that the extension twins in magnesium usually grow quickly and encompass the entire grains. Thus, no evidence of a high volume fraction of extension twins was observed in the final microstructure of ND0ED$_0$ and CA0ED$_0$ samples at the end of the mechanical test.

The samples in the second group (ND90ED$_0$, CA0ND$_0$, and TA90ND$_0$) have similar flow stress with falling strain-hardening rates as shown in Figure 2. Unlike the samples in the first group, the extension twins are not expected to occur in these samples [10, 13, 14, 33]. For ND90ED$_0$ and CA0ND$_0$ samples, the compression direction aligns parallel to the crystallographic c-axis in most grains (Figures 5(a) and 5(d)), and therefore the <a> slips in basal or prismatic slip systems would not be activated. In this case, the second hard pyramidal <c+a> system is the dominant slip mode that can accommodate the plastic strain. Hence, it is not surprising to see high strain-hardening rates in the early portion of the stress-strain curves. The activation of compression twins is an additional possible deformation mechanism reported in the literature for such deformation modes [11, 20, 21]. For TA90ND$_0$ sample in the second group, since the tensile axis TA is parallel to the basal plane of most grains (Figure 5(e)), the prismatic <a> and second pyramidal <c+a> slips have more favorable orientations compared to extension twin and basal slip [34, 35].

It is known that the activation of extension twins in magnesium has significant effects on the texture evolution since tensile twins reorient the crystal by 86° around $1\bar{2}10$ directions. This can be clearly seen in the texture evolution of the samples in the first group (ND0ED$_0$ and CA0ED$_0$). These samples begin the mechanical tests with the loading direction aligned nearly perpendicular to the crystallographic c-axis in most grains (Figures 5(b) and 5(c)). However, the activation of tensile twins after about 0.03 axial strains alters the orientations of the crystallographic c-axis in these samples to become nearly parallel to the loading direction forming the strong basal texture ({0001}-fiber texture) shown in Figures 6(a) and 7(a). On the other hand, the texture of ND90ED$_0$ and CA0ND$_0$ samples in the second group with no active tensile twins did not change significantly (Figures 6(b) and 7(b)) compared to their initial basal texture shown in Figures 5(a) and 5(d). These samples were expected to deform mainly by slip in the second pyramidal <c+a> system. It is interesting to note that even though the initial texture of the samples in the two groups before the mechanical tests is different (the loading direction is perpendicular to the c-axis in ND0ED$_0$ and CA0ED$_0$ samples and parallel to the c-axis in ND90ED$_0$ and CA0ND$_0$ samples), the final texture for these samples is relatively similar. It is also seen that the flow response and strain-hardening rates of the samples in the first group (ND0ED$_0$ and CA0ED$_0$) beyond a small strain of 0.08 (the point at which extension twins deplete or saturate) are roughly similar to the behavior of the ND90ED$_0$ and CA0ND$_0$ samples in the second group. This can be related to the similarity in texture and deformation mechanism, where the texture of the ND0ED$_0$ and CA0ED$_0$ samples at 0.08 strain transforms mainly to a basal texture, which is similar to the initial texture of the ND90ED$_0$ and CA0ND$_0$. The samples with such a fiber texture deform essentially by slip in the second pyramidal <c+a> system.

4.2. Crystal Plasticity Simulation.

In order to examine the effect of texture on the activation of various slip and twin systems, the different experimental deformation modes tested in this study were modeled using crystal plasticity constitutive equations. In this model, the plastic velocity

TABLE 1: Various slip and twin systems used in the crystal plasticity simulations for Mg.

System	Shear direction, m_0				Plane normal, n_0			
Basal <a> slip	0	0	0	1	-2	1	1	0
	0	0	0	1	1	-2	1	0
	0	0	0	1	1	1	-2	0
Prism <a> slip	0	1	-1	0	-2	1	1	0
	1	0	-1	0	1	-2	1	0
	1	-1	0	0	1	1	-2	0
Pyramidal <a> slip	0	1	-1	1	-2	1	1	0
	0	-1	1	1	-2	1	1	0
	1	0	-1	1	1	-2	1	0
	-1	0	1	1	1	-2	1	0
	1	-1	0	1	1	1	-2	0
	-1	1	0	1	1	1	-2	0
Second pyramidal <c+a> slip	1	1	-2	2	-1	-1	2	3
	-1	2	-1	2	1	-2	1	3
	-2	1	1	2	2	-1	-1	3
	-1	-1	2	2	1	1	-2	3
	1	-2	1	2	-1	2	-1	3
	2	-1	-1	2	-2	1	1	3
Extension twin	0	-1	1	2	0	1	-1	1
	1	0	-1	2	-1	0	1	1
	-1	1	0	2	1	-1	0	1
	0	1	-1	2	0	-1	1	1
	-1	0	1	2	1	0	-1	1
	1	-1	0	2	-1	1	0	1

gradient tensor is assumed to have contributions from both slip and deformation twinning [36, 37]:

$$L^P = \sum_{\alpha}^{N^s} \dot{\gamma}^{\alpha}\left(m_0^{\alpha} \otimes n_0^{\alpha}\right) + \sum_{\beta}^{N^{tw}} \dot{f}^{\beta} \gamma^{tw}\left(m_0^{\beta} \otimes n_0^{\beta}\right), \quad (1)$$

where $\dot{\gamma}^{\alpha}$ represents the shearing rate on the slip system α, γ^{tw} describes the constant shear associated with the twin system, and f^{β} denotes its volume fraction in the given crystal. $m_0^{\alpha(\beta)}$ and $n_0^{\alpha(\beta)}$ denote the slip (twin shear) direction and the slip (twin) plane normal to the slip (twin) system $\alpha(\beta)$, respectively, in the initial configuration. N^s and N^{tw} refer to the total number of available slip and twin systems, respectively. The evolution of the plastic shearing rate $\dot{\gamma}^{\alpha}$ and the deformation twin volume fraction $\dot{f}^{\beta}$ are quantified using power-law relations [38, 39]:

$$\dot{\gamma}^{\alpha} = \dot{\gamma}_0 \left|\frac{\tau^{\alpha}}{s_s^{\alpha}}\right|^{1/m} \text{sign}\left(\tau^{\alpha}\right), \quad (2)$$

$$\dot{f}^{\beta} = \frac{\dot{\gamma}_0}{\gamma^{tw}} \left(\frac{\tau^{\beta}}{s_{tw}^{\beta}}\right)^{1/m}, \quad (3)$$

where $\dot{\gamma}_0$ is a reference value of slip rate taken as 0.001 s^{-1} and m is a rate-sensitivity parameter taken as 0.02 and assumed to be the same for both slip and twinning. $\tau^{\alpha} \cdot (\tau^{\beta})$ and $s_s^{\alpha} \cdot (s_{tw}^{\beta})$ denote the resolved shear stress and slip (twin) resistance of the α-slip (β-twin) system, respectively. An extended version of the saturation-type hardening laws has been employed in this work to describe the slip-hardening evolution [36–38]:

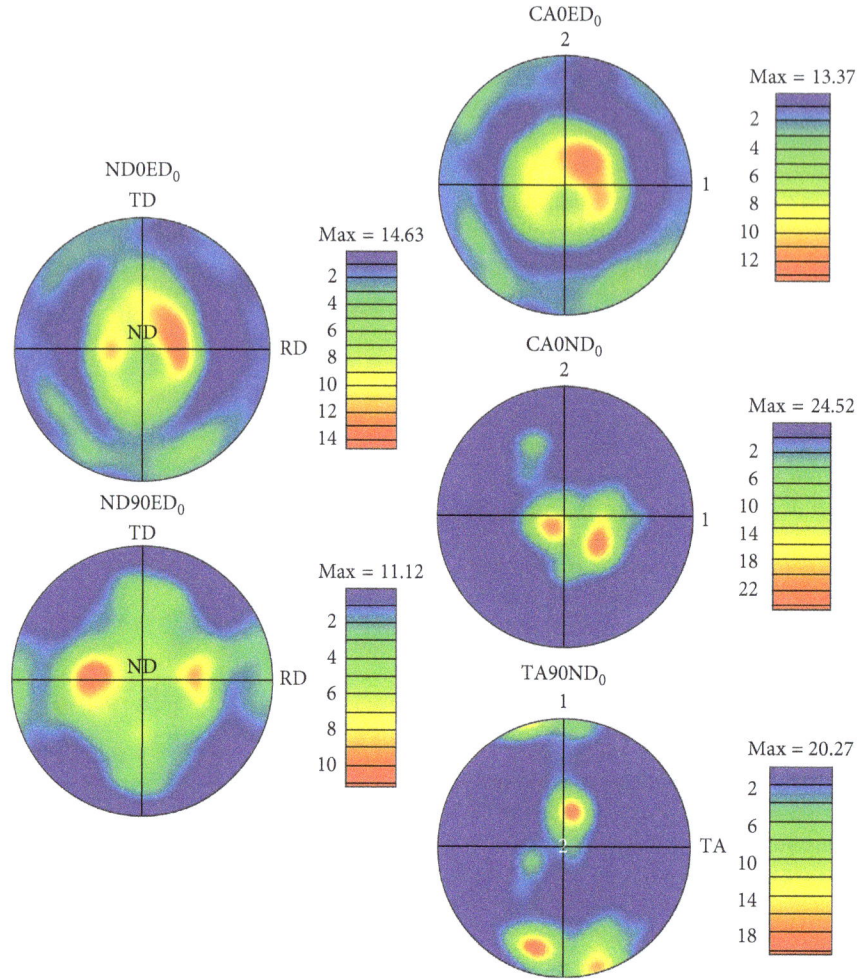

FIGURE 8: Predicted final texture ({0001} pole figures) using crystal plasticity simulations under the five set of mechanical tests performed in this study: plane strain compression (ND0ED$_0$ and ND90ED$_0$), simple compression (CA0ED$_0$ and CA0ND$_0$), and uniaxial tension (TA90ND$_0$).

$$\dot{S}_s^{\alpha} = h_s \left(1 + C \sum_{\beta}^{N^{tw}} f^{\beta} \right) \left(1 - \frac{s_s^{\alpha}}{s_{sat}^{\alpha}} \right)^a \sum_{k}^{N^s} \dot{\gamma}^k,$$

$$s_{sat}^{\alpha} = s_{s0} + s_{pr} \left(\sum_{\beta}^{N^{tw}} f^{\beta} \right)^{0.5}, \tag{4}$$

where the parameters h_s, C, s_{sat}^{α}, a, and s_{pr} aim to capture the phenomenological slip-hardening behavior due to slip and twin interactions. More details about the physical interpretations of these parameters can be found in [36–38]. The twin resistance was assumed to be constant in the current simulation.

A total of 18 slip systems (3 basal <a>, 3 prism <a>, 6 pyramidal <a>, and 6 {10$\bar{2}$2} second pyramidal <c+a>) and 6 {10$\bar{1}$2}$\bar{1}$011 tensile twin systems were considered as listed in Table 1. The simulations were conducted assuming that the slip resistance values for the basal, prismatic, pyramidal <a>, second pyramidal <c+a>, and tensile twin modes were in the ratio of 1 : 50 : 50 : 80 : 7 [30, 40, 41]. The values of the various slip and twin hardening parameters in (4) were

established by fitting the measured stress-strain curves of CA0ND$_0$, CA0ED$_0$, and TA90ND$_0$ samples. These mechanical tests were selected because they produced the most distinct mechanical behavior (Figure 2) since the loading directions align either parallel or perpendicular to the crystallographic c-axis in most grains in these samples. The initial texture in the samples was captured using a set of 1000 discrete crystal orientations selected randomly from the initial measured texture. Five loading conditions were simulated corresponding to plane strain compression, simple compression, and uniaxial tensile tests with the loading direction aligned either along the crystallographic c-axis or along a direction normal to the c-axis. These loading conditions cover the five sets of experiments performed in this study: namely, ND90ED$_0$ (ND ∥ c-axis), ND0ED$_0$ (ND ⊥ c-axis), CA0ND$_0$ (CA ∥ c-axis), CA0ED$_0$ (CA ⊥ c-axis), and TA90ND$_0$ (TA ⊥ c-axis). Figure 8 shows the predicted final textures after an effective strain of about 0.15 for the five set of mechanical tests performed in this study using Taylor-type crystal plasticity simulations. It can be seen that the predicted textures are in reasonable agreement with the measured-final textures shown in Figures 6 and 7. The deviation between the simulated and

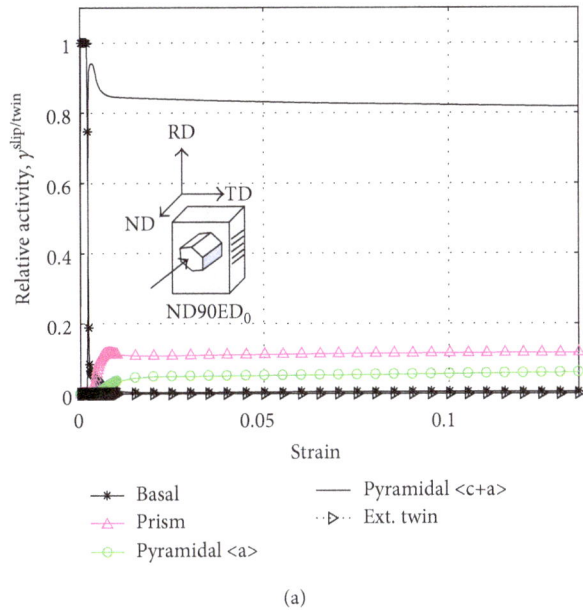

(a)

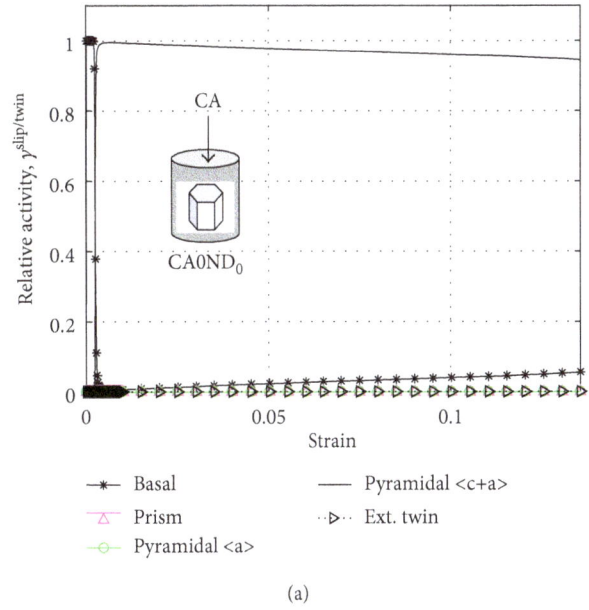

(b)

FIGURE 9: Relative slip and twin activities predicted by crystal plasticity simulation during plane strain compression of pure Mg for the two sample orientations tested in this study: (a) ND90ED$_0$ sample (ND ∥ c-axis) and (b) ND0ED$_0$ sample (ND ⊥ c-axis).

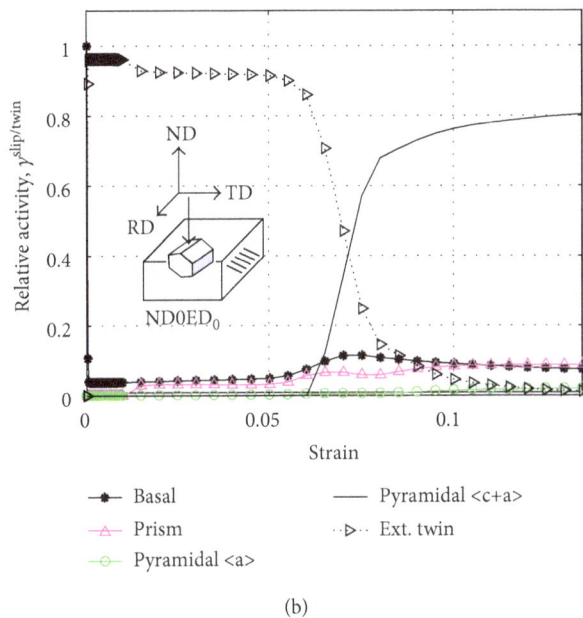

(a)

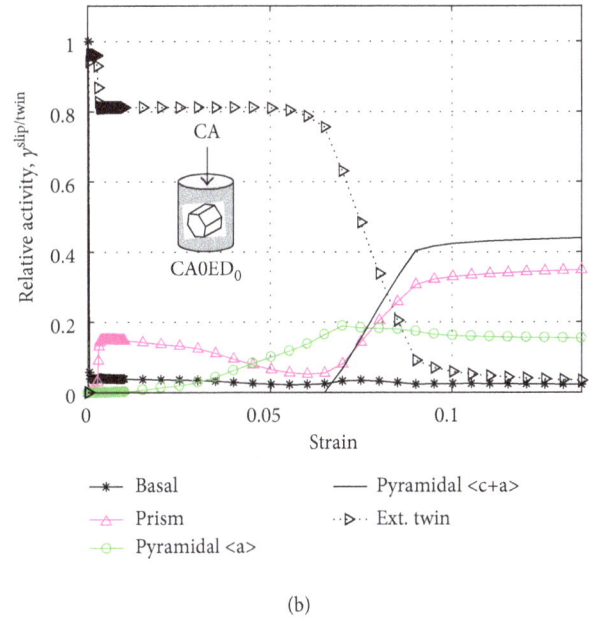

(b)

FIGURE 10: Relative slip and twin activities predicted by crystal plasticity simulation during simple compression of pure Mg for the two sample orientations tested in this study: (a) CA0ND$_0$ (CA ∥ c-axis) and (b) CA0ED$_0$ (CA ⊥ c-axis).

measured textures appears in the pole figure of the plane strain compressed samples where the simulated results do not produce the strong-basal texture obtained in the measured texture (Figure 6).

Our main interest is to capture the most favored (active) slip or twin modes for the various loading conditions performed in this study using crystal plasticity simulation. Figure 9 shows the relative slip and twin activities for plane strain compression of pure Mg with the loading direction aligned either parallel (ND90ED$_0$ sample) or perpendicular (ND0ED$_0$ sample) to the crystallographic c-axis in most grains. The predictions from the crystal plasticity model suggest that much

of the plastic strain in ND90ED$_0$ sample (ND ∥ c-axis) is accommodated by the second pyramidal <c+a> slip. However, the plastic strain in ND0ED$_0$ sample (ND ⊥ c-axis) is initially accommodated by the activation of tensile twins up to a strain of about 0.07, a point at which most grains are oriented such that the c-axis becomes nearly parallel to the loading, beyond that the second hard pyramidal <c+a> slip is activated to accommodate the further deformation. In the case of simple compression, Figure 10 presents the most favored deformation mechanisms predicted by the crystal plasticity simulation. These results are somehow similar to PSC for the two sample orientations tested experimentally in this study: namely,

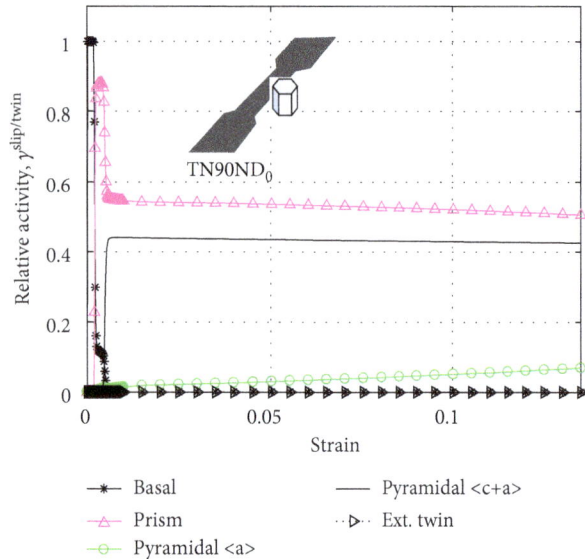

FIGURE 11: Relative slip and twin activities predicted by crystal plasticity simulation during uniaxial tension of pure Mg with the basal plane parallel to the tensile axis, TA90ND$_0$ sample.

$CA0ND_0$ (CA $\parallel$ c-axis) and $CA0ED_0$ (CA $\perp$ c-axis). More specifically, Figure 10(a) shows that the second pyramidal <c+a> slip is the predominant active deformation mechanism in $CA0ND_0$ sample (CA $\parallel$ c-axis) for the entire deformation. However, for $CA0ED_0$ sample (Figure 10(b)), it is seen once again that the tensile twins are initially operated and then followed by the activation of the second pyramidal <c+a> slip (after a strain of about 0.07). Finally, for the uniaxial tensile test ($TA90ND_0$ sample), the crystal plasticity predictions shown in Figure 11 suggest that the plastic strain is most likely accommodated by the prismatic <a> and second pyramidal <c+a> slips. These results are expected for uniaxial tension when the basal plane is parallel to the tensile axis [34, 35].

5. Conclusions

The anisotropic stress-strain responses of strongly textured pure magnesium were investigated at room temperature under three different monotonic deformation paths: (1) plane strain compression; (2) simple compression; and (3) uniaxial tensile tests. The samples in these mechanical tests were oriented such that the loading direction aligned either parallel or normal to the crystallographic c-axis in most of the grains. The main conclusions of this study are summarized below based on our experimental observations and supported by the predicted relative slip and twin activities using crystal plasticity simulations:

(a) The activation of {10$\bar{1}$2} tensile twins and the accompanying texture-hardening effect were associated with the initial increase in strain-hardening rate observed in samples subjected to plane strain compression or simple compression with the loading direction normal to the crystallographic c-axis.

(b) The transition in the stress-strain response from the initial increase in strain-hardening rates to the falling hardening rates was correlated with the exhaustion of deformation twins and the activation of slips along the second pyramidal <c+a> systems.

(c) The slip on the second hard pyramidal <c+a> system was found to be the dominant deformation mechanism that can accommodate the plastic strains in samples subjected to plane strain compression or simple compression with the loading direction parallel to the crystallographic c-axis.

(d) The prismatic <a> and second pyramidal <c+a> slips are the main deformation mechanisms during uniaxial tension when the basal plane is parallel to the tensile axis.

Conflicts of Interest

The authors declare that they have no conflicts of interest.

Acknowledgments

The authors extend their appreciation to the Deanship of Scientific Research at King Saud University for funding this work through RAED Project no. NFG-15-03-11.

References

[1] C. Bettles and M. Gibson, "Current wrought magnesium alloys: strengths and weaknesses," *JOM*, vol. 57, no. 5, pp. 46–49, 2005.

[2] M. K. Kulekci, "Magnesium and its alloys applications in automotive industry," *International Journal of Advanced Manufacturing Technology*, vol. 39, no. 9-10, pp. 851–865, 2008.

[3] W. F. Hosford, *The Mechanics of Crystals and Textured Polycrystals*, Oxford University Press, New York, UK, 1993.

[4] W. F. Hosford and R. M. Caddell, *Metal Forming: Mechanics and Metallurgy*, Cambridge University Press, Cambridge, UK, 2011.

[5] T. Obara, H. Yoshinga, and S. Morozumi, "{11$\bar{2}$2} ⟨11$\bar{2}$3⟩ slip system in magnesium," *Acta Metallurgica*, vol. 21, no. 7, pp. 845–853, 1973.

[6] S. R. Agnew, "2-deformation mechanisms of magnesium alloys," *Advances in Wrought Magnesium Alloys*, pp. 63–104, Woodhead Publishing, Sawston, UK, 2012.

[7] S. Sandlöbes, M. Friák, J. Neugebauer, and D. Raabe, "Basal and non-basal dislocation slip in Mg–Y," *Materials Science and Engineering: A*, vol. 576, pp. 61–68, 2013.

[8] B. A. Bilby and A. G. Crocker, "The theory of the crystallography of deformation twinning," *Proceedings of the Royal Society of London. Series A, Mathematical and Physical Sciences*, vol. 288, no. 1413, pp. 240–255, 1965.

[9] J. W. Christian and S. Mahajan, "Deformation twinning," *Progress in Materials Science*, vol. 39, no. 1-2, pp. 1–157, 1995.

[10] M. R. Barnett, "Twinning and the ductility of magnesium alloys: part I: "tension" twins," *Materials Science and Engineering: A*, vol. 464, no. 1-2, pp. 1–7, 2007.

[11] M. R. Barnett, "Twinning and the ductility of magnesium alloys: part II. "contraction" twins," *Materials Science and Engineering: A*, vol. 464, no. 1-2, pp. 8–16, 2007.

[12] E. Kelley and W. Hosford, "Plane strain compression of magnesium and magnesium alloy crystals," *Transactions of the Metallurgical Society of AIME*, vol. 24, p. 5, 1968.

[13] R. Gehrmann, M. M. Frommert, and G. Gottstein, "Texture effects on plastic deformation of magnesium," *Materials Science and Engineering: A*, vol. 395, no. 1-2, pp. 338–349, 2005.

[14] M. D. Nave and M. R. Barnett, "Microstructures and textures of pure magnesium deformed in plane-strain compression," *Scripta Materialia*, vol. 51, no. 9, pp. 881–885, 2004.

[15] G. T. Gray, "Influence of strain rate and temperature on the structure. Property behavior of high-purity titanium," *Le Journal de Physique IV*, vol. 7, pp. C3–423–C3–428, 1997.

[16] M. R. Barnett, Z. Keshavarz, A. G. Beer, and D. Atwell, "Influence of grain size on the compressive deformation of wrought Mg–3Al–1Zn," *Acta Materialia*, vol. 52, no. 17, pp. 5093–5103, 2004.

[17] E. El-Danaf, S. R. Kalidindi, and R. D. Doherty, "Influence of grain size and stacking-fault energy on deformation twinning in fcc metals," *Metallurgical and Materials Transactions A*, vol. 30, no. 5, pp. 1223–1233, 1999.

[18] A. A. Salem, S. R. Kalidindi, and R. D. Doherty, "Strain hardening of titanium: role of deformation twinning," *Acta Materialia*, vol. 51, no. 14, pp. 4225–4237, 2003.

[19] R. Hill, *The Mathematical Theory of Plasticity*, Oxford University Press, Oxford, UK, 1998.

[20] T. Al-Samman and G. Gottstein, "Room temperature formability of a magnesium AZ31 alloy: examining the role of texture on the deformation mechanisms," *Materials Science and Engineering: A*, vol. 488, no. 1-2, pp. 406–414, 2008.

[21] M. Knezevic, A. Levinson, R. Harris, R. K. Mishra, R. D. Doherty, and S. R. Kalidindi, "Deformation twinning in AZ31: influence on strain hardening and texture evolution," *Acta Materialia*, vol. 58, no. 19, pp. 6230–6242, 2010.

[22] S. Biswas, S. Suwas, R. Sikand, and A. K. Gupta, "Analysis of texture evolution in pure magnesium and the magnesium alloy AM30 during rod and tube extrusion," *Materials Science and Engineering: A*, vol. 528, pp. 3722–3729, 2011.

[23] J. Bohlen, S. B. Yi, J. Swiostek, D. Letzig, H. G. Brokmeier, and K. U. Kainer, "Microstructure and texture development during hydrostatic extrusion of magnesium alloy AZ31," *Scripta Materialia*, vol. 53, no. 2, pp. 259–264, 2005.

[24] N. Dixit, K. Y. Xie, K. J. Hemker, and K. T. Ramesh, "Microstructural evolution of pure magnesium under high strain rate loading," *Acta Materialia*, vol. 87, pp. 56–67, 2015.

[25] S. Kleiner and P. J. Uggowitzer, "Mechanical anisotropy of extruded Mg–6% Al–1% Zn alloy," *Materials Science and Engineering: A*, vol. 379, no. 1-2, pp. 258–263, 2004.

[26] S. J. Liang, Z. Y. Liu, and E. D. Wang, "Microstructure and mechanical properties of Mg-Al-Zn alloy deformed by cold extrusion," *Materials Letters*, vol. 62, pp. 3051–3054, 2008.

[27] J. Victoria-Hernandez, S. Yi, D. Letzig, D. Hernandez-Silva, and J. Bohlen, "Microstructure and texture development in hydrostatically extruded Mg–Al–Zn alloys during tensile testing at intermediate temperatures," *Acta Materialia*, vol. 61, no. 6, pp. 2179–2193, 2013.

[28] J. A. del Valle, F. Carreño, and O. A. Ruano, "Influence of texture and grain size on work hardening and ductility in magnesium-based alloys processed by ECAP and rolling," *Acta Materialia*, vol. 54, no. 16, pp. 4247–4259, 2006.

[29] S. Suwas, G. Gottstein, and R. Kumar, "Evolution of crystallographic texture during equal channel angular extrusion (ECAE) and its effects on secondary processing of magnesium," *Materials Science and Engineering: A*, vol. 471, no. 1-2, pp. 1–14, 2007.

[30] J. Zhang and S. P. Joshi, "Phenomenological crystal plasticity modeling and detailed micromechanical investigations of pure magnesium," *Journal of the Mechanics and Physics of Solids*, vol. 60, no. 5, pp. 945–972, 2012.

[31] G. Zhou, M. K. Jain, P. Wu, Y. Shao, D. Li, and Y. Peng, "Experiment and crystal plasticity analysis on plastic deformation of AZ31B Mg alloy sheet under intermediate temperatures: how deformation mechanisms evolve," *International Journal of Plasticity*, vol. 79, pp. 19–47, 2016.

[32] Y. Gan, W. Song, and J. Ning, "Crystal plasticity analysis of plane strain deformation behavior and texture evolution for pure magnesium," *Computational Materials Science*, vol. 123, pp. 232–243, 2016.

[33] X. Y. Lou, M. Li, R. K. Boger, S. R. Agnew, and R. H. Wagoner, "Hardening evolution of AZ31B Mg sheet," *International Journal of Plasticity*, vol. 23, no. 1, pp. 44–86, 2007.

[34] T. Al-Samman, X. Li, and S. G. Chowdhury, "Orientation dependent slip and twinning during compression and tension of strongly textured magnesium AZ31 alloy," *Materials Science and Engineering: A*, vol. 527, no. 15, pp. 3450–3463, 2010.

[35] C. J. Geng, B. L. Wu, X. H. Du et al., "Stress–strain response of textured AZ31B magnesium alloy under uniaxial tension at the different strain rates," *Materials Science and Engineering: A*, vol. 559, pp. 307–313, 2013.

[36] A. A. Salem, S. R. Kalidindi, and S. L. Semiatin, "Strain hardening due to deformation twinning in α-titanium: constitutive relations and crystal-plasticity modeling," *Acta Materialia*, vol. 53, no. 12, pp. 3495–3502, 2005.

[37] X. Wu, S. R. Kalidindi, C. Necker, and A. A. Salem, "Prediction of crystallographic texture evolution and anisotropic stress–strain curves during large plastic strains in high purity α-titanium using a Taylor-type crystal plasticity model," *Acta Materialia*, vol. 55, no. 2, pp. 423–432, 2007.

[38] S. R. Kalidindi, "Incorporation of deformation twinning in crystal plasticity models," *Journal of the Mechanics and Physics of Solids*, vol. 46, no. 2, pp. 267–290, 1998.

[39] A. Needleman, R. J. Asaro, J. Lemonds, and D. Peirce, "Finite element analysis of crystalline solids," *Computer Methods in Applied Mechanics and Engineering*, vol. 52, no. 1–3, pp. 689–708, 1985.

[40] Y. Gan, W. Song, J. Ning, H. Tang, and X. Mao, "An elastic–viscoplastic crystal plasticity modeling and strain hardening for plane strain deformation of pure magnesium," *Mechanics of Materials*, vol. 92, pp. 185–197, 2016.

[41] J. Zhang, Q. Wei, and S. P. Joshi, "Effects of reinforcement morphology on the mechanical behavior of magnesium metal matrix composites based on crystal plasticity modeling," *Mechanics of Materials*, vol. 95, pp. 1–14, 2016.

Combined Effects of Sustained Loads and Wet-Dry Cycles on Durability of Glass Fiber Reinforced Polymer Composites

Mengting Li,[1] Jun Wang,[1] Weiqing Liu,[1] Ruifeng Liang,[2] Hota GangaRao,[2] and Yang Li[3]

[1]College of Civil Engineering, Nanjing Tech University, Nanjing 211816, China
[2]Department of Civil and Environmental Engineering, West Virginia University, Morgantown, WV, USA
[3]Jiangsu Wei Xin Engineering Consultants Ltd., Nanjing, China

Correspondence should be addressed to Jun Wang; wangjun3312@njtech.edu.cn and Weiqing Liu; wqliu@njtech.edu.cn

Academic Editor: Fabrizio Sarasini

This paper deals with durability of glass fiber reinforced polymer (GFRP) composites under the combined effects of sustained tensile loads and wet-dry (WD) cycles. Two different solutions (distilled water and saltwater) were used to imitate the freshwater and marine environments, respectively. Tensile properties of the unconditioned and conditioned specimens were measured to study the durability of GFRP composites under these 2 effects. The response indicated that both tensile strength and elastic modulus increased initially upon WD cycles, which was attributed to both the postcuring of resin and the sustained tensile stress allowing for fastec cure. Further exposure to WD cycles in distilled water or saltwater led to a steady decrease in tensile strength and modulus. WD cycles of saltwater and distilled water have similar effects on the degradation of the tensile properties for unstressed specimens. However, the elastic modulus and elongation at rupture of stressed specimens under WD cycles of saltwater decreased more than those specimens under WD cycles of distilled water. Moreover, increase of sustained loads led to a decrease in tensile strength. Based on Arrhenius method, a prediction model which accounted for the effects of postcure processes was developed. The predicted results of tensile strength and elastic modulus agree well with those obtained from the experiments.

1. Introduction

The use of fiber reinforced polymer (FRP) composites in the construction of waterfront and marine infrastructures has been increasing rapidly, due to their advantage of light weight, high tensile strength, and better reinforce to aqueous corrosion. Because of economic benefits in construction, glass fiber reinforced polymer (GFRP) composites are becoming prevalent in civil engineering structures. FRP structures are usually exposed to harsh in-service environment, such as moisture, corrosive media, sustained stress, temperature, and ultraviolet radiation. Cyclic moisture effect near the splash-zone of marine structures is one of the important environmental factors to deteriorate the properties and behavior of polymer matrix composites [1]. Moreover, the environmental attack is combined with sustained loading prone to aggravate the deterioration of GFRP. Therefore, better understanding of the long-term behavior of GFRP under the combined effects of load and moisture cycles is helpful to promote its applications.

In recent decades, several researchers have investigated durability issues of FRP reinforced concrete (RC) structures subjected to wet-dry (WD) cycles. Soudki et al. [2] studied the behavior of CFRP strengthened RC beams subjected to WD cycles with 3% NaCl. Their test results displayed that the critical chloride levels reached the reinforced steel bars after 200 WD cycles and visible cracks and stains appeared after 300 WD cycles. Belarbi and Bae [3] evaluated the combined effects of freeze-thaw (FT) cycles and WD cycles on the long-term behavior of RC columns strengthened with CFRP and GFRP sheets. Their studies indicated that, after 300 FT cycles, the failure load and axial stiffness of both CFRP and GFRP strengthened RC columns increased slightly due to the matrix hardening effects at extremely low temperatures. The combined FT and WD cycles have insignificant effects on mechanical properties of CFRP wrapped RC columns, while the failure loads of GFRP wrapped RC columns decreased remarkably. These results are in good agreement with former findings reported by Toutanji and EI-Korchi [4]. Yin et al. [5] investigated the mechanical properties and durability of RC columns strengthened by hybrid carbon and E-glass fiber/epoxy textile. It showed that, under the coupled effects of WD cycles and sustained stress, the bearing capacity and ductility of the columns and the reinforcement effects of textile decreased more rapidly as the sustained compression stress increased. Moreover, exposure to chloride WD cycles generated corrosive products ($C_3 \cdot CaCl_2 \cdot 10H_2O$) on the interface between textile and concrete.

One of the major challenges to be addressed for FRP wrapped concrete is the long-term durability of externally bonded FRP [6, 7]. A number of experimental programs have been conducted to examine the behavior of FRP-concrete interface subjected to WD cycles. Silva and Biscaia [8] investigated the durability of RC beams reinforced with GFRP strips under different conditions (temperature cycles, WD cycles, complete immersion in water, and salt fogging cycles, for up to 1000 h). Their test results showed that the temperature cycles (−10 to +10°C) were the most severe condition on the bending strength, followed by the salt fog cycles, while total immersion in water was the most severe condition on the pull-off strength, followed by WD cycles and salt fogging cycles. For CFRP bonded concrete columns, WD cycles (one-week wetting at 95% RH and 30–32°C followed by one-week drying at 20–23°C) were the most severe condition on fracture energy compared with temperature cycles (30 to 40°C) and outdoor environment [9]. Kim et al. [10, 11] investigated the interfacial behavior of CFRP composite sheets bonded to concrete blocks subjected to freezing-wet-dry cycles and WD cycles, respectively. The results showed that the effects of freezing-wet-dry cycles were more detrimental than those of WD cycles on the behavior of CFRP-concrete interface, especially when the temperature was below −20°C. Biscaia et al. [12] examined the interface of GFRP and concrete subjected to salt fog cycles and WD cycles with saltwater. A 27% reduction of cohesion was found after 416 salt fog cycles and 8% reduction of the friction angle was found after 60 WD cycles due to the water attack at the interface. Wang and Amidi [13] applied environment-assisted subcritical debonding testing (EASD) program to evaluate the long-term durability of the CFRP-to-concrete interface subjected to the coupled effects of mechanical load, humidity, and temperature. They found that silane coupling agent could improve the durability of the CFRP-to-concrete interface due to introducing covalent bonds between the epoxy and concrete. Lipatov et al. [14] reported that the adding zirconia to basalt fibers contributed to improving the durability of fibers in alkali solution, and basalt fibers with 5.7 wt% ZrO_2 had the best alkali resistance properties.

With the development of novel FRP structural members in recent years [15], some researchers start to pay attention to the durability of new FRP construction [16, 17]. Toutanji and Saafi [18] proposed a new hybrid concrete column consisting of an exterior PVC-FRP shell with a concrete core. The compressive test showed that the PVC-CFRP confined concrete columns performed well when subjected to FT and WD cycles, whereas the PVC-GFRP and PVC-AFRP confined concrete columns experienced significant reductions in both

strength and ductility after 400 FT and 400 WT cycles. Chen et al. [19] evaluated the durability of FRP reinforcing bars for concrete structures from accelerated aging tests. It showed that continuous immersion in solutions with a pH value of 13.6 resulted in greater degradation than exposure to WD cycling for GFRP bars, while FT cycles had insignificant effects on the strength of GFRP bars.

Most of FRP structures are designed to resist some level of sustained load in-service engineering. The combination of environmental attacks with sustained stress can aggravate the loss of mechanical properties of FRP composites [20, 21]. Myers et al. [22] investigated the coupled effects of environmental attacks (FT cycles, moisture, high temperature cycling, and indirect ultraviolet radiation exposure) and sustained loads on the performance of interface between FRP and concrete. It is observed that the bond property of specimens under sustained load (40% of ultimate load) degraded greater than that of unstressed specimens in the same environment. Mahato et al. [23] studied the tensile behavior of GFRP composites under the coupled effects of temperature and loads. The tensile strength of the GFRP composites was observed to increase with the enhancement of loading speeds, and exposure in different temperature environments (25°C to 110°C) resulted in little change in glass transition temperatures (T_g). The authors [24] investigated GFRP durability under combined effects of moisture and sustained loads and found that the tensile properties of GFRP decreased significantly as the sustained loading increased under salt water for 360 days, whereas GFRP specimens with and without sustained loads had little changes in tensile properties under tap water.

Although many focused on single or coupled effects of environmental attacks under sustained loads in several studies, the durability of FRP composites under the combined effects of WD cycles and sustained loads has received scant attention. The coupled effects cannot be simplified by simply superposing the responses under different environmental attacks, and the mechanisms of degradation under single environmental attack are far different when compared with those under combined environmental attacks. This paper is concerned with investigating the long-term behavior of GFRP composites subjected to the combined effects of tensile loads and WD cycles. Microstructural and physical observations were performed to examine the conditioned and unconditioned samples. Moreover, a theoretical model was developed to predict the tensile properties of GFRP composites based on Arrhenius method.

2. Experimental Program

2.1. Materials. The test specimens were reinforced with bidirection (0°/90°) E-glass plain woven fabric provided by Sinoma Nanjing Fiberglass Research & Design Institute. The fabric had an average thickness of 0.5 mm and a unit weight of 500 g/m². Isophthalic polyester resin (Synolite 0593-I-3) supplied by DSM China was used as matrix. 3 wt% cyclo-hexanone peroxide and 2 wt% cobalt isocaprylate-styrene were used as the initiator and acceleration employed for isophthalic polyester resin curing.

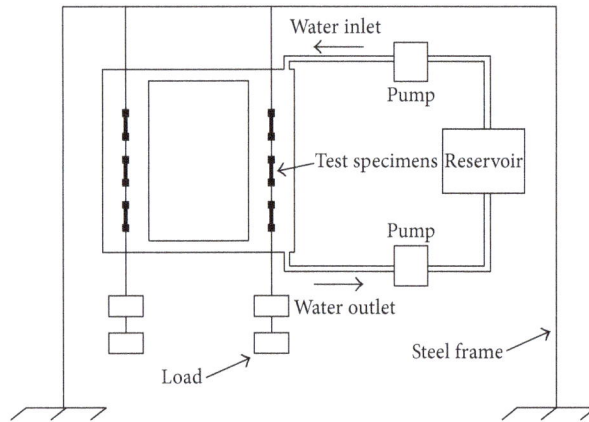

FIGURE 1: Schematic of the wet-dry exposure set-up.

FIGURE 2: Tensile test set-up.

A flat 1 m × 2 m panel with a thickness of 2 mm was fabricated using resin infusion under flexible tooling (RIFT2) method and cut to coupons of 250 mm × 20 mm × 2 mm size. In addition, a surface layer of resin was applied to seal the specimens. The fiber volume fraction for the GFRP composites was about 60%. The specimens were cured at room temperature for 3 days, and then were cured in an oven at 60°C for 3 hours.

2.2. Apparatus. Figure 1 shows the environmental exposure set-up. The specimens were exposed to two environments: (1) WD cycles of distilled water; (2) WD cycles of saltwater with a salinity content of about 3.5%. Each cycle consisted of full immersion in water for 12 hours followed by air drying for 12 hours at room temperature. The numbers of cycles for the GFRP composites were 15, 30, 60, 90, 180, and 360, respectively. The applied stresses chosen for the specimens were 0%, 20%, and 40% of the tensile strength of the GFRP composites, respectively. Three duplicate specimens were connected in series and were tensioned by the weights of concrete. Both ends of the specimens were reinforced by 50 mm × 20 mm × 5 mm aluminum tab attachments. The water used to immerse the specimens was controlled by pumps. In the drying condition, the extra water was drawn back into the reservoir, while in the wetting condition, the water was poured into the specimen containers.

2.3. Tensile Tests. After conditioning, tensile tests were conducted, in accordance with ASTM D638-10 [25] to collect the tensile strength, elastic modulus, and tensile elongation at the rupture of GFRP composites under different cycles. A universal test machine with 50 kN capacity was used for all specimens, as shown in Figure 2. The tensile load was applied through a stiff steel clamp and was recorded via a load cell mounted directly above the top steel clamp. Before testing, the average cross-sectional area of each specimen was measured. The strain values of the specimens were measured by an extensometer with 25.4 mm gage length which was connected to the middle of the specimens. Each specimen was stressed to about 50% of the tensile strength to obtain the elastic modulus, and then the extensometer was removed before

TABLE 1: Mechanical properties of GFRP under WD cycles.

Specimens	WD cycles	Tensile strength (MPa)	Elastic modulus (GPa)	Elongation (mm)
WWD0	0	416.75	24.70	1.813
	15	457.40	26.77	1.003
	30	448.47	26.95	1.123
	60	431.72	25.87	0.834
	90	420.36	25.43	0.725
	180	410.85	23.89	0.534
	360	377.67	20.86	0.515
SWD0	0	416.75	24.70	1.813
	15	480.32	26.60	1.370
	30	442.68	27.08	1.014
	60	436.58	26.57	0.739
	90	422.58	25.74	0.452
	180	411.04	24.42	0.474
	360	368.66	21.85	0.467

specimen was stressed to failure. The failure load was used to calculate the tensile strength of the GFRP composites in MPa.

2.4. Microstructural Observations. Electron microscopy (EM) observations and image analysis were performed to examine microstructure of specimens before and after conditions with DM4000M (Leica, Germany). The samples included unconditioned specimens and conditioned specimens after 60 and 180 cycles under different sustained loads. All specimens were first cut and polished before observations. These observations were made on surfaces and sections parallel to the fiber layers for potential degradation of the glass fibers, resin, or interfaces.

3. Results and Discussion

3.1. Specimens without Sustained Load. Tensile tests were conducted on GFRP composites subjected to 0, 15, 60, 90, 180, and 360 WD cycles of both distilled water and saltwater. Table 1 shows the mechanical properties of GFRP under WD

(a)

(b)

(c)

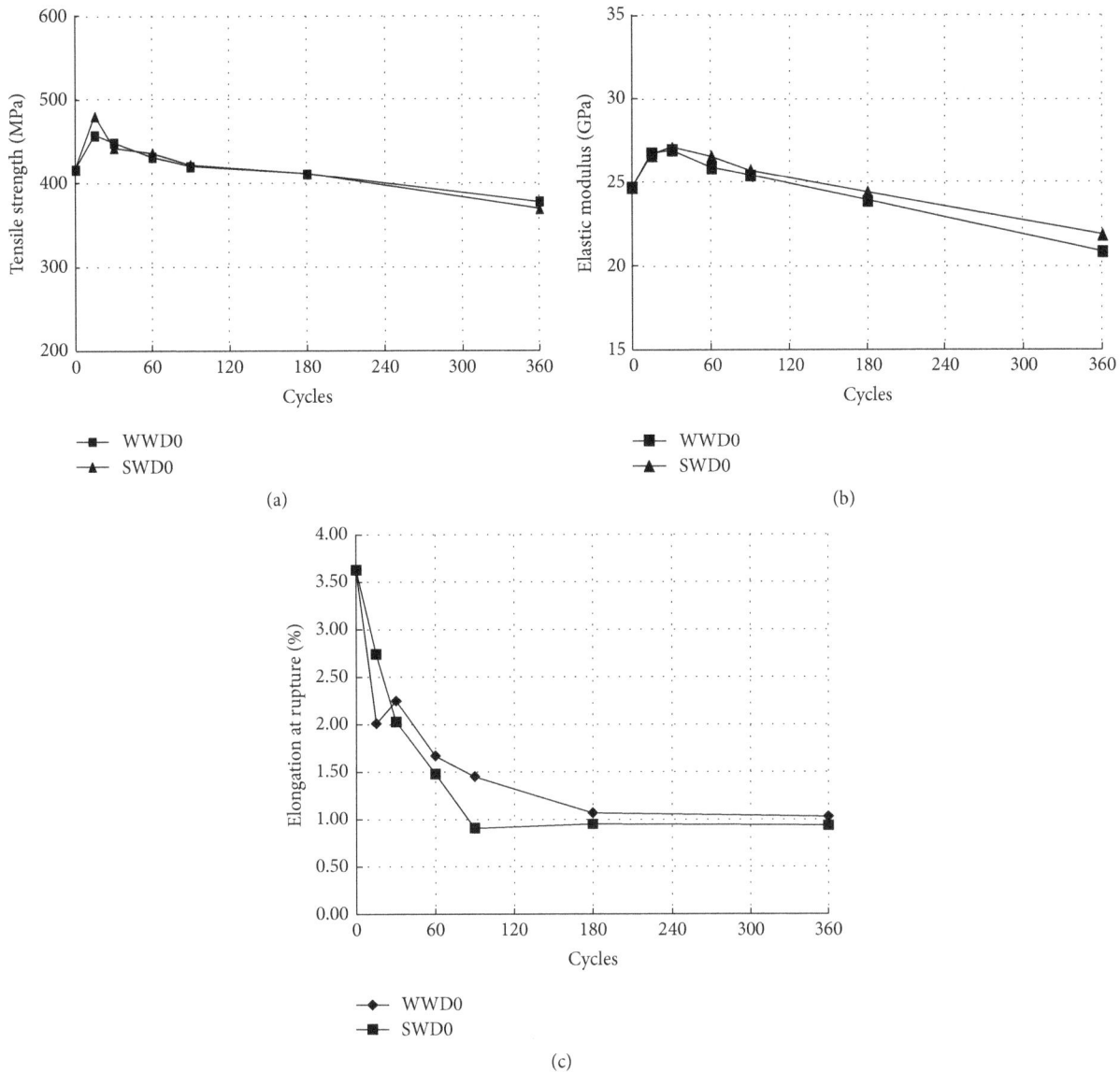

FIGURE 3: Mechanical properties of unstressed specimens exposed to WD cycles: (a) tensile strength; (b) elastic modulus; (c) elongation at rupture. *Note.* WWD0 is unstressed specimen exposed to WD cycles of distilled water; and SWD0 is unstressed specimen exposed to WD cycles of saltwater.

cycles. The reduction in the tensile strength, modulus, and elongation at rupture of aged specimens versus the number of cycles is presented in Figure 3. Both the tensile strength and elastic modulus have an initial increase after 15 WD cycles, followed by a continued decline with the increasing number of cycles. Although the specimens were cured at room temperature for 3 days and then cured in an oven at 60°C for 3 hours, the resin was not polymerized completely. Immersion in water led to the occurrence of postcuring of polymer, yielding an initial increase in tensile strength and modulus [24, 26, 27]. The differential scanning calorimetry (DSC) measurement of the identical samples confirmed that the resin partially cured before conditions [24].

Further exposure to WD cycles led to a steady decrease in tensile strength and modulus. After 360 cycles, the loss

of tensile strength is 9% and 12% for WWD0 and SWD0, respectively, and the loss of modulus is 16% and 12% for WWD0 and SWD0, respectively, as shown in Figures 3(a) and 3(b). However, the elongation at rupture decreased initially after 90 cycles and almost remained unchanged from 90 cycles to 360 cycles. After 360 cycles, the loss of elongation at rupture is 72% and 74% for WWD0 and SWD0, respectively. The durability of unstressed specimens under WD cycles of distilled water has negligible differences with that of specimens under WD cycles of saltwater. The sharp decrease of elongation at rupture indicates that GFRP composites become brittle in exposure of WD cycles.

Micro observations were conducted to examine the physical damage of GFRP composites due to exposure to WD cycles. Figure 4 shows the typical micrographs of the

(a)

(b)

(c)

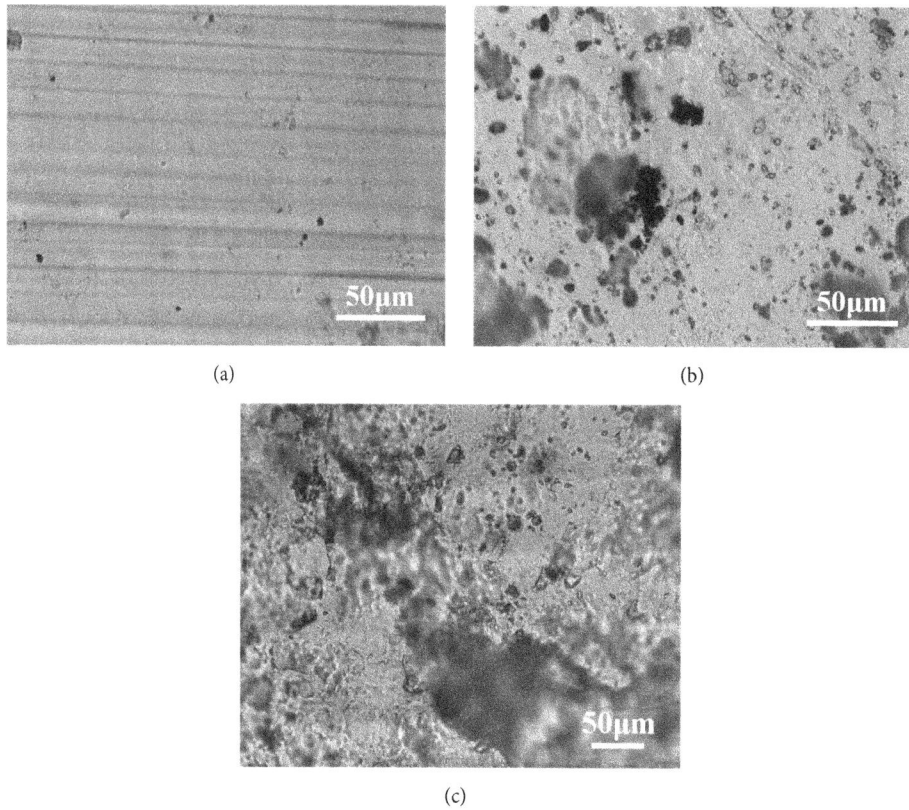

FIGURE 4: Micrographs of the surfaces (×50) of the unstressed specimens: (a) unconditioned specimen; (b) SWD0 after 60 WD cycles; and (c) WWD0 after 180 WD cycles.

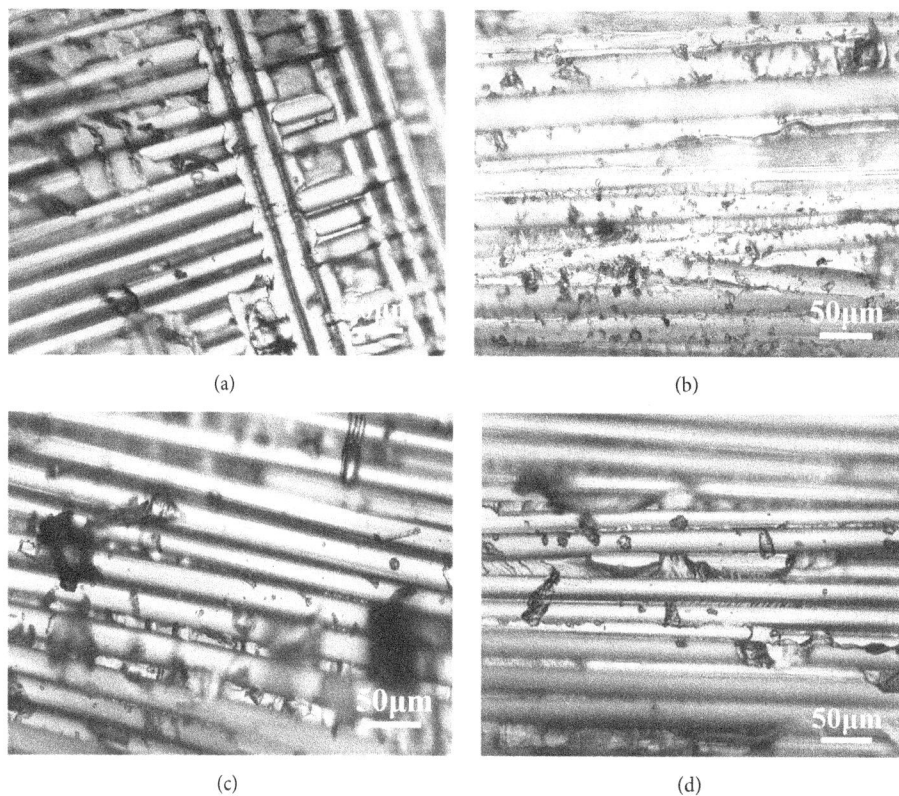

(a)

(b)

(c)

(d)

FIGURE 5: Micrographs of the longitudinal section (×50) of the unstressed specimens: (a) unconditioned specimen; (b) WWD0 after 60 WD cycles; (c) SWD0 after 60 WD cycles; and (d) WWD0 after 180 WD cycles.

TABLE 2: Mechanical properties of GFRP under the combined effects of sustained loads and WD cycles.

Specimens	Load level	WD cycles	Tensile strength (MPa)	Elastic modulus (GPa)	Elongation (mm)
WWD2	20%	0	416.75	24.70	1.813
		15	440.85	26.87	0.958
		30	487.13	27.17	1.235
		60	508.28	28.10	0.929
		90	431.74	24.56	0.684
		180	403.64	22.65	0.507
		360	345.90	20.61	0.484
WWD4	40%	0	416.75	24.70	1.813
		15	435.91	25.89	1.258
		30	474.93	28.27	1.017
		60	532.71	29.75	0.873
		90	442.15	24.00	0.557
		180	390.90	22.05	0.485
		360	329.12	19.59	0.443
SWD2	20%	0	416.75	24.70	1.813
		15	470.66	24.29	0.703
		30	478.37	25.10	1.055
		60	499.34	27.36	0.980
		90	477.48	26.85	0.796
		180	398.63	22.28	0.505
		360	347.23	19.89	0.486
SWD4	40%	0	416.75	24.70	1.813
		15	478.71	25.87	0.942
		30	493.56	27.87	1.071
		60	534.80	28.56	0.971
		90	465.53	25.35	0.654
		180	390.89	21.98	0.431
		360	328.69	18.90	0.423

surfaces of unconditioned and conditioned specimens. It is observed that the void content of the polyester resin increases gradually with the number of WD cycles in both distilled water and saltwater. Voids and porosities not only lead to water diffusion but also lead to stress concentration. Once water molecules diffuse into a polyester matrix, they would make hydrogen-bond with the cross-linked polymer at the polar ester linkages [28]. As shown in Figure 5, hydrolysis of isophthalic polyester resins occurred due to exposure to WD cycles. Moisture absorption test of the identical samples showed that the moisture uptake content M_t decreased with

the immersion time after M_t reached its peak value, which confirmed the hydrolysis of isophthalic polyester resins [24]. Moreover, chemical reaction between glass fibers and water usually cause alkali elements to leach out [29]. In addition, for specimens exposed to WD cycles of saltwater, the spatial distribution of Cl from NaCl correlates with that of Si from the E-glass fibers, yielding the aggravation of deterioration in tensile property of GFRP composites.

3.2. *Specimens with Sustained Loads.* Table 2 shows the mechanical properties of GFRP under combined effects of

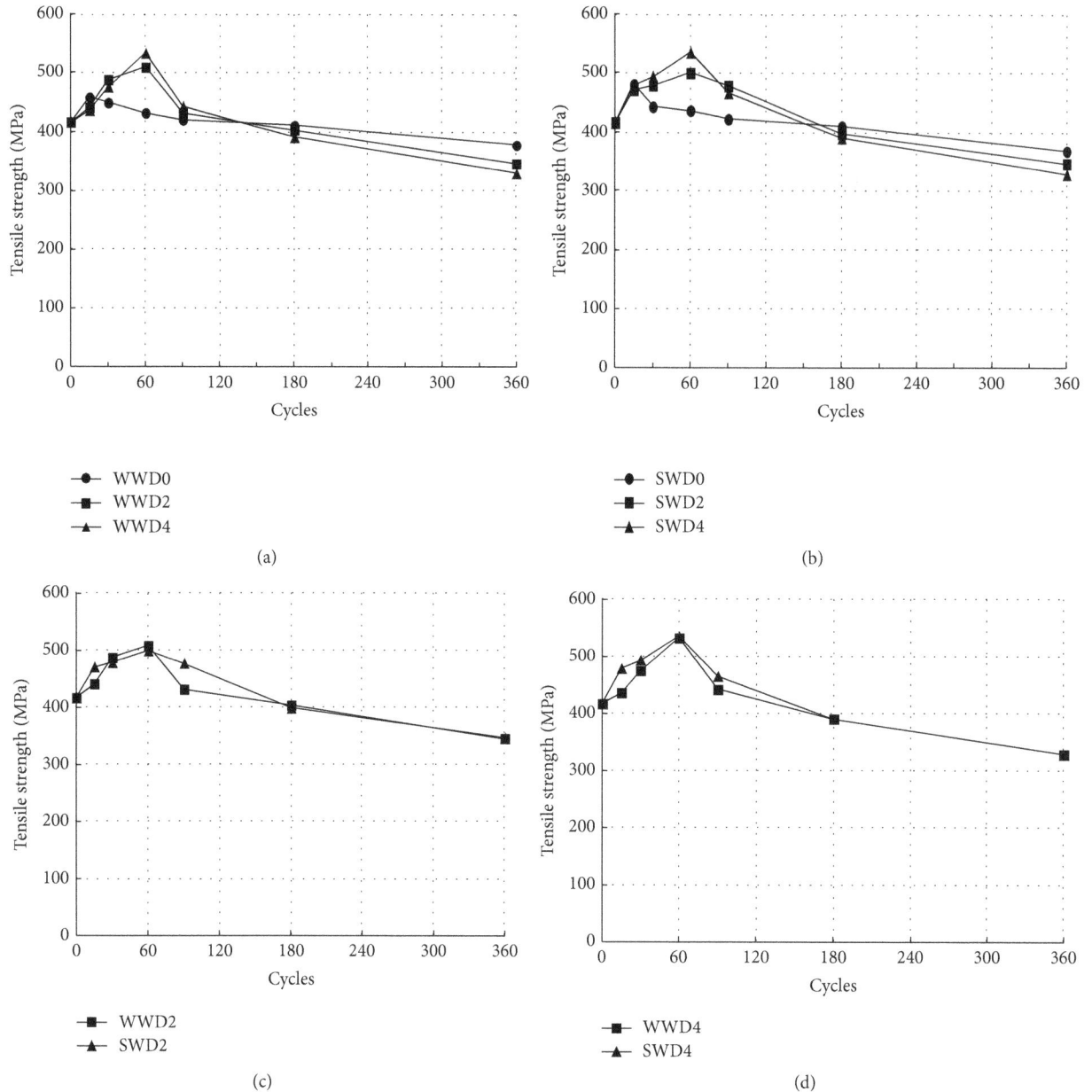

FIGURE 6: Tensile strength of stressed specimens (a) subjected to WD cycles of distilled water and different loads; (b) subjected to WD cycles of saltwater and different loads; (c) subjected to WD cycles and 20% ultimate tensile strength; (d) subjected to WD cycles and 40% ultimate tensile strength. *Note.* WWD2 and WWD4 are specimens exposed to WD cycles of distilled water and subjected to 20% and 40% ultimate tensile strength, respectively, and SWD2 and SWD4 are specimens exposed to WD cycles of saltwater and subjected to 20% and 40% ultimate tensile strength, respectively.

sustained loads and WD cycles. The specimens with sustained loads have an initial increase in both tensile strength and elastic modulus after 60 cycles, as shown in Figures 6 and 7. After 60 cycles, the tensile strength increased by 22%, 20%, 28%, and 28% for WWD2, SWD2, WWD4, and SWD4, respectively. Meanwhile, after 60 cycles, the elastic modulus increased by 14%, 11%, 20%, and 16%, for WWD2, SWD2, WWD4, and SWD4, respectively. These results indicate that the initial increases of tensile strength and elastic modulus of the specimens with sustained loads are greater than those of

specimens without sustained loads, which may be attributed to the combined effects of postcuring of resin and the sustained tensile stress. Similar behavior was also observed by Abdel-Magid et al. [30]. The tensile stress aligns the fibers in the stress direction and endows the material with more efficiency in carrying the load after short period of prestress.

From 60 cycles to 360 cycles, the tensile strength and elastic modulus of stressed specimens decreased steadily. After 360 cycles, the tensile strength decreased by 17%, 17%, 21%, and 21% for WWD2, SWD2, WWD4, and SWD4,

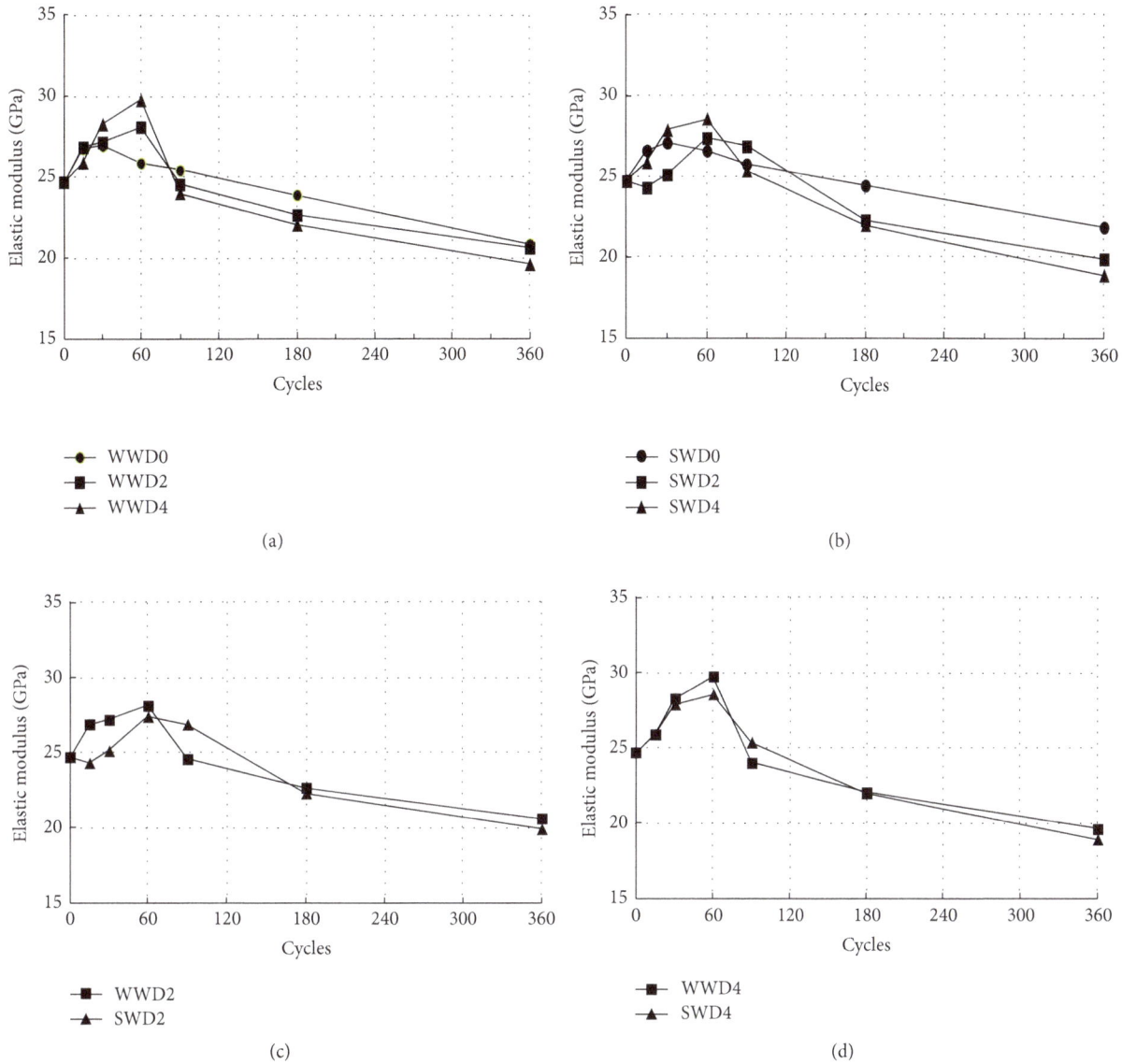

FIGURE 7: Elastic modulus of stressed specimens (a) subjected to WD cycles of distilled water and different loads; (b) subjected to WD cycles of saltwater and different loads; (c) subjected to WD cycles and 20% ultimate tensile strength; (d) subjected to WD cycles and 40% ultimate tensile strength.

respectively, and the elastic modulus decreased by 17%, 19%, 21%, and 23% for WWD2, SWD2, WWD4, and SWD4, respectively. Meanwhile, after 360 cycles, the loss of elongation at rupture is 73%, 73%, 76%, and 77% for WWD2, SWD2, WWD4, and SWD4, respectively, as shown in Figure 8. It is obvious that the increase of sustained loads leads to a decrease in tensile property. Moreover, the effect of WD cycles of saltwater on elastic modulus and elongation at rupture is greater than that of WD cycles of distilled water. Both water diffusion and stress corrosion are associated with the degradation of resin and the interface between fiber and matrix. The previous research of authors has proven that the effect of saltwater immersion on tensile property is greater than that of tap water due to the chemical reaction between leaked Cl ions from

saltwater and metal ions (Si, Na, and K) from E-glass fibers [24].

The micrographs of the surfaces of stressed specimens showed that the voids of polyester resins progressed in both distilled water and saltwater, as shown in Figure 9. Hydrolysis of isophthalic polyester resins was also observed under the combined effects of WD cycles and sustained loads, as shown in Figure 10.

4. Prediction of Long-Term Behavior

Several models have been developed to predict the long-term responses of FRP composites based on Arrhenius concept which assumed that the single dominant deterioration

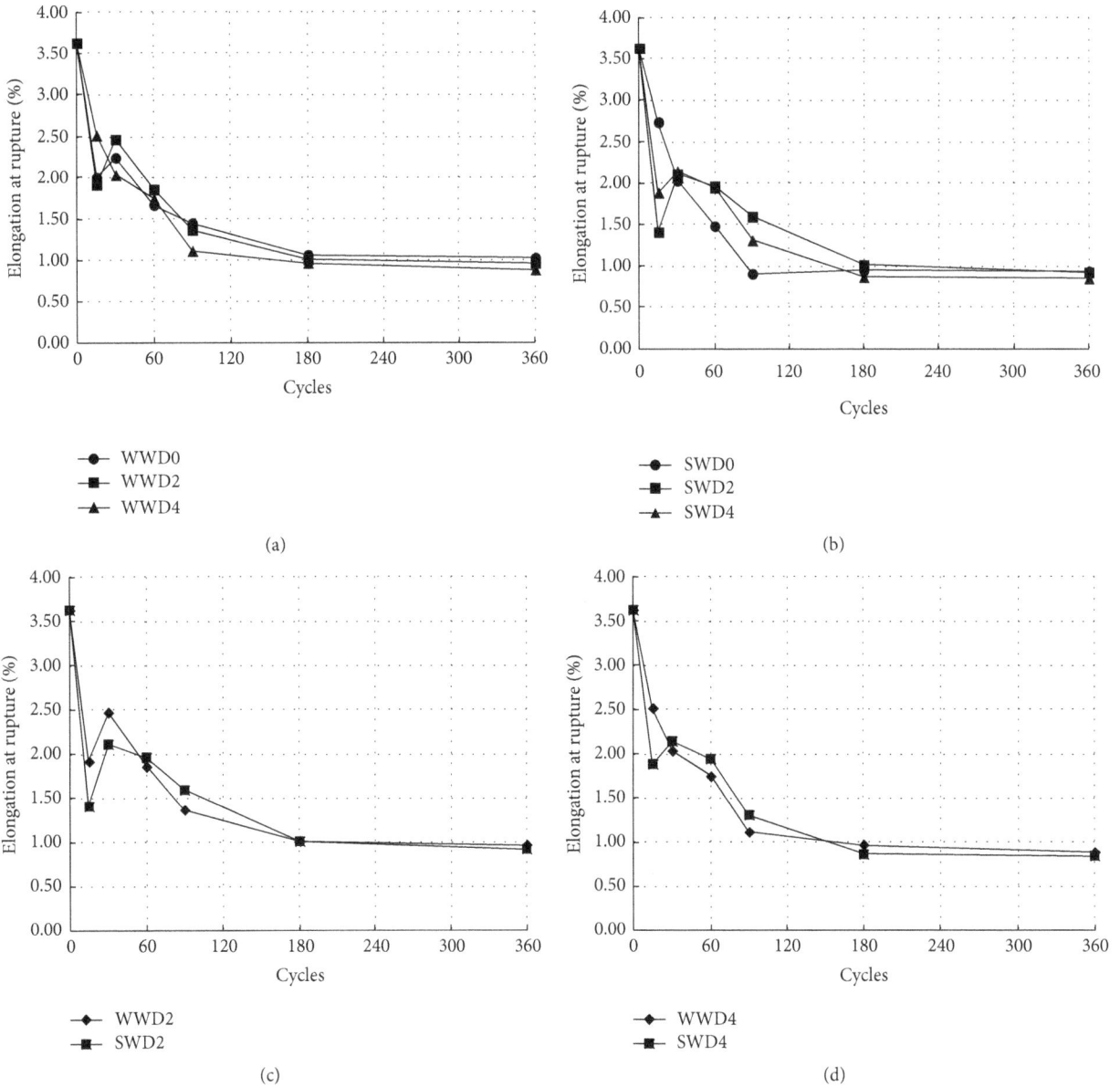

FIGURE 8: Elongation at rupture of stressed specimens (a) subjected to WD cycles of distilled water and different loads; (b) subjected to WD cycles of saltwater and different loads; (c) subjected to WD cycles and 20% ultimate tensile strength; (d) subjected to WD cycles and 40% ultimate tensile strength.

mechanism of the material will not change with time and temperature in corrosive environment [8, 31].

The Arrhenius equation for predicting the mechanical property of FRP composites is given by [8]

$$P(t) = \frac{P_0}{100}[A\ln(t) + B], \qquad (1)$$

where $P(t)$ and P_0 are the performance attributes of FRP composites at exposure time t and zero time of exposure, respectively, A is degradation rate, and B is a material constant associated with postcure progression of polymer during exposure. A value of $B > 100$ relates to the initial

dominance of postcure processes in environmental exposure, while a value of $B = 100$ indicates full polymerization of the resin. Equation (1) is only valid for $t > 0$. In the case of $t = 0$, the percent retention of $P(t)$ is assumed to be 100%.

For exposure in WD cycles, (1) can be transformed into

$$P(x) = \frac{P_0}{100}[A\ln(x) + B], \qquad (2)$$

where x is the number of WD cycles.

The tensile strength and elastic modulus of FRP composites tend to increase gradually with time in the initial postcuring processes and then become stable, regardless of

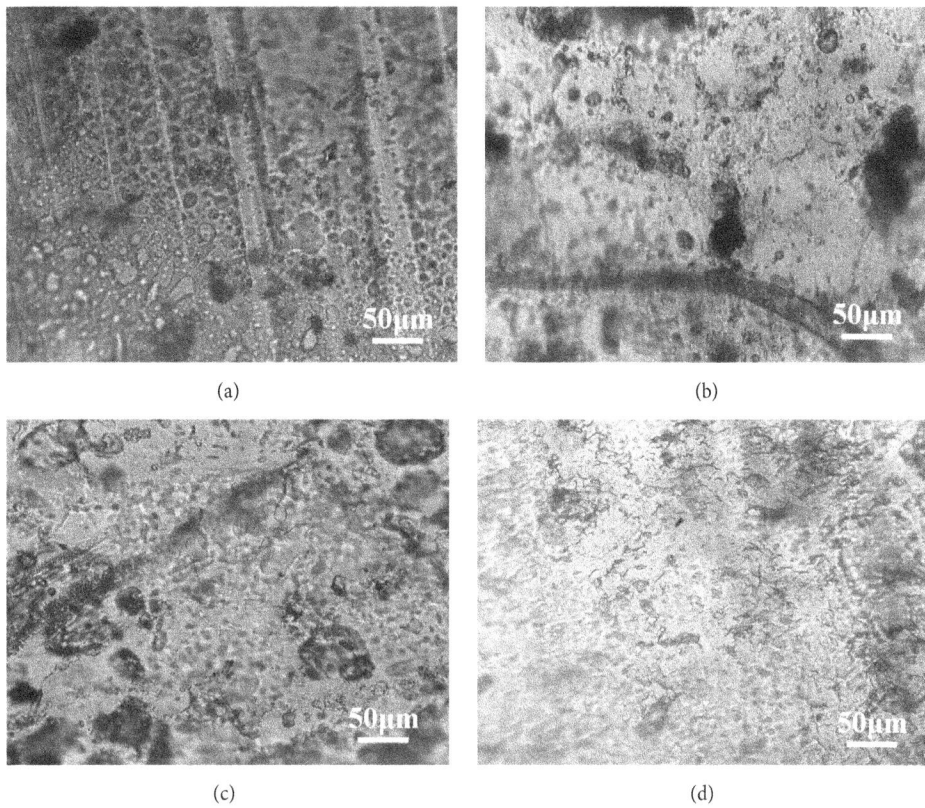

FIGURE 9: Micrographs of the surfaces (×50) of stressed specimens: (a) SWD2 after 60 WD cycles; (b) SWD2 after 180 WD cycles; (c) WWD2 after 180 WD cycles; and (d) SWD4 after 180 WD cycles.

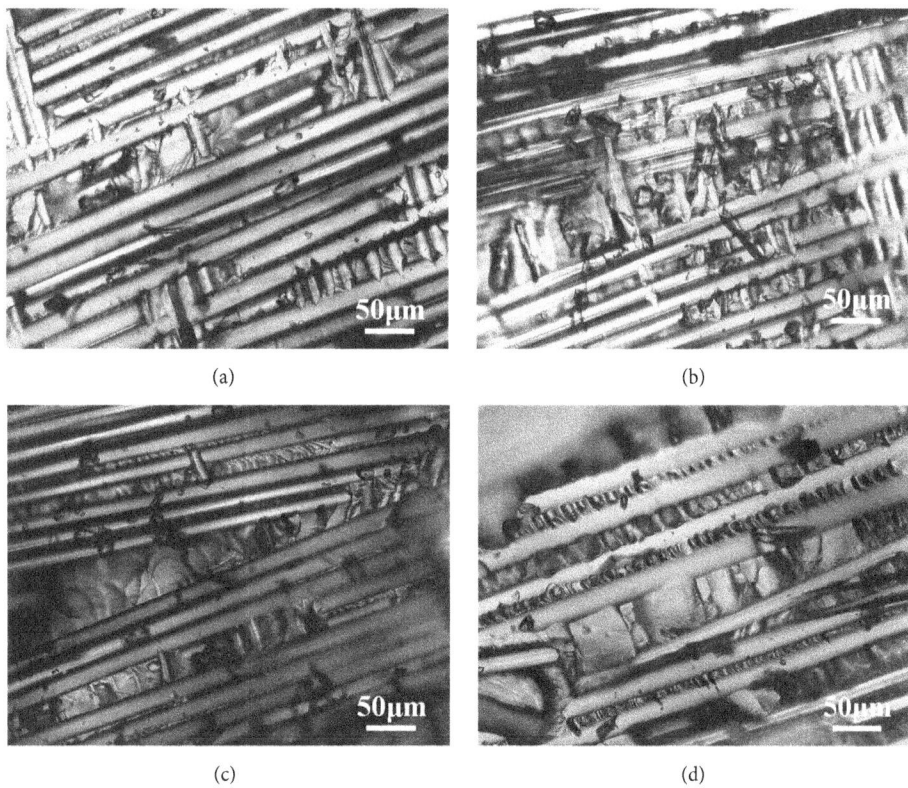

FIGURE 10: Micrographs of the longitudinal section (×50) of stressed specimens: (a) SWD2 after 60 WD cycles; (b) SWD2 after 180 WD cycles; (c) WWD2 after 180 WD cycles; and (d) WWD4 after 180 WD cycles.

(a) WWD2

(b) WWD4

(c) SWD2

(d) SWD4

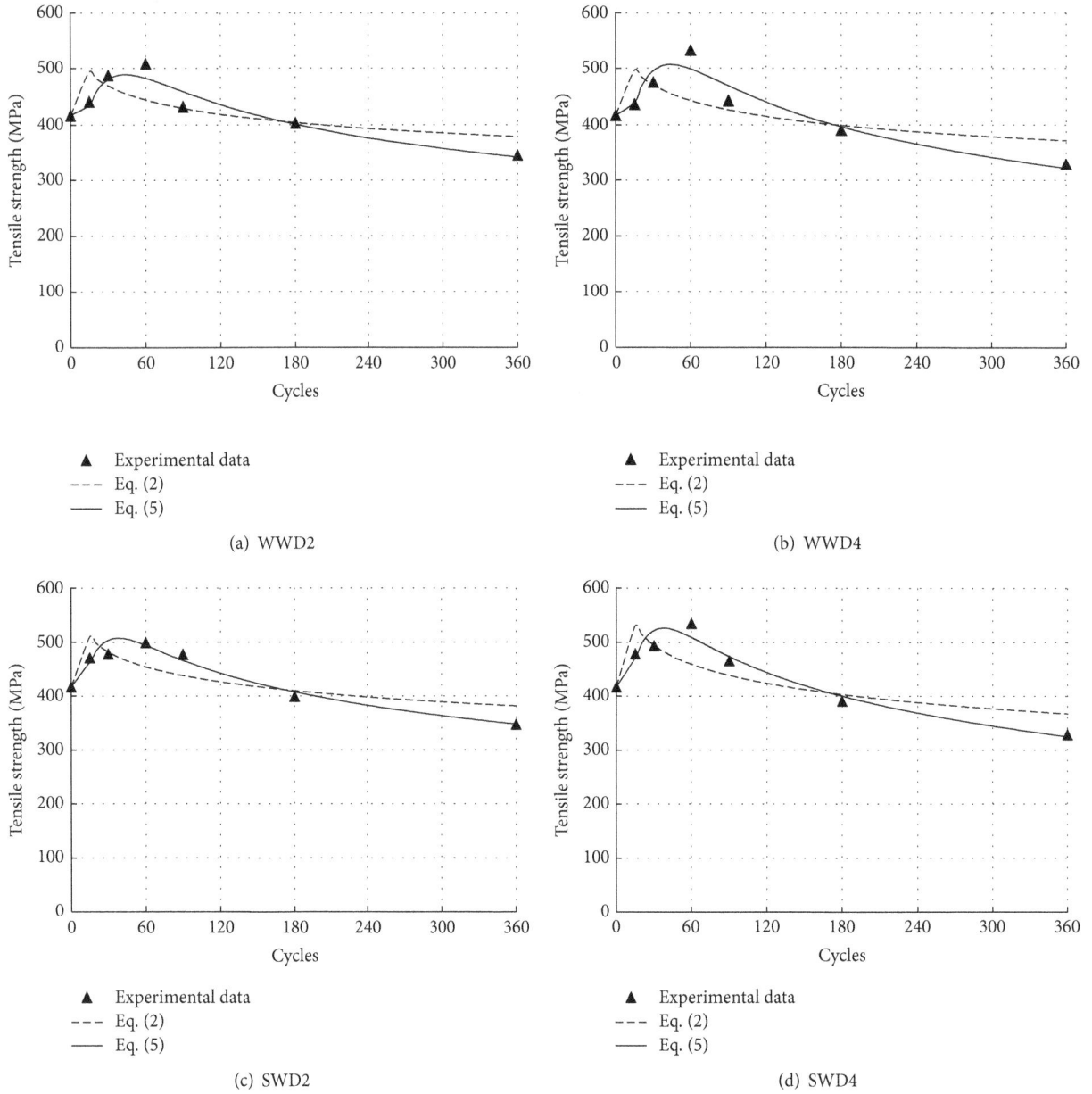

FIGURE 11: Comparison of prediction of tensile strength.

the effects of environmental attacks. Thus, the effects of postcuring on the mechanical property of FRP composites can be described as a two-parameter exponential equation:

$$f(t) = \alpha \left(1 - \beta^{-t}\right), \qquad (3)$$

where $f(t)$ is the performance attribute of FRP composites due to postcuring, and α and β are regression constants.

Considering the combined effects of postcure processes and exposure environments, the prediction model can be given by

$$P(t) = \frac{P_0}{100} \left[A \ln(t) + 100\right] + \alpha \left(1 - \beta^{-t}\right). \qquad (4)$$

For exposure in WD cycles, (3) can be transformed into

$$P(x) = \frac{P_0}{100} \left[A \ln(x) + 100\right] + \alpha \left(1 - \beta^{-x}\right), \qquad (5)$$

where x is the number of WD cycles.

The predictive equations of tensile strength and elastic modulus according to (5) and (2) are given in Tables 3 and 4, respectively. For unstressed specimens, the predicted results from (2) and (5) agree well with those obtained from experiments. However, for specimens with sustained loads, (5) provides more accurate prediction results than (2), as shown in Figures 11 and 12.

(a) WWD2

(b) WWD4

(c) SWD2

(d) SWD4

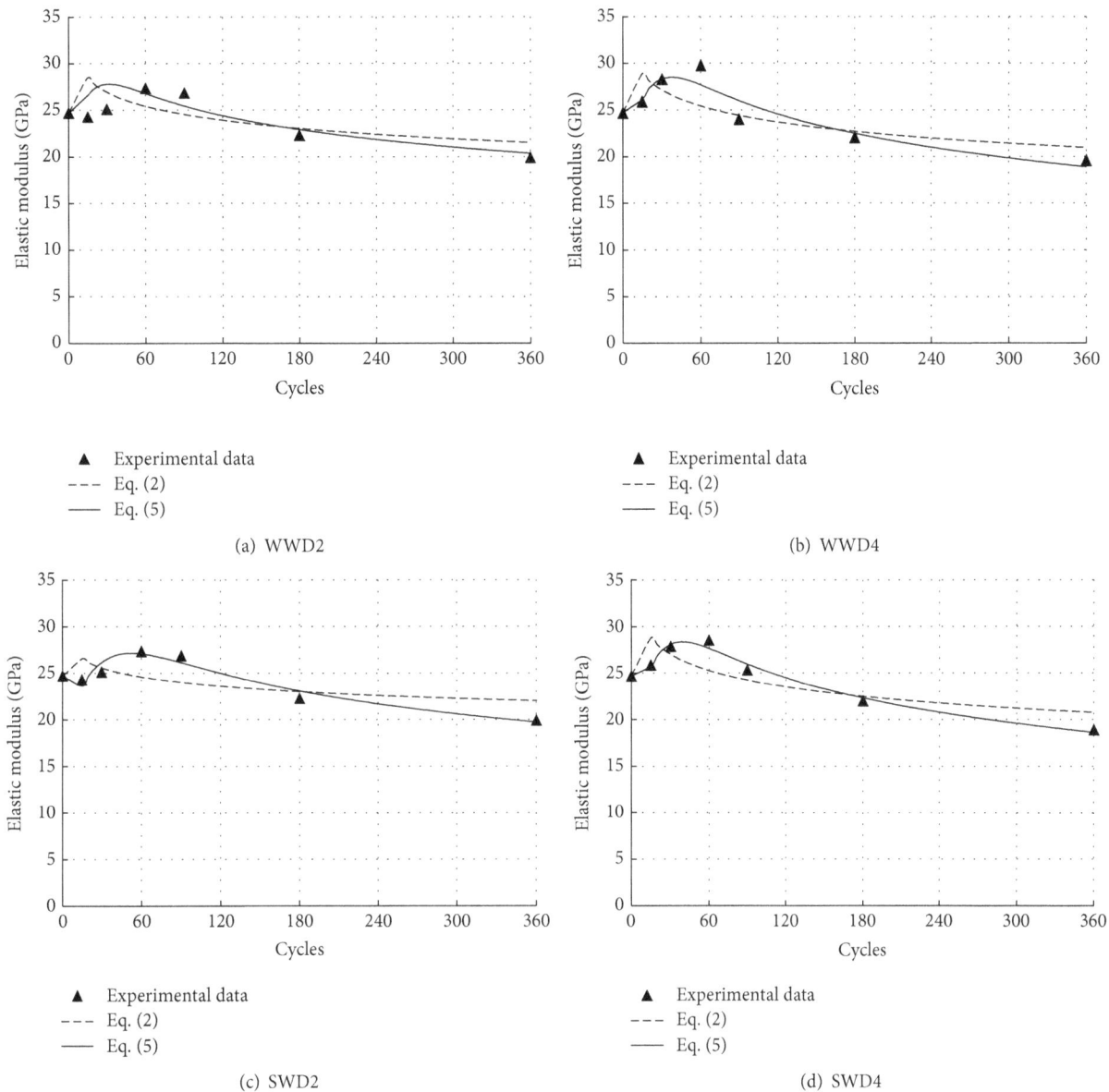

FIGURE 12: Comparison of prediction of elastic modulus.

5. Conclusions

The durability of GFRP composites exposed to WD cycles of both distilled water and saltwater was investigated under three different sustained loading levels. The tensile properties of both the unconditioned and conditioned specimens were tested to study the long-term behavior of GFRP composites under different WD cycles. Based on Arrhenius concept, a model was developed to predict the tensile property of GFRP. In addition, the electron microscopy (EM) was used to characterize the aging effect on GFRP composites. The results obtained from this study are summarized as follows:

(1) The tensile strength and elastic modulus of all conditioned specimens increased initially and then decreased with the increase of exposure cycle numbers. The coupled effects of postcuring and alignment of the fibers in the stress direction yield greater increases in tensile strength and elastic modulus of stressed specimens than those of unstressed specimens. Further exposure to WD cycles led to a steady decrease in tensile strength and modulus. For unstressed specimens, the degradation of the tensile property of specimens exposed to distilled water was almost equal to specimens exposed to saltwater after 360 cycles. However, for stressed specimens, the effect of WD cycles of saltwater on elastic modulus and elongation at rupture is greater than that of WD cycles of distilled water. Moreover, increasing sustained loads led to a decrease in tensile property.

(2) The elongation at rupture decreased sharply with the increasing number of cycles in both distilled water

TABLE 3: Predictive equations for tensile strength.

Exposure environment	Load level	Equation	R-square
Wet-dry cycles in distilled water	0%	$4.1675[-6.614\ln(x) + 100] + 128.63(1 - 1.16^{-x})$	0.97
		$4.1675[-5.794\ln(x) + 126.74]$	0.99
	20%	$4.1675[-20.479\ln(x) + 100] + 427.57(1 - 1.06^{-x})$	0.92
		$4.1675[-8.590\ln(x) + 141.59]$	0.51
	40%	$4.1675[-25.814\ln(x) + 100] + 538.77(1 - 1.06^{-x})$	0.91
		$4.1675[-9.379\ln(x) + 144.51]$	0.42
Wet-dry cycles in saltwater	0%	$4.1675[-4.912\ln(x) + 100] + 103.00(1 - 1.21^{-x})$	0.83
		$4.1675[-7.430\ln(x) + 134.38]$	0.94
	20%	$4.1675[-20.604\ln(x) + 100] + 436.32(1 - 1.07^{-x})$	0.96
		$4.1675[-9.687\ln(x) + 148.475]$	0.63
	40%	$4.1675[-25.888\ln(x) + 100] + 543.50(1 - 1.07^{-x})$	0.95
		$4.1675[-12.266\ln(x) + 160.37]$	0.62

Note. R-square is the coefficient of determination. It denotes the proportion of the variance in the dependent variable which is predictable from the independent variables.

TABLE 4: Predictive equations for elastic modulus.

Exposure environment	Load level	Equation	R-square
Wet-dry cycles in distilled water	0%	$0.247[-11.092\ln(x) + 100] + 12.77(1 - 1.091^{-x})$	0.97
		$0.247[-7.347\ln(x) + 132.63]$	0.84
	20%	$0.247[-14.902\ln(x) + 100] + 17.33(1 - 1.08^{-x})$	0.93
		$0.247[-8.835\ln(x) + 139.158]$	0.75
	40%	$0.247[-20.873\ln(x) + 100] + 24.57(1 - 1.068^{-x})$	0.88
		$0.247[-10.062\ln(x) + 144.14]$	0.57
Wet-dry cycles in saltwater	0%	$0.247[-10.324\ln(x) + 100] + 12.49(1 - 1.08^{-x})$	0.98
		$0.247[-6.07\ln(x) + 128.83]$	0.78
	20%	$0.247[-19.638\ln(x) + 100] + 23.57(1 - 1.049^{-x})$	0.85
		$0.247[-5.789\ln(x) + 123.24]$	0.34
	40%	$0.247[-21.73\ln(x) + 100] + 25.5(1 - 1.065^{-x})$	0.98
		$0.247[-10.22\ln(x) + 144.13]$	0.63

and saltwater, which indicated the GFRP composites became brittle in exposure of WD cycles. The sustained loads have insignificant influences on the elongation at rupture.

(3) A modified model which considered the effects of postcure processes was developed to predict the tensile strength and elastic modulus of GFRP composites subjected to WD cycles. It is shown that the model can capture the degradation trend of specimens under the combined effects of WD cycles and sustained loads.

Conflicts of Interest

The authors declare that they have no conflicts of interest.

Acknowledgments

The financial support from the National Natural Science Foundation of China (Grant 51578283), Jiangsu Province Science and Technology Support Program of Construction Industry Modernization (Grant 2017-13), and Key Program of the National Natural Science Foundation of China (Grant 51238003) is greatly appreciated.

References

[1] C. Tuakta and O. Büyüköztürk, "Conceptual model for prediction of FRP-concrete bond strength under moisture cycles," *Journal of Composites for Construction*, vol. 15, no. 5, pp. 743–756, 2011.

[2] K. Soudki, E. El-Salakawy, and B. Craig, "Behavior of CFRP strengthened reinforced concrete beams in corrosive environment," *Journal of Composites for Construction*, vol. 11, no. 3, pp. 291–298, 2007.

[3] A. Belarbi and S.-W. Bae, "An experimental study on the effect of environmental exposures and corrosion on RC columns with FRP composite jackets," *Composites Part B: Engineering*, vol. 38, no. 5-6, pp. 674–684, 2007.

[4] H. A. Toutanji and T. El-Korchi, "Tensile durability of cement-based frp composite wrapped specimens," *Journal of Composites for Construction*, vol. 3, no. 1, pp. 38–45, 1999.

[5] S.-P. Yin, C. Peng, and Z.-Y. Jin, "Research on Mechanical Properties of Axial-Compressive Concrete Columns Strengthened

with TRC under a Conventional and Chloride Wet-Dry Cycle Environment," *Journal of Composites for Construction*, vol. 21, no. 1, Article ID 04016061, 2017.

[6] V. M. Karbhari and M. A. Abanilla, "Design factors, reliability, and durability prediction of wet layup carbon/epoxy used in external strengthening," *Composites Part B: Engineering*, vol. 38, no. 1, pp. 10–23, 2007.

[7] Z. Wu, Y. J. Kim, H. Diab, and X. Wang, "Recent developments in long-term performance of FRP composites and FRP-concrete interface," *Advances in Structural Engineering*, vol. 13, no. 5, pp. 891–903, 2010.

[8] M. A. G. Silva and H. Biscaia, "Environmental effects on bond of GFRP external reinforcement to RC beams," in *Proceeding of the 8th International Symposium on Fiber Reinforced Polymer Reinforcement for Concrete Structures, FRPRCS-8*, University of Patras, Patras, Greece, July 2007.

[9] M. I. Kabir, B. Samali, and R. Shrestha, "Fracture Properties of CFRP-Concrete Bond Subjected to Three Environmental Conditions," *Journal of Composites for Construction*, vol. 20, no. 4, Article ID 4016010, 2016.

[10] Y. J. Kim, M. Hossain, and Y. Chi, "Characteristics of CFRP-concrete interface subjected to cold region environments including three-dimensional topography," *Cold Regions Science and Technology*, vol. 67, no. 1-2, pp. 37–48, 2011.

[11] Y. J. Kim, M. Hossain, and J. Zhang, "A probabilistic investigation into deterioration of CFRP-concrete interface in aggressive environments," *Construction and Building Materials*, vol. 41, pp. 49–59, 2013.

[12] H. C. Biscaia, M. A. G. Silva, and C. Chastre, "An experimental study of GFRP-to-concrete interfaces submitted to humidity cycles," *Composite Structures*, vol. 110, pp. 354–368, 2014.

[13] J.-L. Wang and S. Amidi, "Subcritical debonding of FRP-to-concrete bonded interface under synergistic effect of load, moisture, and temperature," *Mechanics of Materials*, vol. 92, pp. 80–93, 2016.

[14] Y. V. Lipatov, S. I. Gutnikov, M. S. Manylov, E. S. Zhukovskaya, and B. I. Lazoryak, "High alkali-resistant basalt fiber for reinforcing concrete," *Materials and Corrosion*, vol. 73, pp. 60–66, 2015.

[15] P. Feng, L. Hu, P. Qian, and L. Ye, "Compressive bearing capacity of CFRP-aluminum alloy hybrid tubes," *Composite Structures*, vol. 140, pp. 749–757, 2016.

[16] B. Benmokrane, F. Elgabbas, E. A. Ahmed, and P. Cousin, "Characterization and Comparative Durability Study of Glass/Vinylester, Basalt/Vinylester, and Basalt/Epoxy FRP Bars," *Journal of Composites for Construction*, vol. 19, no. 6, Article ID 04015008, 2015.

[17] J. Xu, H. Kolstein, and F. S. K. Bijlaard, "Moisture diffusion and hygrothermal aging in pultruded fibre reinforced polymer composites of bridge decks," *Materials and Design*, vol. 37, pp. 304–312, 2012.

[18] H. Toutanji and M. Saafi, "Durability studies on concrete columns encased in PVC-FRP composite tubes," *Composite Structures*, vol. 54, no. 1, pp. 27–35, 2001.

[19] Y. Chen, J. F. Davalos, I. Ray, and H. Y. Kim, "Accelerated aging tests for evaluations of durability performance of FRP reinforcing bars for concrete structures," *Composite Structures*, vol. 78, no. 1, pp. 101–111, 2007.

[20] P. Boer, L. Holliday, and T. H.-K. Kang, "Independent environmental effects on durability of fiber-reinforced polymer wraps in civil applications: a review," *Construction and Building Materials*, vol. 48, pp. 360–370, 2013.

[21] J. U. N. Wang, H. O. T. A. Gangarao, R. Liang, and W. Liu, "Durability and prediction models of fiber-reinforced polymer composites under various environmental conditions: A critical review," *Journal of Reinforced Plastics and Composites*, vol. 35, no. 3, pp. 179–211, 2016.

[22] J. J. Myers, S. S. Murthy, and F. Micelli, "Effect of combined environmental cycles on the bond of FRP sheets to concrete," in *Proceedings of the 2001 International Conference-Composites in Construction*, Porto, Portugal, October 2001.

[23] K. K. Mahato, K. Dutta, and B. C. Ray, "High-temperature tensile behavior at different crosshead speeds during loading of glass fiber-reinforced polymer composites," *Journal of Applied Polymer Science*, vol. 134, no. 16, Article ID 44715, 2017.

[24] J. Wang, H. GangaRao, R. Liang, D. Zhou, W. Liu, and Y. Fang, "Durability of glass fiber-reinforced polymer composites under the combined effects of moisture and sustained loads," *Journal of Reinforced Plastics and Composites*, vol. 34, no. 21, pp. 1739–1754, 2015.

[25] ASTM D638-10, "Standard Test Method for Tensile Properties of Plastics," in *Proceedings of the American Society for Testing and Materials*, January 2010.

[26] M. S. Sciolti, M. Frigione, and M. A. Aiello, "Wet lay-up manu factured FRPs for concrete and masonry repair: influence of water on the properties of composites and on their epoxy components," *Journal of Composites for Construction*, vol. 14, no. 6, pp. 823–833, 2010.

[27] B. Ghiassi, G. Marcari, D. V. Oliveira, and P. B. Lourenço, "Water degrading effects on the bond behavior in FRP-strengthened masonry," *Composites Part B: Engineering*, vol. 54, no. 1, pp. 11–19, 2013.

[28] S. P. Sonawala and R. J. Spontak, "Degradation kinetics of glass-reinforced polyesters in chemical environments: Part II. Organic solvents," *Journal of Materials Science*, vol. 31, no. 18, pp. 4757–4765, 1996.

[29] M. A. Abanilla, Y. Li, and V. M. Karbhari, "Durability characterization of wet layup graphite/epoxy composites used in external strengthening," *Composites Part B: Engineering*, vol. 37, no. 2-3, pp. 200–212, 2005.

[30] B. Abdel-Magid, S. Ziaee, K. Gass, and M. Schneider, "The combined effects of load, moisture and temperature on the properties of E-glass/epoxy composites," *Composite Structures*, vol. 71, no. 3-4, pp. 320–326, 2005.

[31] P. Marru, V. Latane, C. Puja, K. Vikas, P. Kumar, and S. Neogi, "Lifetime estimation of glass reinforced epoxy pipes in acidic and alkaline environment using accelerated test methodology," *Fibers and Polymers*, vol. 15, no. 9, pp. 1935–1940, 2014.

Functional Modification Effect of Epoxy Oligomers on the Structure and Properties of Epoxy Hydroxyurethane Polymers

Victor Stroganov,[1] Oleg Stoyanov,[2] Ilya Stroganov,[2] and Eduard Kraus ⓘ[3]

[1]Kazan State University of Architecture and Engineering, Zelenaya 1, 420043 Kazan, Russia
[2]Kazan National Research Technology University, Karl Marx 68, 420015 Kazan, Russia
[3]SKZ-German Plastic Center, Friedrich-Bergius-Ring 22, 97076 Wuerzburg, Germany

Correspondence should be addressed to Eduard Kraus; e.kraus@skz.de

Academic Editor: Ana María Díez-Pascual

We introduce different ways to solve the actual fragility problem of the epoxy-amine polymers by curing epoxidian oligomers with aliphatic amines without additional heat input. The pathways are the oligomer-oligomeric modification of epoxy resins-epoxy oligomers (EO), with their conversion to oligoethercyclocarbonates (OECC) by carbonization with carbon dioxide. The cocuring of these oligomers as a result of aminolysis competing reactions is "epoxide-amine" (forming a network polymer) and "cyclocarbonate-amine" (forming the linear hydroxyurethane, extending the internodal chains). Formation of internal and intermolecular hydrogen bonds was established on hydroxycarbonates (HA) and linear polyhydroxyurethanes (PHU) model compounds by IR and NMR spectroscopy. The results of the hydrogen bond system formation processes explain the change in the relaxation and physicomechanical properties of hard polymers modified by the epoxy-amine compositions (OECC), containing aromatic and aliphatic links. This paper presents a possible OECC modificator, the optimal EO:OECC ratio and its influence on the cross-link frequency, the polarity, the fragment and chain flexibilities and, as a consequence, the possible stiffness regulation for selected epoxy polymers. Thus, the causes of the increase in deformation-strength and adhesion characteristics were established by a factor of 1.5 to 3.0 due to an increase in cohesive strength (as a result of the combined network operation with covalent and physical bonds), as well as reduction of residual stresses (by adding the aliphatic fragments as additional relaxants), and reducing the defectiveness of the boundary layers (polymer-substrate).

1. Introduction

Materials based on epoxy oligomers such as epoxy polymers and composites with epoxy polymeric matrix possess a unique complex of valuable technological and operational properties. Characteristics like high adhesion to most materials, low shrinkage during curing, high strength, low creep under load, good chemical, biological stability, and electrical insulation ensure their successful and effective use in various applications and industries. The production development and the use of epoxy-based compositions and materials expand very fast. The new types of resin-oligomers, hardeners, active diluents, and compositions are emerging. These trends are undoubtedly associated with increasing interest in the physicochemistry of epoxy polymers, their structural organization, and connection with properties, which is reflected in monographic literature and reviews, published in collections and journals. Accordingly, the need for development of new composite materials is also very high. Relevant in this context are the novel possibilities of modifying epoxy polymers [1–9].

2. Theory

A feature of high molecular substances is their diverse nature and complex structure, which can not be described in one way. Conventionally, four interrelated structural levels can be distinguished: molecular, topological, supramolecular, and dispersive colloidal [10–18]. The main reason for this interrelation is the influence of the chemical structure of grid fragments (atoms and atomic groups) on the realizing possibility for each known type of the intermolecular interactions (physical bonds grid). In the chemical structure of polymers, the elemental composition of structural units, the type and position of functional groups, and the configuration and conformation of polymer chains characterize their molecular level.

The topological structure characterizes the cohesion and branching in a reticular polymer, which can be represented as a macroscopic molecule combining a system of cyclic structures with a system of different internode chains distributed throughout the volume. Such a molecule can be represented as an infinite graph [16]. Within the framework of this representations, the concept of nodes (γ_c), chains (n_c) or the average static mass of the chain interstitial segment (M_c) is the main characteristics of the topological structure, under the assumption of ideal networks (tetrahedral, cubic, etc.). The presence of topological defects (grid irregularity, chain inefficiency, closed loops, and unbound sol fraction) is a characteristic difference of real structures in reticular polymers from ideal ones. It has been shown that for epoxy-amine polymers, the main types of chemical defects are "free ends," and topological ones are chemical assemblies that have different connectivity with the wire mesh [19]. The abovementioned topological elements of the structure (I–V) can be conditionally classified: 3- and 4-related nodes (I, II) forming a grid; 2-connected nodes (III) of linear fragments, 1-bound (IV) nodes of free ends, and sol-fraction connections (V). These defects cause a decrease of degree of conversion for the reactive groups to the values of $\alpha_{\text{lim}}^{\text{top}} = 92\text{–}93\%$ (topological limit) at theoretical limit values $\alpha_{\text{lim}}^{\text{theor}} = 97\text{-}98\%$ for three-dimensional grid models. It is generally accepted that these characteristics determine the basic properties of network polymers [3, 20, 21].

The supramolecular structure is determined by the nature of intermolecular interactions of chemical structural elements, lateral active chains, and the degree of ordering in their mutual arrangement. The nature of the intermolecular interaction in epoxy polymers is determined by the presence of polar atoms, groupings, and their interaction in the topological grid fragments (I–IV). The following types of intermolecular interactions are distinguished in polymers [22]: dispersion, inductive, dipole, and hydrogen bonds. These named interactions form a labile spatial grid (with a stable network of covalent bonds). In particular, in densely cross-linked epoxy polymers, the physical grid of hydrogen bonds exceeds the concentration of chemical assemblies by 2–4 times, reaching densities up to 4×10^{21} 1/cm^3 [23, 24]. The role of the hydrogen bond between the oxygen atoms of the oxirane ring and the hydrogen for hydrogen bonding of the hydroxyl group during the aminolysis of epoxy compounds has been considered in a number of works. The results of calculating the distribution of hydrogen-bonded groups and the types of H-bonds are considered in detail in [25]. Based on the literature data, the authors believe that the defining influence on the topological structure formation of the network polymers has the formation of an intramolecular hydrogen bond between the epoxy and hydroxyl groups of the following type [26]:

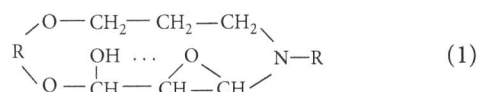

$$
\begin{array}{c}
\text{O—CH}_2\text{—CH}_2\text{—CH}_2 \\
R \qquad \text{OH} \cdots \quad \text{O} \qquad \text{N—R} \qquad (1) \\
\text{O—CH—CH—CH}
\end{array}
$$

Formation of an intracellular bond system with hydrogen bonds leads to the strengthening and ordering the spatial network of cured epoxy oligomers [27]. Pochan et al. [28] associate relatively high values of crystallization temperature T_c and destructive loading of polymers with intermolecular hydrogen bonds formed by hydroxyl groups:

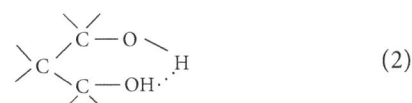

$$
\begin{array}{c}
\text{C—O} \\
\text{C} \qquad\qquad \text{H} \qquad (2) \\
\text{C—OH}\cdots
\end{array}
$$

Despite the contradictory nature of these researchers, the prevailing view is that the influence of intermolecular interactions is multifaceted. This approach explains and causes wide possibilities for modifying the properties of epoxy-amine network polymers, also due to covalent and physical bonds.

The presented data make possible the representation of the epoxy polymer structure as a superposition of two grids: a covalent bonds network characterized by inhomogeneity at the molecular and topological levels and a thermofluctuation physical bonds grid caused by the realization of all possible types of intermolecular interactions groups and grid fragment atoms. The presented model assumes the possibility of regulating the molecular mobility of network polymers due to a change in the chemical structure (the nature of the node, internodal fragments and their kinetic flexibility, polarity of groups, etc.). Numerous studies on the effect of the chemical structure and properties of epoxy polymers show that changing the initial oligomers nature, hardeners, and their ratio can affect the structure and physicomechanical properties of epoxy polymers and materials based on them.

By analyzing the properties of epoxy polymeric material complexes, the "brittleness" of epoxide-amine polymers obtained with amine curing without additional heat input (such compositions are often used in practice in various technologies) should be noted as a considerable disadvantage. This drawback is due to an inhibition of the aminolysis reaction, a decrease in the degree of reactive group conversion $\alpha_{\text{lim}}^{\text{top}}$, and as a result, a significant reduction in the level of cohesive (σ_c, ε_c) and adhesive strength (σ_a, τ_a). It is known that aliphatic chains increase the flexibility of densely cross-linked epoxy-amine polymers. Their conformational set and kinetic flexibility are practically independent from location of these chains (e.g., in the oligomer molecule, in the linear polymer chain, or in the mesh polymer). This provision is very important since it indicates the fundamental influence of the initial molecule structure of oligomers and hardeners on the cured polymer properties. Among the main factors regulating the molecular level of the structure, the reactivity of the initial components (the presence of functional groups in the oligomer, the hardener, and their potential activity in various types of reactions), the temperature-time regimes for process realization, the mechanisms of polyreactions themselves, (polymerization, polycondensation, and polyaddition), and so on were presented [29].

Proceeding from the theoretical propositions above, to solve the epoxy-amine polymers' "stiffness" problems, following purposes of this article have been formulated:

(i) The analysis of theoretical bases for the formation of epoxy-amine polymer networks.

(ii) The process study of epoxy oligomers functional modification with different chemical structures and their carbonization into cyclic carbonates.

(iii) The properties of epoxy hydroxyurethane polymers formed as a result of the competition of aminolysis oligomers reactions (EO and OECC) at their simultaneously curing.

It can be assumed that a partial change in the network structure of epoxy-amine polymers (due to the introduction of linear hydroxyurethane fragments obtained on the basis of OETC with different structures containing highly polar hydroxyurethane groups) will provide the opportunity to regulate intermolecular interactions (a physical bonds network) and the molecular mobility of modified polymers in vitreous and elastomeric states. Such a functional modification will make it possible to purposefully apply the relaxation and physicomechanical properties of epoxy-amine polymers and understand the reasons for their "rigidity."

3. Materials and Methods

3.1. Materials. The epoxy oligomer ED-20 was used in the compositions studied. ED-20 contains 21.5% epoxy, with molecular weight 400 to 450 g/mol and oligomerization degree $n = 0.2$. Following OECC modifiers containing fragments differing in structural rigidity have been used: aromatic (based on ED-20), aliphatic with methyl, chloromethyl, and cyclic fragments in the side chain, and an amine hardener, the diethylenetriamine (DETA). OECC were obtained by the interaction of various epoxy oligomers with CO_2. The structure can be represented by the following general formula:

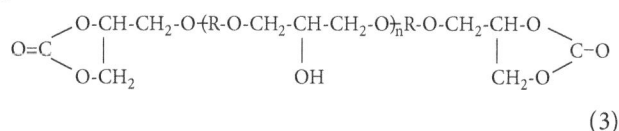

$$O=C \overset{\displaystyle O-CH-CH_2-O\!\!\left(\!R-O-CH_2-CH-CH_2-O\!\right)_{\!\!\overline{n}}R-O-CH_2-CH-O}{\underset{\displaystyle O-CH_2 \qquad\qquad OH \qquad\qquad CH_2-O}{\Big|\qquad\qquad\quad\Big|\qquad\qquad\quad\Big|}} \overset{}{\diagdown} C-O$$

(3)

where R the is residues of alcohol, phenol, or carboxylic acid and n is the degree of oligomerization. The characteristics of the OECC are shown in Table 1.

It has been established that relatively close reactivity values are observed in the reactions of CO_2 with glycidyl ethers of phenols, alcohols, and acids. Consequently, the conditions for the synthesis of cyclic carbonates based on glycidyl ethers [30] have been determined: the ratio of reagents, temperature, type and amount of catalyst, mass exchange conditions, and others.

3.2. Synthesis of the OECC. The epoxy oligomers (1 mol) were placed in a reactor equipped with a stirring device, a heating element, and a coil to cool the reaction mass. The mixture was heated to the prescribed temperature of 120 to 140°C, the required amount of catalyst was introduced $C_c \approx 0.025$ mol/kg, and after 2 min a sample was taken to determine the epoxy groups initial concentration. Subsequently the reactor was sealed and carbon dioxide was added ($V_{CO_2} = 4.5$ l/h with stirring $n = 300$ rpm and $P = 0.3$ MPa). The reaction was monitored by changing the concentration of the epoxy groups by the sampling method (1 ml of the reaction mixture was taken, quenched, and analyzed).

Various catalysts for the reaction of epoxide compounds with carbon dioxide can be found in the literature today [31]. Catalytic systems containing halide salts found actually the greatest application among them. The catalytic activity of haloanions increases in the series as follows: I > Br > Cl. For alkali metal halides, the catalytic activity increases with increasing cation radius. The localization of the alkali metal accelerates the reaction to a greater extent. Thus, the catalytic system "KCl and crown ether of dibenzo-18-crown-6" exhibit greater activity (reaction rate constant $K = 1.28 \times 10^{-2}$ kg·mol^{-1}·s^{-1}) as KCl ($K = 9.56 \times 10^{-3}$ kg·mol^{-1}·s^{-1}). It has been established that an increase in the reactor volume does not adversely affect the quality of the cyclocarbonates obtained.

3.3. Saponification Number Determination of Cyclocarbonate Groups. The analyzed product between 0.1 and 0.2 g was weighed (±0.0002) in a 100 ml flask and dissolved in 10 ml of acetone. Subsequently 5 ml of 0.7 M sodium hydroxide solution was poured in. The resulting mixture was heated for 30 minutes in a boiling water bath with a reflux condenser. After that, the flask was lifted, slightly cooled, and disconnected from the refrigerator. 20 ml of distilled water and 10 ml of a 10% solution of barium chloride (previously neutralized by phenolphthalein) were added to the mixture. The flask was closed with stopper, mixed, and cooled. Excess sodium hydroxide was titrated with 0.1 M hydrochloric acid in the presence of phenolphthalein until the pink color disappeared.

In parallel, a blank test was carried out under the same conditions, but without analyzing the product. The saponification number (X) in of KOH/g product was calculated by the formula: $X = (V_0 - V) \cdot K \cdot 5.6/m$, where V_0 and V are the volumes of the hydrochloric acid solution consumed for titration in the blank test and the sample of the product, respectively. K is the correction factor to 0.1 M HCl solution. 5.6 is the amount of potassium hydroxide corresponding to exactly 1 ml of 0.1 M HCl solution, and M is the weight of the sample of the product. For the result analysis, the average of three parallel definitions was taken, which discrepancy was no more than 20 mg KOH/g with a confidence level of 0.95.

3.4. Mass Fraction Determination of Epoxy Groups (Epoxy Number). To determine the residual epoxy number, 1 g of CC sample was used. The reaction products (OECC) were dissolved in chloroform, and DCC/ED-20 was solved in

TABLE 1: Characteristics of OECC based on epoxy oligomers.

The chemical nature of epoxy oligomers (epoxy number (%))	Radicals, reflecting the nature of the OECC, $R=$	Conditional abbreviation of OECC	Content of epoxy groups (%) (residual)	Saponification number, (mg KOH/g)		Viscosity at 25°C (Pa·s)	T_m (°C)	Density, 20°C (g/cm³)	Appearance
				Determined	Calculated				
Epoxidated oligomer ED-20 (21.6%) based on bisphenol-A		DCC/ED-20	0.10	452	459	—	55	—	A hard brittle mass of yellow color (brown-white powder after grinding)
Diglycidyl oligomer based on diethylene glycol DEG-1 (26.8%)	-CH$_2$-CH$_2$-O-CH$_2$-CH$_2$-	DCC/DEG-1	0.17	552	560	1.12	—	1.4762	Low-viscosity liquid of yellowish color
Diglycidyl oligomer based on dipropylene glycol (25.5%)		DCC/DPG	0.16	530	535	1.85	—	1.4722	Liquid of yellowish color
Diglycidyl oligomer based on 1,1-di (oxymethyl)-3-cyclohexene (25.0%)		DCC/DOC	0.06	555	560	26.10	—	1.4960	Viscous liquid of cherry color
Epoxy oligomer based on epichlorohydrin (polyepicholorhydrin) -E-181 (25.9%)		DCC/E-181	0.10	524	522	17.04	—	1.4906	Viscous liquid of cherry color

dioxane. The epoxy number determining was carried out according to GOST 12497-78.

3.5. Calculation of Amine Hardener Amount for CC-Compositions.
The hardener was always taken in a stoichiometric amount under consideration of its interaction with epoxy and cyclocarbonate groups. The amount of hardener (X) per 100 g of the mixture of oligomers was determined according to the following formula:

$$X = \alpha \times k_{EO} \times M_{EG} + \beta \times kcc \times \% \, CC, \qquad (4)$$

where α and β are the mass fractions of oligomers in the mixture, K_{EO} and K_{CC} are the stoichiometric coefficients of the hardener with respect to epoxy and cyclocarbonate oligomers, CC is the mass fraction of cyclocarbonate groups, and M_{EG} is the mass fraction of epoxy groups.

3.6. Strength and Adhesion Properties.
For evaluating the mechanical properties, the following values were determined: the tensile strength (σ_p) and the elongation (ε_p) at break according to DIN EN ISO 527 and the tensile shear strength of adhesive joints (τ) in accordance with DIN EN 1465 with uniform separation ($\sigma_{p.o}$) in accordance with ISO 4587. The data of physical and mechanical tests were processed using the software Statgrafica.

3.7. Thermomechanical Analysis (TMA).
TMA was carried out under uniaxial compression at stress of 1.5 MPa and a temperature rise rate of 2.5°C/min. Samples of cylindrical shape with diameter 10 mm and height 10 mm were used. The glass transition temperature (T_g) and the transition temperature to the high-elastic state (T_m) were determined by conventional methods (by the tangents intersection).

3.8. Topological Grid Parameters.
The most important topological grid parameter is the interstitial chain fragment molecular mass (M_c). It was determined from Wall's formula [12] that $M_c = 3 \cdot \rho \cdot R \cdot T \cdot \varepsilon_e/\sigma$, where ρ is the density of the polymer, T is the absolute temperature, R is the gas constant, ε_e is the relative deformation in the highly elastic state, and $\sigma = N/S$ is the stress applied to the sample. The effective density of the polymer network nodes (γ_c) was determined from the following relation: $\gamma_c = 2 \cdot \rho \cdot N_0/(3 \cdot M_c)$, where N_0 is the Avogadro number.

3.9. IR Spectroscopy.
IR spectra of oligomeric and polymeric systems were recorded on a two-beam Carl Zeiss UR-20 spectrophotometer with the detection range from 400 to 4000 cm^{-1}. The following operating parameters were used: target program nr. 4, scanning speed was 160 cm^{-1} for normal spectrum, and 32 cm^{-1} for frequency refinement spectrum. The registration scale was 20 min/100 cm^{-1}. The spectra of the polymer samples were taken in the condensed state in the film with the thickness $\delta = 10$ to 20 μm between the KBr plates. At elevated temperatures, a special thermocuvette was used. The solutions spectra were recorded in KBr and KRS-5 cuvettes with a thickness of 0.07 to 3.00 mm.

For investigating the curing processes of polymer systems, the cuvette windows were covered with a fluoroplastic film with thickness $\delta = 10 \, \mu$m. The spectra were recorded in the regions 800 to 1000 cm^{-1} and 1600 to 1900 cm^{-1}. In these ranges, an uncompensated absorption of the fluoroplastic film could be considered. The fraction of unreacted functional groups (epoxy and cyclocarbonate) was determined by normalizing the optical density of the corresponding band at time of the first measurement t_c (1 min after the oligomers were mixed with the curing system).

3.10. Dielectric Loss Method.
Measurement of the dielectric parameters was carried out in the frequency range from 10^3 to 10^6 Hz and the temperature range from −180 to +250°C on samples in the disk form with a diameter of 50 mm and a height of 2 and 3 mm. The samples were previously covered with aluminum foil.

3.11. Nuclear Magnetic Resonance Method.
Two methods are most effective for research in solids and viscous liquids (oligomers and polymers): pulsed NMR and NMR of wide lines. The first is based on the study of magnetic relaxation at various temperatures, estimated by spin-lattice (T_1) and spin-spin (T_2) relaxation. The second is based on the shape study of the line and its temperature dependence. The study was performed on a laboratory coherent NMR relaxometer at a frequency of 17 MHz. The decay curves of the transverse magnetization (DCTM) were recorded by the Carr–Parcell–Meibum–Gil method from the free induction decay [32]. The measurement was carried out under isothermal conditions, as well as with a stepwise temperature rise in the range from 20 to 220°C with an isothermal holding time of 15 minutes. The molecular mobility was estimated from the times of transverse spin-spin relaxation T_2. In the general case, the free induction decay is described by a function as a superposition of several terms:

$$f(t) = P_a \cdot \exp\left(-\frac{t}{T_{2a}}\right) + P_b \cdot \left(-\frac{t}{T_{2b}}\right) + P_c \cdot \exp\left(-\frac{t}{T_{2c}}\right), \qquad (5)$$

where P_a, P_b, and P_c are the relative proton nuclei fractions that relax with transverse relaxation times T_{2a}, T_{2b}, and T_{2c}.

"Phases a" is formed by nuclei with longer relaxation times and "phase b and c" with shorter ones, respectively. The times T_{2b} and T_{2c} were determined by successively subtracting the values of the longer relaxation component from the values of the experimental curves. The population of the "phases" (the number of protons entering this "phase") P_a, P_b, and P_c was calculated from the contribution to the initial amplitude of the signal by extrapolating the lines to the zero line (the line passing through the point of excitation of the oscillograph). Measurements by the wide-line method [33] were carried out on a laboratory NMR spectrometer at a frequency of 16 MHz.

4. Results and Discussion

By the development of new polymeric materials, the modification of epoxy polymers with urethanes is successfully used

[34, 35]. The most promising direction here is the oligomer modification containing urethane groups in the chain and epoxy groups at the ends. One of the obstacles for its wide use is the difficulty to obtain the urethane-containing materials using conventional isocyanate technology, which also has its drawbacks such as toxicity of isocyanates, the complexity of their production, storage, processing, and the possibility of side reactions in the presence of even small amounts of water [31, 35]. Some improvement in processing conditions for epoxy compositions provides the use of blocked isocyanates [36]. Among the known nonisocyanate methods [37, 38] for the preparation of urethane-containing compounds, the urethane-forming reaction "cyclocarbonate-amine" deserves attention. Cyclic carbonates (CC) are a relatively new and poorly studied class of compounds that causes the urgency of the work analysis by methods of their production, reactivity, and methods for modifying polymers [31].

4.1. Oligomer-Analogous Transformations of Epoxy Oligomers. Highlighting the main stages of CC synthesis, it should be noted that the carbonatization of glycidyl ethers can be represented as a chemisorption process described by the following equation:

$$\alpha = 1 - e^{-kT}, \qquad (6)$$

where α is the degree of reactive groups conversion, T is the time, and K is the adsorption coefficient, which is a function of the gas content (φ), the rotational speed of the stirrer (n), the fluid viscosity (ν_L), and the apparatus diameter (D). The quantitative interrelation of these parameters is established for the process under study [30] by the following equation:

$$K = 1.66 \left(\frac{\nu_L}{D^2} \right)^{0.088} \cdot \varphi^{0.34} \cdot n^{0.912}. \qquad (7)$$

By these studies, the possibility of obtaining oligomeric CC at both excess and atmospheric pressure was established. Moreover, the process of carbonization at atmospheric pressure can take place not only in solution, but also in the mass with a sufficiently high rate to high degrees of transformation (α) in the range from 120 to 140°C and concentration of catalyst $(C_2H_5)_4$ NJ tetraethylammonium iodide in an amount of no more than 1×10^{-2} mol/kg. The reaction is described by second-order kinetic equation [39]:

$$\frac{dx}{d\tau} = K \cdot XY, \qquad (8)$$

where X is the concentration of epoxide groups and Y is the catalyst concentration (C_c). In the synthesis of the CC, the mass of the reaction mixture (m) increases with increasing conversion (α) and the catalyst concentration (Y) decreases accordingly. Dependence of the change in the mass of the reaction mixture (m) and α is related by the following relation: $m = n_{EO} \times (M_{ER} + 44 \cdot f \cdot \alpha)$, where M_{ER} is the molecular weight of epoxy resins (ER), f is the functionality of the ER, and n_{EO} is the number of the epoxy oligomer moles in initial time. The obtained experimental data on the interaction of α-oxides with CO_2 allowed determining the conditions for the oligoethercyclocarbonate preparation, based on epoxy oligomers with different structures (Table 1). The stable quality of the OECC is probably similar due to the oligomeramino conversion of epoxide groups to cyclocarbonate ones. The presence of such a transformation is confirmed by IR spectroscopy data (the peak presence at 1800 cm^{-1} corresponding to stretching vibrations >C=O groups in the CC) and chemical analysis (coincidence of the calculated and determined saponification number). After carbonatization, the molecular-weight distribution of epoxy oligomers is practically unchanged inherited by cyclocarbonate oligomers. This was clearly confirmed for the epoxidian oligomer and the corresponding cyclic carbonate DCC/ED-20. It was satisfactorily confirmed also for other epoxy oligomers.

4.2. Structuring of Epoxyurethane Mesh Polymers. The polymer formation process based on epoxy-cyclocarbonate amine curing compositions was determined by the conditions of two basic competing reactions: epoxide-amine, with the formation of a network structure and cyclocarbonate-amine, with the formation of linear hydroxyurethane fragments. The variety in the resulting epoxyurethane polymer properties cannot be excluded under the conditions determining the formation of a single polymeric network: a common curing agent (an aliphatic amine) and a close reactivity of epoxy and cyclocarbonate oligomers. By varying the ratio of components and the structure of EC oligomers, it is possible to regulate the cross-link density, polarity, and flexibility of the grid chains formed by chemical bonds. In addition, the modification of OECC epoxy-amine compositions leads to the formation of intermolecular hydrogen bonds involving urethane groups. That can affect the molecular mobility and the level of physical and mechanical properties of polymers (the contribution of a physical bonds network). The totality of the modification processes can be represented step by step as follows.

4.3. Aminolysis of Cyclic Carbonates and Their Curing with Epoxy Oligomers. Kinetic studies of the CC aminolysis were performed on the example of the interaction of 1-tetrahydrophenylcarboxy-2,3-propylene carbonate (obtained on the basis of phenyl glycidyl ether (PGE), the content of epoxy groups was 0%, the determined saponification number was 582, and the calculated saponification number was 577, $T_m = 93$°C, white powder) with benzylamine in chlorobenzene [40]. It is established that the investigated process proceeds by two parallel flows: noncatalytic and catalyzed by two amine molecules. The mechanism of the process can be represented in the following form:

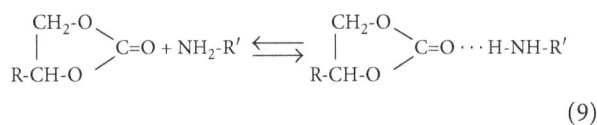

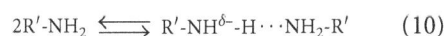

$$(9)$$

$$2R'\text{-}NH_2 \rightleftharpoons R'\text{-}NH^{\delta-}\text{-}H \cdots NH_2\text{-}R' \qquad (10)$$

In the first stage of the process, a formation of associates is possible: a hydrogen-bonded complex of benzylamine with a CC and two amine molecules as a result of self-association. Further, the catalytic reaction develops with the opening of the cyclocarbonate ring, obviously, through an intermediate cyclic transition state formed by the interaction of the activated amine in the associate and cyclocarbonate in the associate:

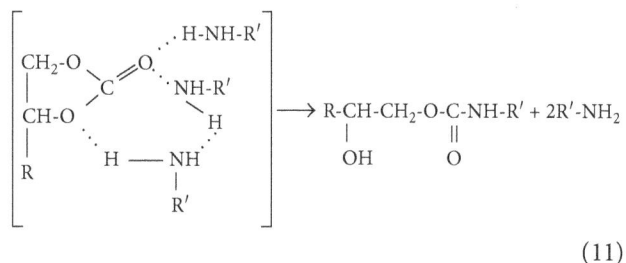

$$
\begin{bmatrix}
\begin{array}{c}
CH_2\text{-}O \\
| \\
\quad\quad C=O \\
CH\text{-}O \\
| \\
R \quad\quad H\text{---}NH \\
\quad\quad\quad | \\
\quad\quad\quad R'
\end{array}
\begin{array}{c}
\cdot\cdot H\text{-}NH\text{-}R' \\
\cdot O \\
\cdot\cdot NH\text{-}R' \\
H
\end{array}
\end{bmatrix}
\longrightarrow
\begin{array}{c}
R\text{-}CH\text{-}CH_2\text{-}O\text{-}C\text{-}NH\text{-}R' + 2R'\text{-}NH_2 \\
\quad\quad\quad | \quad\quad\quad || \\
\quad\quad\quad OH \quad\quad O
\end{array}
$$

$$(11)$$

Analogous assumptions about the formation possibility of a cyclic transition compound were also expressed in [41], but with the participation of two amine molecules. In the case of noncatalytic aminolysis, the cyclocarbonate is attacked by the carbonyl carbon atom by one amine molecule. The probability of a cyclic transition state is confirmed by the low activation energy in the catalytic reaction [42], calculated approximately at two temperatures. With a decrease in temperature and an increase in the amine concentration, the catalytic flow contribution increases to the overall process of the CC aminolysis (Table 2).

By real curing conditions of oligomers EO and OECC (when the process is carried out in "mass"), higher reaction rates should be expected, since the amine concentration under these conditions is 4-5 g/l, which confirms the validity and possibility of using the CC as reactive epoxy modifiers of amine curing compositions.

The curing process of the epoxy and cyclocarbonate oligomer was studied by IR spectroscopy. Comparative studies were performed on aromatic (DCC/ED-20) and aliphatic oligoethercyclocarbonates (DCC/DEG-1 and DCC/E-181) (Table 3). When curing the (ED-20 + DCC/DEG-1 + DETA) and (ED-20 + DCC/ED-20 + DETA) compositions, a redistribution of the intensities of the absorption bands of $920\,cm^{-1}$ (epoxy groups), $1802\,cm^{-1}$ (cyclocarbonate Group), 1700 and $1715\,cm^{-1}$ (carbonyl groups of urethane fragments) could be considered in IR spectra. This indicates the occurrence of simultaneous reactions over epoxide and cyclocarbonate groups.

In the aminolysis study on model compounds, it was shown that the reaction rate for cyclocarbonate is higher than for epoxy. This conclusion can be confirmed for OECC (with a content of 20 to 30 %): after 5 minutes, the reaction rate is high (this is indicated by the intense peak of urethane carbonyl), and after 60 minutes, the conversion (α) is about 60 %. With an OECC content of more than 30 %, the consumption rate of the cyclocarbonate groups decreases and that of the epoxide groups increases (slopes of the curves in Figure 1). This may be a catalytic effect consequence of the hydroxyurethane groups formed.

Further, as a result of the predominant epoxy group interaction, the composite system is depleted by the primary amine. The limiting degree of the CC group transformation

TABLE 2

Value	60°C	80°C	E (kJ/mol)
C_o (l·mol^{-1}·min)	5.58×10^{-1}	9.95×10^{-4}	28.3
C_v (l·mol^{-1}·min)	9.92×10^{-3}	1.14×10^{-2}	6.8

Note. C_o is the noncatalytic constant of the CC bimolecular interaction on the PGE basis and C_v is the catalytic rate constant.

decreases since the interaction with the secondary CC amines at 20 to 22°C is very slow. This conclusion is confirmed by the fact that with an excess of amine (1.2 to 1.3 from stoichiometry), the degree of reactive group conversion increases sharply, and after 5 to 8 h (for cyclocarbonate) and 16 to 20 h (for epoxy), changes in the intensity of the characteristic bands almost do not occur. After 24 h, α is 90 to 95% (for cyclocarbonate) and 80 to 90% (for epoxy) groups. The noted signs of inhibition due to the network polymer solidifying [43] are also retained when the OECC is modified, reaching $\alpha \approx 70$ to 75%. The properties of unmodified epoxy-amine polymers stabilize after 5 to 7 days, but they do not reach the level of polymers characteristic for highly cured polymers (22°C with 24 h and 100°C with 10 h) (Table 3).

For systems containing DCC-DEG-1, this difference is insignificant, which is quite convincing evidence of the effect of modification.

It should be noted that with polymer characteristics improvement (σ_p, ε_p, and T_c), the adhesive properties also increase (T_B and $\sigma_{p.o}$). This result is worthy of note, since Lipatov et al. noted [44, 45] that epoxy-amine systems have a low adhesive strength as a result of the weak boundary layer formation due to the selective sorption of epoxy polymers on high-energy hard surfaces. This, as a consequence, leads to a violation of the stoichiometry of the components and the lack of solidification of the composition in the boundary layer.

To compare the adhesive strength in the boundary layers of the systems (Table 4), the IR absorption spectroscopy in absorption (1, 2) and ATR arrangement (1', 2') was used (Figure 2). It can be considered that for the unmodified system, the degree of reactive epoxy groups conversion (α) was 72% and in the boundary layer 36% (high free surface energy-element KRS-5). For systems modified with 20% DCC/DEG-1, the values α for epoxy groups are relatively close to 72% and 62%, respectively. These results make it possible to understand not only the reasons for the increase in adhesion strength as a result of the OECC modification, but also the previously described aminolysis features of the CC and EO. As noted above, in the first minutes of mixing oligomers with an amine hardener, a significant amount of urethane groups are formed in the system, which are capable to blocking the active centers of the substrate solid surface. It prevents the selective sorption of EO and weak boundary layers formation after the composition is applied.

4.4. The Contribution of Hydroxyurethane Fragments to the Epoxy Polymer Properties. The topological structure studies, formed by epoxy-amine mesh modification with cyclic carbonate containing hydroxyurethane fragments were

TABLE 3: Physicomechanical properties of polymers obtained under different conditions of curing epoxy-amine compositions.

Composition	Curing mode	Physical and mechanical properties				
		σ_p (MPa)	ε_p (%)	τ_B (MPa)	$\sigma_{p.o}$ (MPa)	T_c (°C)
ED-20 + DETA	7 d at $(22 \pm 2°C)$	20.7	0.6	4.6	8.8	46
	1 d at $(22 \pm 2°C)$ and 10 h at 100°C	72.5	2.5	12.5	28.0	108
ED-20 + DCC/DEG-1 + DETA	7 d at $(22 \pm 2°C)$	75.2	5.2	15.8	30.5	42
	1 d at $(22 \pm 2°C)$ and 10 h at 100°C	88.3	4.4	22.8	50.0	68

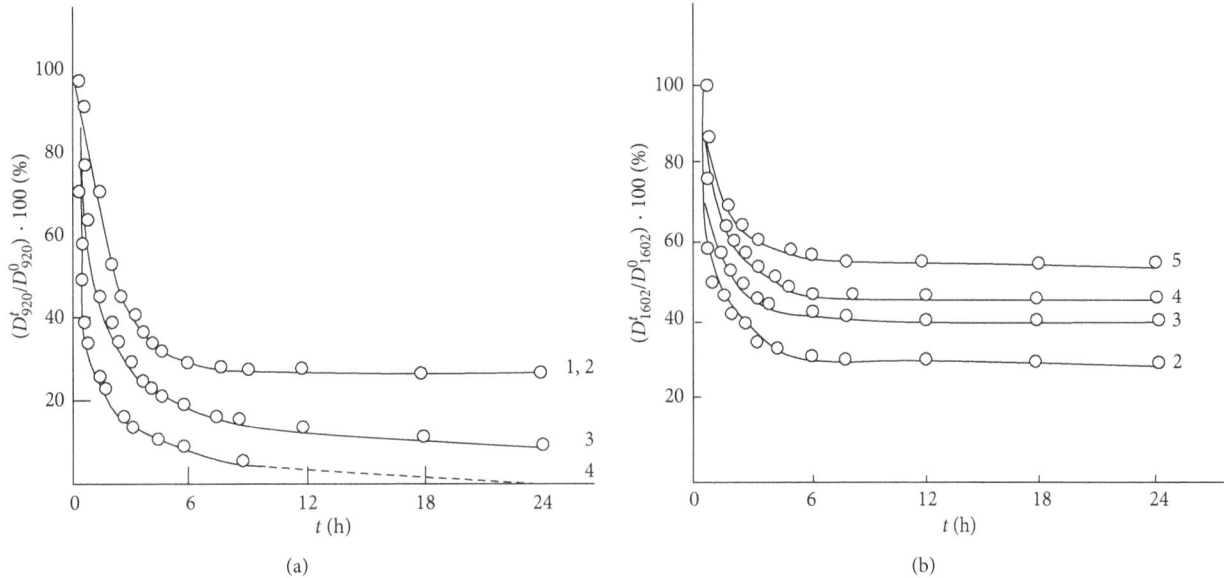

FIGURE 1: Changes in the content of epoxide (a) and cyclocarbonate groups (b) during the curing of ED-20 + DCC/DEG-1 + DETA compositions as a function of the DCC/DEG-1 content with 0% (1), 20% (2), 40% (3), 50% (4), and 60% (5).

TABLE 4: Influence of modifiers (20% OECC) and ED-20 + DETA composition curing at the second moment of NMR absorption.

Composition	Second moment of NMR absorption	
	Curing 7 d at 22°C	Postcuring 10 h at 100°C
ED-20 + DETA	4.50	2.50
ED-20 + DCC/ED-20 + DETA	4.50	5.20
ED-20 + DCC/DEG-1 + DETA	1.68	2.70

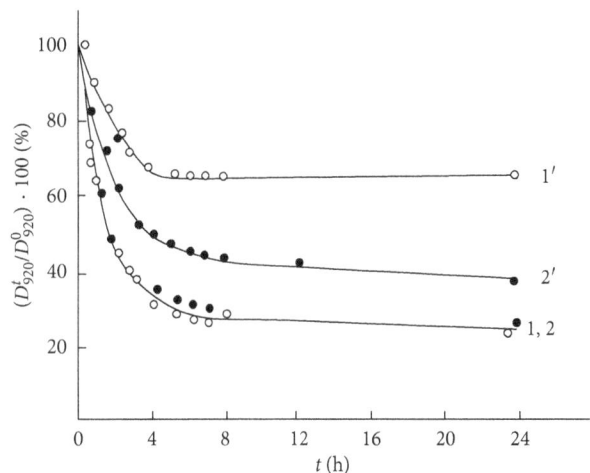

FIGURE 2: The change in the content of unreacted epoxy groups during the curing of ED-20-DETA (1, 1′) and ED-20 + DCC/DEG-1 + DETA (2, 2′) compositions, determined by IR transmission spectroscopy (1, 2) and ATR-IR (1′, 2′).

performed by IR spectroscopy and a number of relaxation methods. As noted above, the disadvantage of unmodified epoxy-amine compositions is their high stiffness (brittleness), which results in low cohesive strength, especially for cured compositions without heat input (Table 3). The cohesive strength depends on both the density of the chemical bonds network and the intermolecular interaction forces (a grid of physical bonds) in the glassy state. The increase in rigidity and heat resistance (to a greater extent for polymers cured at $T < 22°C$) mainly depends on intermolecular interactions (IMI) in chains and packing of aromatic nuclei. Based on these provisions, it was of interest to determine the manifestation and relative level of these factors in the initial epoxy-amine system, and

to follow the changes that occur during the modified compositions curing.

Controlling the optical density and the integrated intensity of the complex deformation vibration in benzene ring bands (wavenumbers of $1612\,\mathrm{cm}^{-1}$ and $1584\,\mathrm{cm}^{-1}$), which are sensitive to changes in the universal intermolecular interaction of aromatic nuclei, the changes occurring during glass transition of the systems were followed. During the ED-20 + DETA compositions curing process, an increase in the integrated intensity of the spectral contour in the frequency range of 1570 to $1650\,\mathrm{cm}^{-1}$ was observed. This is proportional to the change in the optical density of $1612\,\mathrm{cm}^{-1}$ band (D_{1612}), which indicates the enhancement of the aromatic nuclei IMI.

The D_{1612} values by curing for 3 d at $22°\mathrm{C}$ increase from 0.681 to 0.724, and after curing for 8 h at $100°\mathrm{C}$ they decrease to 0.685. These results indirectly indicate a change in the stiffness and molecular mobility of the polymer structure elements, which is also confirmed by data determined from the second magnetic moments values of NMR absorption (M_2). It is known [32] that the larger the value of M_2 indicates the lower molecular mobility. A polymer based on an unmodified epoxy-amine composition cured at $22°\mathrm{C}$ is characterized by a high level of M_2 values with $4.5\,E^2$, which decreases after postcuring at $100°\mathrm{C}$ (Table 4). According to these results, the additional curing should help increase the chemical bonds number and further increase the rigidity of the polymer. Can this fact be explained?

For more rigid epoxy-amine systems containing slow-moving polyhedra fragments, it was shown that by amine curing without additional heat input, linear polymer chains are predominantly formed in the composition as a result of the predominant interaction of more active primary amino groups with epoxy groups EO (with the example of bisphenol A diglycidyl ether). The resulting linear chains are capable of denser packaging, in particular aromatic nuclei (in the case of adamantanes, bulk cycloaliphatic fragments), which determines the high rigidity of the polymer. The postcure at $T > T_c$ not only leads to an increase in the cross-linking frequency (over the secondary amino groups), but also to the destruction of the formed ordered structures [46] and, consequently, to a polymer rigidity reduction.

For example, when the ED-20 + DETA composition is modified with an aromatic OECC, the structure of the newly formed polymer is characterized by the presence of urethane groups and a lower cross-linking frequency (η_c). However, despite a slight decrease in η_c, the rigidity of the limit-cured polymer is much higher ($M_2 = 5.2\,E^2$) than for an unmodified polymer. This unambiguously demonstrates the contribution of urethane groups to an increase in stiffness and a decrease in molecular mobility. Comparing with the stiffness of modified aliphatic OECC, it can be seen that the level for polymers cured at $22°\mathrm{C}$ with $M_2 = 1.68\,E^2$ is much lower, and after the hardening, the M_2 value is close to the level of the highly cured unmodified polymer. This is obviously a consequence of the combined effect of high-polar urethane groups (decreased mobility) and flexible diethylenic fragments (increased mobility). According to the values of M_2, the optical density of the band is $1612\,\mathrm{cm}^{-1}$ and the values of D_{1612} increase during the curing process from 1.145 to 1.205 and 1.252 (after the postcuring), which

corresponds to the increase in rigidity of the system. Thus, the cumulative effect is that the introduction of an aliphatic modifier into the composition increases the system deformation reserves and reduces the overall level of its rigidity (from $M_2 = 4.50\,E^2$ to $M_2 = 1.68\,E^2$), which provides a high level of cohesive strength of the polymers also curing without additional heat (Table 3).

The evidence for the urethane group role for increasing the system rigidity can be confirmed by an experiment with the blocking of these groups by lithium chloride (4% solution in dimethylformamide, taken in the stoichiometric ratio to the calculated number of urethane groups). The composition as well as the cured polymer remained transparent, when combined with LiCl and after the addition of the hardener. The Li^+ and Cl^- ions block the >C=O and NH-groups formed during the curing, excluding (in part or in whole) the intra- and intermolecular interactions. The data obtained clearly illustrate the effect of polymer hardening due to physical interactions of urethane groups (Table 5). The performed experiment indicates that the hydrogen bonds in the studied epoxy polyurethane combinations have a significant influence not only on the processes of polymer formation but also on their macroscopic properties. For unmodified epoxy-amine polymers, the greater contribution of hydrogen bonds to the macroscopic properties of polymers should be expected in the temperature range below the β transition [28]. The linear homo- and copolymers of styrene and methacrylates showed [47] that the β-transition "loosens up" the hydrogen bonds and leads to their partial destruction. The hydrogen bonds shift T_c to higher temperatures, preventing large-scale molecular motion.

The molecular mobility in the range from -100 to $+200°\mathrm{C}$ has been studied by the dielectric relaxation method. It was established that for the investigated polymers, two different transitions are the low-temperature transition in the range from -70 to $+100°\mathrm{C}$, corresponding to the processes of dipole-group β-relaxation, and high-temperature transition by $T > +100°\mathrm{C}$, corresponding to dipole-segmental α-relaxation (Figure 3). It can be seen from the relaxation curves that the β-relaxation peak intensity decreases with an increase in the OECC concentration (partial degeneration of the β-transition was observed). It can be assumed that the intra- and intermolecular hydrogen bonds of the hydroxyl groups on the urethane group carbonyl ($-\mathrm{OH}\cdots\mathrm{O=C}<$) prevent the internal rotation. The defrosting of these movements obviously occurs with the onset of the polymer melting. An increase in the M_2 NMR absorption values, the degeneration of the β-transition, and a sharp decrease in the deformation-strength characteristics for the ED-20 + DCC/ED-20 + DETA system indicate a decrease in the molecular mobility in the glassy state. In the highly elastic state, its increase is obviously associated with the rotation of hydrogen-bonded hydroxyurethane fragments during "defrosting" of the aromatic nuclei movements (with the α-relaxation process). This is manifested by an increase in the intensity of the α-transition peak on the dielectric relaxation curves and the appearance of a second (longer) NMR relaxation time at $T > T_c$. Modification of aliphatic DCC/DEG-1 (graph 3 in Figure 3) differs from the

TABLE 5: Effect of blocking of urethane groups by LiCl on the epoxy polymer properties.

Composition	Properties of polymers				
	σ_p (MPa)	ε_p (%)	τ_B (MPa)	$\sigma_{p.o}$ (MPa)	T_c (°C)
ED-20 + DETA with 20% DCC/DEG-1	88.3	4.4	22.8	50.0	68
ED-20 + DETA with 20% DCC/DEG-1 and 4% LiCl	62.4	5.1	20.5	32.2	38

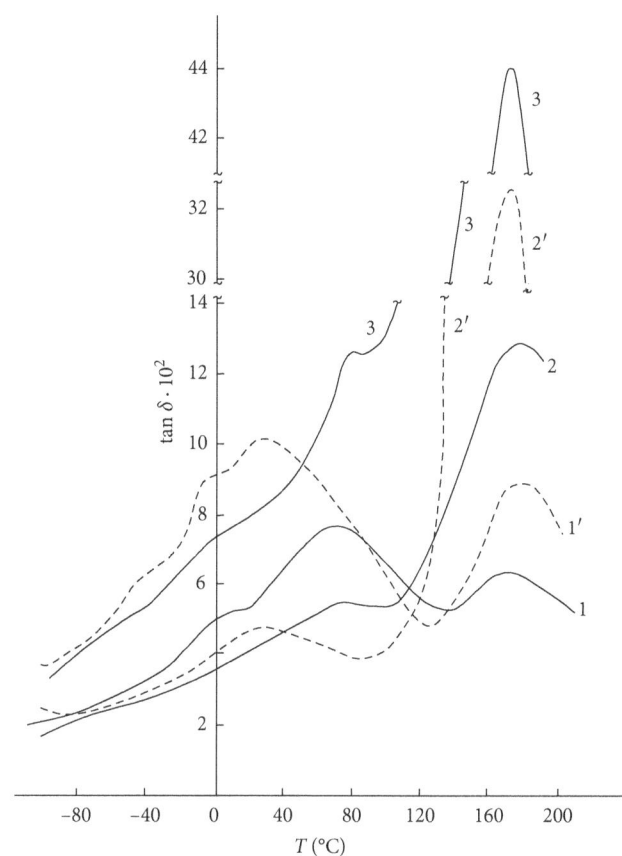

FIGURE 3: The temperature dependence of the dielectric loss angle tangents at a frequency of 10^6 Hz (1–3) and 10^5 Hz (1′, 2′) for polymers obtained on the basis of the modified ED-20 + DETA (1, 1′), modified 20% DCC/ED-20 (2, 2′), and 20% of DCC/DEG-1 (3).

considered variant in the presence of mobile diethylene glycol units in the polymer structure, increasing its molecular mobility in both glassy and highly elastic states.

A characteristic feature of most epoxy-amine compositions is the two-component decrease in magnetization during their curing, with the appearance of the transverse relaxation times T_{2a} and T_{2b} (Figure 4) in the initial stage of the induction period. The isolated relaxation times T_{2a} and T_{2b} decrease monotonically and are combined in one short time T_2 at a level of 10 to 20 μs during the reaction proceeding. This is characteristic for rigid polymers with frozen segmental mobility. However, the yield of T_2 values at this level does not mean the completion of the structure

formation processes, as evidenced by the high values of P_a in the range of 0.4 to 0.5, which are retained by the glass transition of the polymers (Figures 4 and 5).

High values of P_a indicate the intermolecular interactions enhancement, as well as the molecular mobility limitation. Analyzing the data of NMR spectroscopy, some peculiarities for polymer system behavior (before gelation) should be noted. First, the induction period decreases, and at 20% to 30% of the OECC, it is already absent, which can be explained by the reaction acceleration due to the realization of the OECC catalytic aminoalkylation reaction. Secondly, the time for achieving the glassy state is shortened and the rate of in the dynamic rigidity increase for the system is raised. This is apparently not only the catalytic process consequence, but also a consequence of polarity increase in the polymer chains due to the formation of urethane groups. A similar picture was observed in other epoxyurethane systems. Thirdly, the values of P_a significantly decrease up to 0.30 (Figure 4).

A comparison on the transverse relaxation times of ultimately cured polymers and temperature dependence data shows that unmodified epoxide-amine polymers have only one time T_2 in a wide temperature range form +22 to +200°C. The appearance of the time T_{2a} is obviously associated with the formation of hydroxyurethane fragments. The molecular mobility changes with increasing temperature (transition to a highly elastic state) in accordance with the polymer structure, for example, the beginning and completion of the increase in T_2 level for unmodified (Figure 6, graph 1) and modified with aromatic OECC (Figure 6, graph 2) polymers differ from polymers modified with aliphatic OECC (Figure 6, graph 3). Moreover, in the case of modification with aliphatic OECC, the time T_{2a} was detected much earlier (at 120°C), and the changes in the levels of T_{2a} and T_{2b} occur simultaneously. This is typical for a non-uniform structure, but a uniform polymer network. Obviously, the time T_{2a} corresponds to elongated internode chains containing hydroxyurethane fragments. The molecular motion in them is initiated by the β-relaxation process by "loosening" the hydrogen bonds formed by urethane and hydroxyl groups.

5. Discussion

The formation processes study of the developed system for the hydrogen bonds (network of physical bonds), a decrease in molecular mobility in the glassy state and an increase in the highly elastic state, allows to understand the causes of the change in the relaxation and physicomechanical properties of polymers due to the "discharge" of the chemical bonds network by the modification of hard epoxy-amine compositions by oligoethercarbonates, containing aromatic and aliphatic links. The choice of the modifying by OECC and the change in the ratio of components allow influencing the frequency of cross-linking, the polarity, the flexibility of fragments and chains and, as a consequence, the rigidity of epoxy polymers and adhesives. For example, the level of polymer hardness modified by aliphatic OECC (Table 4) is much lower ($M_2 = 1.62\,E^2$) than unmodified or modified by

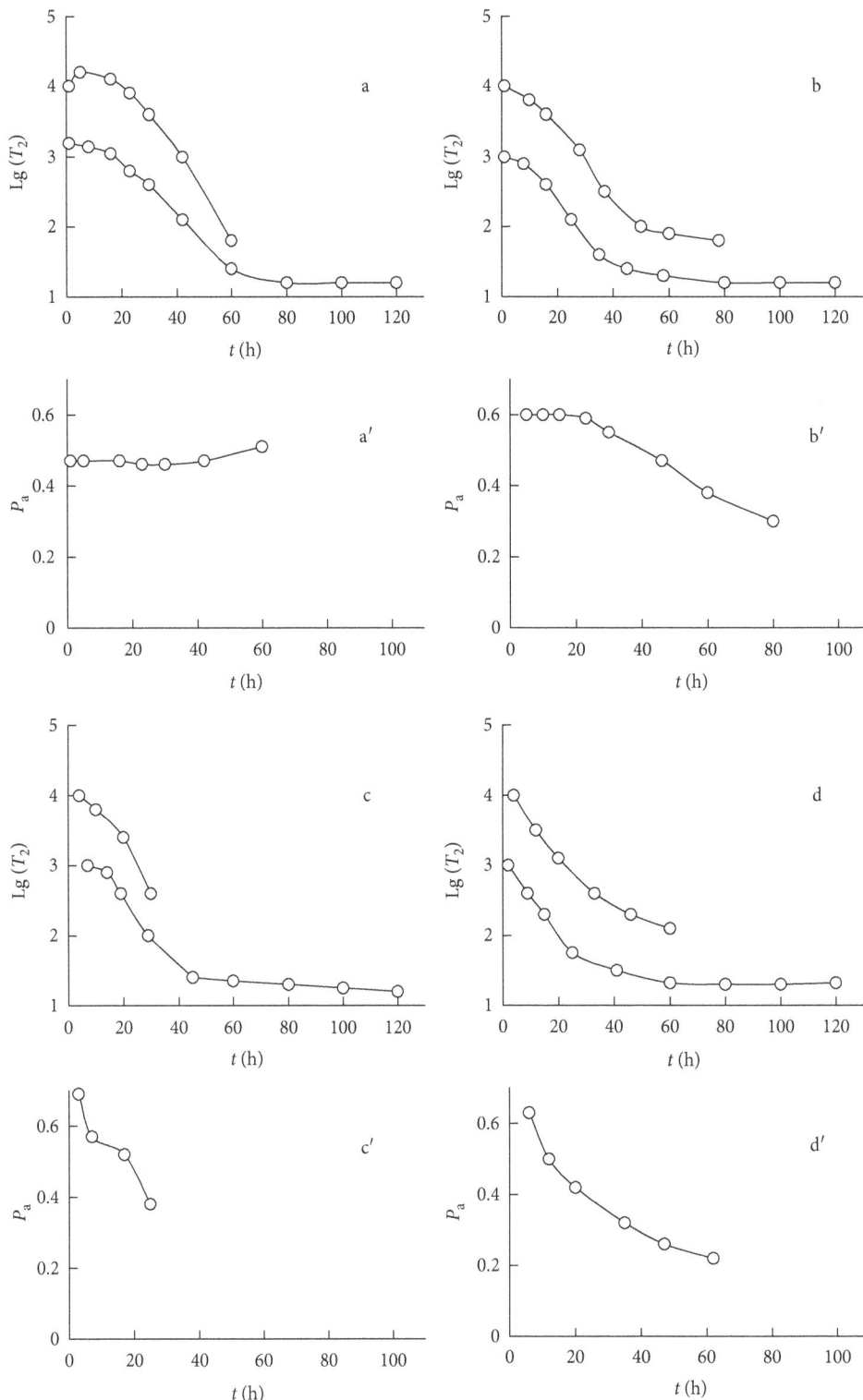

FIGURE 4: Change in the transverse relaxation times (a–d) and the protons population of the mobile "phase" (a–d) during the curing of ED-20 + DCC/DEG-1 + DETA with the content of DCC/DEG-1: 0% (a, a'), 10% (b, b'), 20% (c, c'), and 50% (d, d').

aromatic OECC. When the polymer is postcured, the combined effect of urethane groups is realized: the manifestation and contribution of physical bond network (reduced mobility) and flexible diethylenic fragments (increased mobility).

The observed changes are evidently due to the rotation of hydrogen-bonded hydroxyurethane fragments during defreezing of the aromatic nuclei movements (in the α-relaxation process). This is manifested in an intensity increase of the α-relaxation transition peak (Figure 3, graph 2, 2') and the

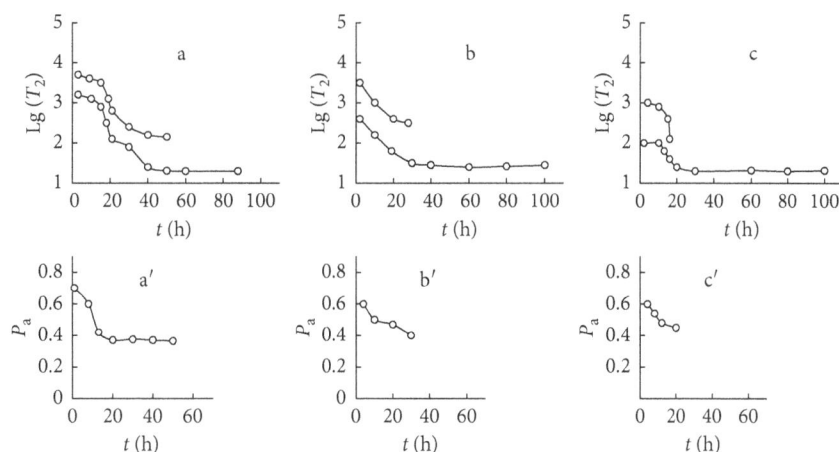

FIGURE 5: Change in the transverse relaxation times (a–c) and the population of the mobile "phase" protons (a′–c′) during the curing of ED-20 + DCC/ED-20 + DETA composition with a content of DCC/ED-20: 10% (a, a′), 20% (b, b′) and 30% (c, c′).

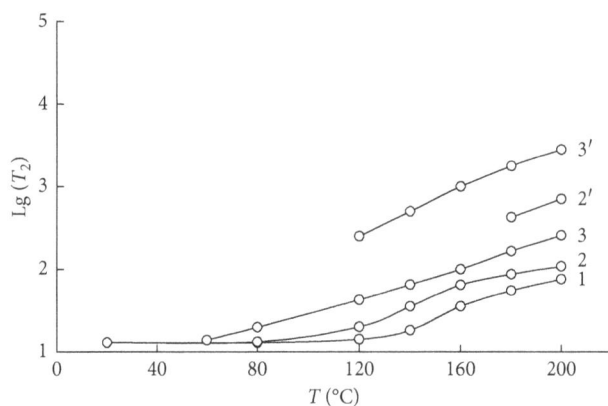

FIGURE 6: Temperature dependences of transverse relaxation times for polymers, based on ED-20 + DETA (1), ED-20 + DCC/ED-20 + DETA (2, 2′) ED-20 + DCC/DEG-1 + DETA (3, 3′) 20% OECC.

TABLE 6: Mechanical properties of polymers based on epoxy compositions modified by OECC and cured DETA.

OECC type	σ_p (MPa)/ε_p (%) of polymers by ratio OE : OECC				
	100 : 0	90 : 10	80 : 20	70 : 30	60 : 40
Aromatic	72.5/2.5	42.0/0.8	20.0/0.3	8.0/—	—
Aliphatic	72.5/2.5	82.5/2.3	88.3/4.4	77.0/2.8	55.0/5.0

appearance of the second (longer) NMR relaxation time at $T > T_c$ (Figure 6). The introduction of aromatic DCC/ED-20 (solid and as evidenced by the presence on the wide-angle X-ray diffractogram of only amorphous halos, amorphous product) leads to a sharp reduction in the deformation reserves of the polymer (despite the reduction in the cross-linking frequency), which is accompanied by a drop in the physicomechanical characteristics of the polymer: $\sigma_p = 20.0$ MPa, and $\varepsilon_p = 0.3\%$ (Table 6), that is, significantly lower than for the unmodified ED-20 + DETA (Table 6, Figure 7(a)) with $\sigma_p = 75.2$ MPa and $\varepsilon_p = 2.5\%$. A similar manifestation of macroscopic properties is observed when aliphatic OECC are used in the concentration range from 15 to 30% (Figures 7–9).

The widely used in practice epoxy composition modification with aliphatic epoxy oligomers (e.g., DEG-1) is less effective. Comparison of the absolute indicators level in Figures 7(b) and 10 clearly demonstrates a more significant contribution of the urethane component in the application of aliphatic OECC. The use of different OECC structures (DCC/DPG, DCC/COC, and DCC/E-181) gives similar dependences in physicomechanical properties,

which differ in the positions of the maxima (Figures 8, 9, and 11).

A number of examples on the practical application convincingly confirm the OECC modification effectiveness of epoxy-amine compositions and the perspectives of their application in solving a number of problems in polymer materials science. For example, for technologies of adhesive bonding parts with large tolerances, in honeycomb structures, lightweight products, and so on, operated in the temperature range from −150 to +200°C, fast-setting foam-adhesives have been developed. These adhesives have a relative low density (0.45 g/cm³) and higher strength (1.5 times) and adhesion (2 times) compared to the known foam-adhesives VK-9V and CW2513, HM, and DY050 (manufacturer Ciba Geigy).

A low viscosity composition based on a mixture of aliphatic and aromatic EO, aliphatic OECC, and a mixture of amine- curing agent for the reinforced concrete structures repair was developed. Due to the elimination of the selective sorption effect for the composition components, it was possible to ensure the reliability of products (water pipes with a diameter of 2000 mm and a length of 6000 mm), which is evaluated under hydraulic tests at a pressure of 1.0 MPa. This composition combines low viscosity (0.6 MPa·s) with high adhesion and deformation characteristics: for steel and glass-ceramic up to 27.0 MPa (concrete breaks at lower loads), σ_p up to 50 MPa, $\varepsilon_p = 5\%$, which is comparable or superior to the analog Araldite K-79 Kit (manufacturer Ciba Geigy). A number of the "Vicor-UP"-type compositions have been developed for corrosion protection of chemical equipment operating under conditions of 5 to 30% mineral acids solutions (hydrochloric, sulfuric, and phosphoric acids) at

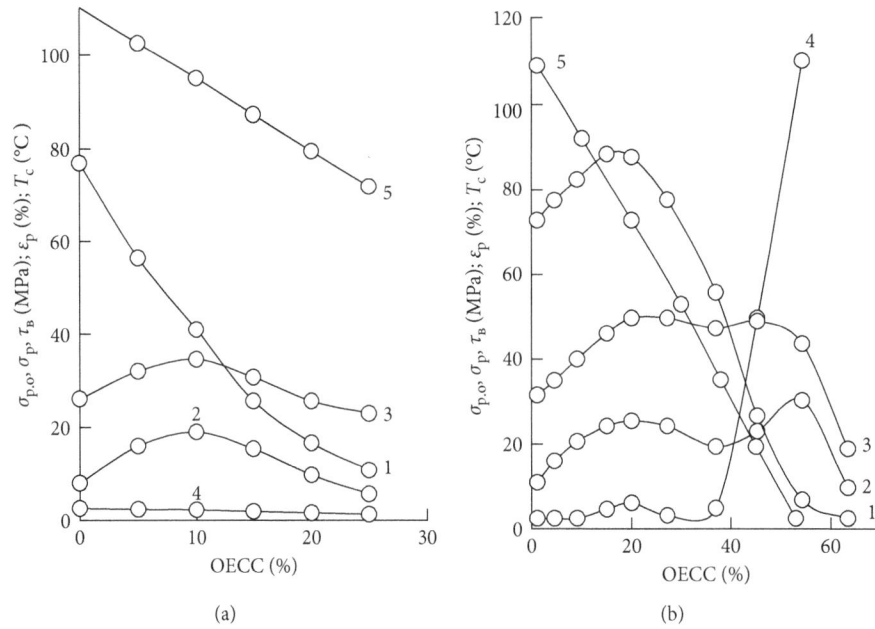

FIGURE 7: Dependence of the epoxyurethane polymer properties on the basis of ED-20 + DCC/ED-20 + DETA (a) and ED-20 + DCC/DEG-1 + DETA (b) on the OECC modifying concentration, σ_p (1), τ_B (2), $\sigma_{p.o}$ (3), E_p (4), and T_c (5).

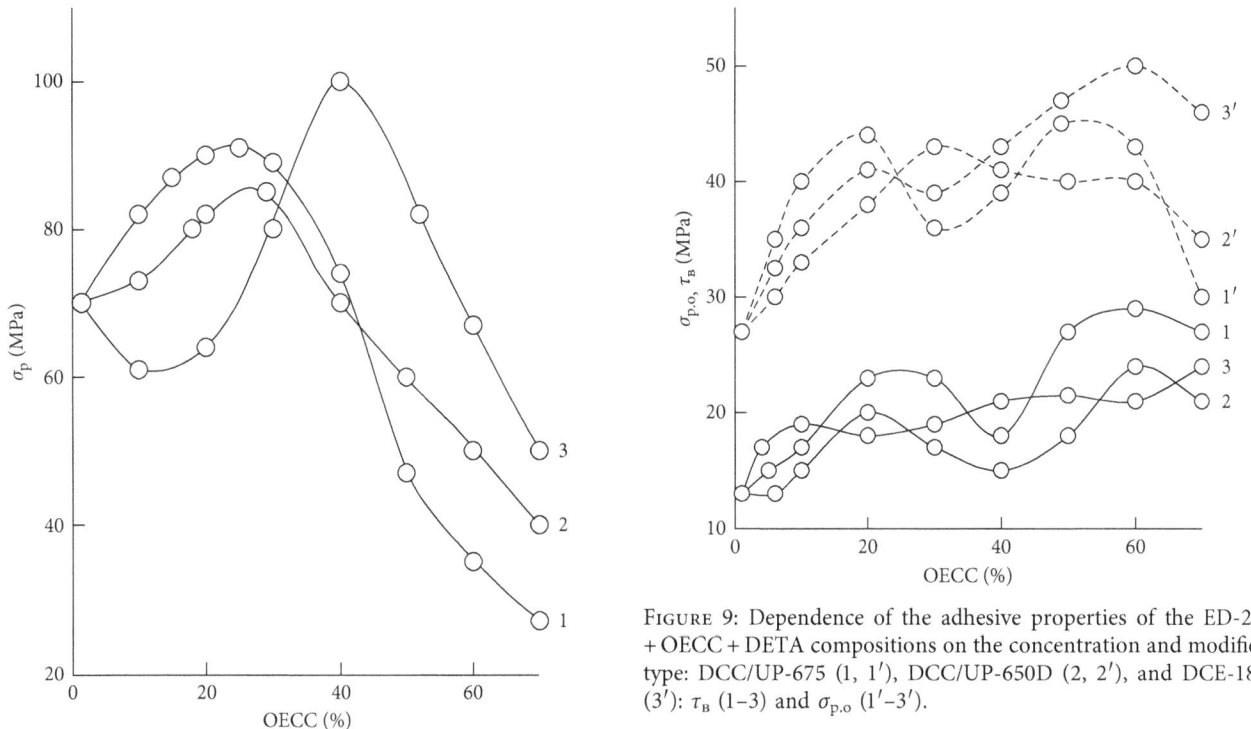

FIGURE 8: Dependence of the tensile strength of polymers obtained on the basis of ED-20 + OECC + DETA compositions, on the concentration and modifier type DCC/UP-675 (1), DCC/UP-650D (2), and DCC-181 (3).

FIGURE 9: Dependence of the adhesive properties of the ED-20 + OECC + DETA compositions on the concentration and modifier type: DCC/UP-675 (1, 1′), DCC/UP-650D (2, 2′), and DCE-181 (3′): τ_B (1–3) and $\sigma_{p.o}$ (1′–3′).

products have also been developed. These applications testify the wide possibilities for OECC as modifiers in epoxy-amine compositions in various technologies and prospects for the research and development in this direction.

6. Conclusions

One of the promising directions of epoxy-amine network polymers, in order to eliminate their "hardness," is the

+120°C, as well as for cold-drying technology. The composition and technology of polymer-sand mandrels obtaining with an increased (by 1.5 times) strength, by reducing thickness and mass, in the technology of manufacturing

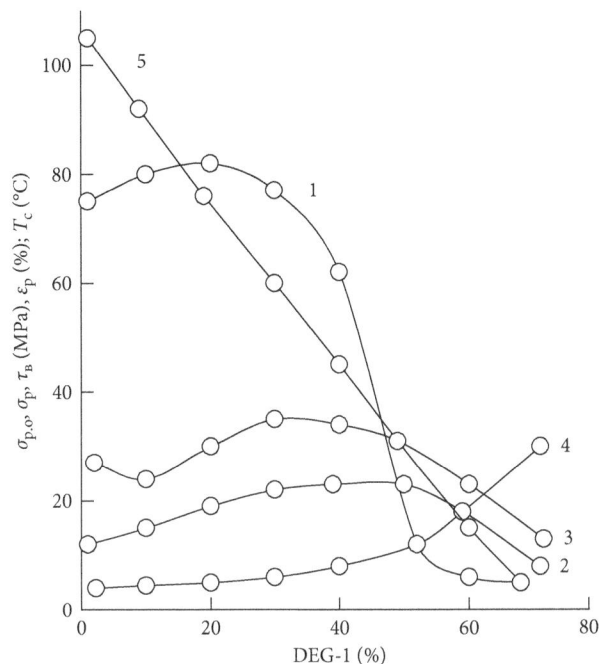

FIGURE 10: The effect of the DEG-1 content in the composition of ED-20 + DEG-1 + DETA on the properties of polymers: σ_p (1), τ_B (2), $\sigma_{p.o}$ (3), ε_p (4), and T_c (5).

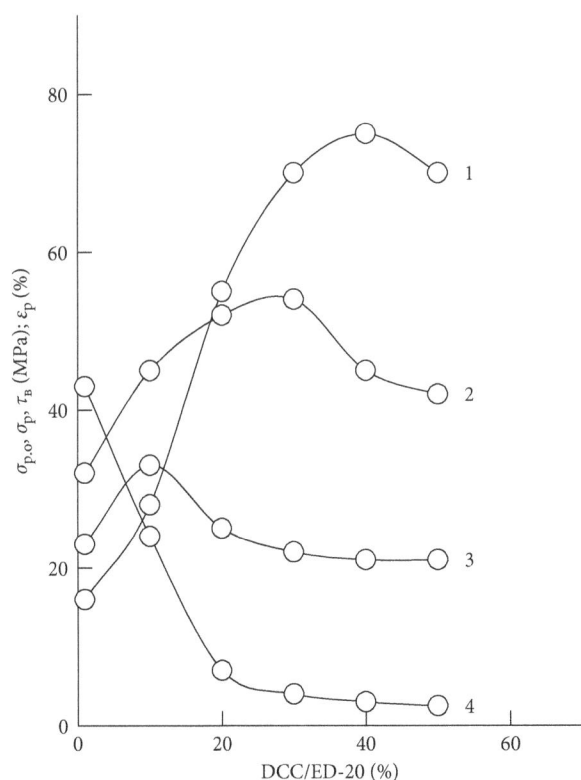

FIGURE 11: Dependence of the mechanical and adhesion properties of polymers obtained on the basis of the E-181 + DCC/ED-20 + DETA compositions on the modifier concentration: σ_p (1), τ_B (2), $\sigma_{p.o}$ (3), and ε_p (4).

preparation of oligoethercyclocarbonates (OECC) and their use in joint curing with epoxy oligomers. The resulting polymers contain in the network structure additional linear hydroxyurethane fragments. These "relaxators" are capable of manifesting intermolecular interactions that affect the relaxation properties and molecular mobility. The last causes an increase in adhesion and elastic-deformation characteristics and opens additional opportunities in the development of new materials and technologies in the polymer material science.

Conflicts of Interest

The authors declare that they have no conflicts of interest.

Acknowledgments

The work was carried out within the framework of State Assignment number 10.4763.2017/8.9. The authors would like to thank the companies and employees of KSUAE, KNRTU, and SKZ for the supporting work, which have made a significant contribution to the implementation of these results.

References

[1] B. Erman and J. E. Mark, *Structure and Properties of Rubberlike Networks*, Oxford University Press, New York, NY, USA, 1997.

[2] R. F. Stepro, *Polymer Networks. Principles of Their Formstion, Structure and Properties*, Springer, Luxemburg, Belgium, 1998.

[3] Y. Osada and A. R. Khokhlov, *Polymer Gels and Networks*, Marcel Dekker, New York, NY, USA, 2002.

[4] A. S. Lipatov, T. T. Alekseeva, L. A. Sorochinskaya, and G. V. Dudarenko, "Confinement effects on the kinetics of formation of sequential semi-interpenetrating polymer networks," *Polymer Bulletin*, vol. 59, no. 6, pp. 739–747, 2008.

[5] S. Goswami and D. Chakrabarty, "Sequential interpenetrating polymer networks of novolac resin and poly(n-butyl methacrylate)," *Journal of Applied Polymer Science*, vol. 102, no. 4, pp. 4030–4039, 2006.

[6] M. Patri, C. V. Reddy, C. Narasimhan, and A. B. Samui, "Sequential interpenetrating polymer network based on styrene butadiene rubber and polyalkyl methacrylates," *Journal of Applied Polymer Science*, vol. 103, no. 2, pp. 1120–1126, 2007.

[7] L. V. Karabanova, L. M. Sergeeva, and A. V. Svyatyna, "Heterogeneity of glass transition dynamics in polyurethane-poly(2-hydroxyethyl methacrylate) semi-interpenetrating polymer networks," *Journal of Polymer Science Part B: Polymer Physics*, vol. 45, no. 8, pp. 963–975, 2007.

[8] J. F. Fu, L. Y. Shi, S. Yuan, Q. D. Zhong, D. S. Zhang, and Y. Chen, "Morphology, toughness mechanism, and thermal propertiesof hyperbranched epoxy modified diglycidyl ether of bisphenol A (DGEBA) interpenetrating polymer networks," *Polymers for Advanced Technologies*, vol. 19, pp. 1597–1607, 2008.

[9] A. Martinelli, L. Tighzert, L. D'Ilario, I. Francolini, and A. Piozzi,

"Poly(vinyl acetate)/polyacrylate semi-interpenetrating polymer networks. II. Thermal, mechanical, and morphological characterization," *Journal of Applied Polymer Science*, vol. 111, no. 6, pp. 2675–2683, 2009.

[10] V. I. Irzhak and S. M. Mezhikovski, "Kinetics of oligomer curing," *Russian Chemical Reviews*, vol. 77, no. 1, pp. 77–104, 2008, in Russian.

[11] A. A. Askadski and V. I. Kondrashenko, *Computer Material Science of Polymers*, Scientific World, Moscow, Russia, 1999, in Russian.

[12] D. W. Van Krevelen and K. T. Nijenhuis, *Properties of Polymers*, Elsevier, Amsterdam, Netherlands, 2009.

[13] D. R. Wentzel and W. Oppermann, "Orientation relaxation of linear chains enclosed in a network studied by birefringence measurements," *Colloid and Polymer Science*, vol. 275, no. 3, pp. 205–213, 1997.

[14] I. T. Smith, "The mechanism of the crosslinking of epoxide resins by amines," *Polymer*, vol. 2, pp. 95–108, 1961.

[15] B. A. Rozenberg, "Epoxy resins and composites II," *Advances in Polymer Science*, vol. 75, pp. 113–165, 1986.

[16] A. M. Elyashevich, "Computer simulation of network formation processes, structure and mechanical properties of polymer networks," *Polymer*, vol. 20, no. 11, pp. 1382–1388, 1979.

[17] P. J. Flory, *Principles of Polymer Chemistry*, Cornell University Press, New York, NY, USA, 1953.

[18] V. M. Lanzov, V. F. Stroganov, and L. A. Abdrahmanova, "Interrelation of kinetic and structural-topological heterogeneity of molecules in polycondensation epoxy-amine network," *High-Molecular Compounds*, vol. 31, pp. 409–413, 1989, in Russian.

[19] V. I. Irzhak, *Architecture of Polymers*, in Russian, Science, Moscow, Russia, 2012.

[20] K. Dušek and M. Duškova-Smrckova, "Network structure formation during crosslinking of organic coating systems," *Progress in Polymer Science*, vol. 25, no. 9, pp. 1215–1260, 2000.

[21] V. I. Irzhak, "Methods of description of the polycondensation kinetics and the structures of the polymers formed," *Russian Chemical Reviews*, vol. 66, no. 6, pp. 541–552, 1997.

[22] V. Bellenger, J. Verdu, and J. Francillette, "Infra-red study of hydrogen bonding in amine-crosslinked epoxies," *Polymer*, vol. 28, no. 7, pp. 1079–1086, 1987.

[23] E. Morel, V. Bellenger, and J. Verdu, "Structure-water absorption relationships for amine-cured epoxy resins," *Polymer*, vol. 26, no. 11, pp. 1719–1724, 1985.

[24] P. J. Bell, "Mechanical properties of a glassy epoxide polymer: effect of molecular weight between crosslinks," *Journal of Applied Polymer Science*, vol. 14, no. 7, pp. 1901–1906, 1970.

[25] R. E. Cuthrell, "Macrostructure and environment-influenced surface layer in epoxy polymers," *Journal of Applied Polymer Science*, vol. 11, no. 6, pp. 949–952, 1967.

[26] T. Hirai and D. E. Kline, "Dynamic mechanical properties of nonstoichiometric, amine-cured epoxy resin," *Journal of Applied Polymer Science*, vol. 16, no. 12, pp. 3145–3157, 1972.

[27] D. M. Brewis, J. Comyn, and J. R. Fowler, "An aliphatic amine cured rubber modified epoxide adhesive: 2 further evaluation," *Polymer*, vol. 18, no. 9, pp. 951–954, 1977.

[28] J. M. Pochan, R. J. Gruber, and D. F. Pochan, "Dielectric relaxation phenomena in a series of polyhydroxyether copolymers of bisphenol-a engcopped polyethelene glycol with epichlorhydrin," *Journal of Polymer Science: Polymer Physics Edition*, vol. 19, no. 1, pp. 143–149, 1981.

[29] H. Batzer and S. A. Zahir, "Studies in the molecular weight distribution of epoxide resins. IV. Molecular weight distributions of epoxide resins made from bisphenol A and epichlorohydrin," *Journal of Applied Polymer Science*, vol. 21, no. 7, pp. 1843–1857, 1977.

[30] V. Besse, F. Camara, C. Voirin, R. Auvergne, S. Caillol, and B. Boutevin, "Synthesis and applications of unsaturated cyclocarbonates," *Polym Chem*, vol. 4, no. 17, pp. 4545–4561, 2013.

[31] V. F. Stroganov, V. N. Savchenko, and S. I. Omelchenko, *Cyclocarbonates and Their Use for the Synthesis of Polymers*, Institute of Technical and Economic Research, Moscow, Russia, 1984, in Russian.

[32] A. C. Lind, "An NMR study of inhomogeneities in epoxy resins," *American Chemical Society, Division of Polymer Chemistry*, vol. 21, pp. 241-242, 1980.

[33] D. W. Larsen and J. H. Strange, "Diglycidyl ether of bisphenol-A with 4,4'-methylenedianiline: a pulsed NMR study of the curing process," *Journal of Polymer Science Part A-2: Polymer Physics*, vol. 11, no. 7, pp. 1453–1459, 1973.

[34] T. I. Kadurina, V. A. Prokopenko, and S. I. Omelchenko, "Curing of epoxy oligomers by isocyanates," *Polymer*, vol. 33, no. 18, pp. 3858–3864, 1992.

[35] Z. S. Petrović, Z. Zavargo, J. H. Flyn, and W. J. Macknight, "Thermal degradation of segmented polyurethanes," *Journal of Applied Polymer Science*, vol. 51, no. 6, pp. 1087–1095, 1994.

[36] A. D. Wicks and Z. W. Wicks, "Blocked isocyanates III: part B: uses and applications of blocked isocyanates," *Progress in Organic Coatings*, vol. 41, no. 1–3, pp. 1–83, 2001.

[37] J. Guan, Y. Song, Y. Lin et al., "Progress in study of non-isocyanate polyurethane," *Industrial and Engineering Chemistry Research*, vol. 50, no. 11, pp. 6517–6527, 2011.

[38] W. Zhijun, C. Wang, C. Ronghua, and Q. Jinqing, "Synthesis and properties of ambient-curable non-isocyanate polyurethanes," *Progress in Organic Coatings*, vol. 119, pp. 116–122, 2018.

[39] M. A. Levina, V. G. Krasheninnikov, and M. V. Zabalov, "Nonisocyanate polyurethanes from amines and cyclic carbonates: kinetics and mechanism of a model reaction," *Polymer Science Series B*, vol. 56, no. 2, pp. 139–147, 2014.

[40] V. F. Stroganov and I. V. Stroganov, "Peculiarities of structurization and properties of nonisocyanate epoxyurethane polymers," *Polymer Science Series C*, vol. 49, no. 3, pp. 258–263, 2007.

[41] J. Tabushi and R. Oda, "Kinetic study of the reaction of ethylene carbonate and amines," *Nippon Kagaki Zasshi*, vol. 84, no. 2, pp. 162–167, 1963.

[42] V. F. Stroganov, V. N. Savchenko, and G. D. Tizkij, "Aminolysis of 1-phenoxy-2,3-propylene carbonate benzylamine in chlorobenzene," *Journal of Organic Chemistry*, vol. 24, pp. 501–504, 1988, in Russian.

[43] Y. Smirnov, B. Komarov, P. Kushch, T. Ponomareva, and V. Lantsov, "Structural and kinetic features of formation of high-strength epoxy-amine cross-linked polymers by combined polycondensation-polymerization process," *Russian Journal of Applied Chemistry*, vol. 75, no. 2, pp. 265–275, 2002.

[44] Y. S. Lipatov, "Interfacial regions in the phase-separated interpenetrating networks," *Polymer Bulletin*, vol. 58, no. 1, pp. 105–118, 2007.

[45] Y. S. Lipatov, R. A. Veselovsky, and Y. K. Znachkov, "Some properties of glues based on interpenetrationg polymeris networks," *Journal of Adhesion*, vol. 10, no. 2, pp. 157–161, 1979.

[46] V. F. Stroganov, V. M. Mihalchuk, and V. M. Lanzov, "Study of molecular mobility during the curing of diphenylolpropane-1,3-bis (aminomethyl) adamant digymondyl ether system," *Russian Academy of Sciences*, vol. 291, pp. 908–912, 1986, in Russian.

[47] V. A. Bershtein, N. N. Peschanskaya, J. L. Halary, and L. Monnerie, "The sub-Tg relaxations in pure and anti-plasticized model epoxy networks as studied by high resolution creep rate spectroscopy," *Polymer*, vol. 40, no. 24, pp. 6687–6698, 1999.

18

Microstructure and Damping Property of Polyurethane Composites Hybridized with Ultraviolet Absorbents

Jiang Chang,[1,2] Bing Tian (iD),[1] Li Li,[1] and Yufeng Zheng[1]

[1]Institute of Materials Processing and Intelligent Manufacturing, College of Materials Science and Chemical Engineering, Harbin Engineering University, Harbin 150001, China
[2]College of Light Industry and Textile Engineering, Qiqihar University, Qiqihar 161006, China

Correspondence should be addressed to Bing Tian; tianbing@hrbeu.edu.cn

Academic Editor: Marino Lavorgna

This article investigated the microstructure and damping property of TPU composites with different contents of ultraviolet absorbents. It can be found that the ultraviolet absorbents formed fiber-shaped precipitations in the TPU matrix. The UV-328 was randomly distributed in the matrix and exhibited a weak interfacial bonding with the matrix. In comparison, the UV-329 was well embedded in the matrix and formed a relatively better interfacial bonding and compatibility with the TPU matrix. The damping factor tanδ of both 328-composites and 329-composites had been reduced gradually with increasing content of ultraviolet absorbents at the glass transition temperature range due to the fact that the ultraviolet absorbents were in the crystalline state which decreased the volume content of the viscoelastic TPU matrix. But the tanδ increased at the temperature range of higher than the glass transition temperature, which should be related to the dominance of interfacial frictions between the ultraviolet absorbents and the matrix on the energy absorption.

1. Introduction

Polyurethane has been widely investigated in the past few years due to its comprehensive properties of high hydrolysis resistance, wear resistance, good flexibility, and damping ability [1–4]. The high damping property of polymers is mainly observed during glass transition process, in which the mechanical energy caused by vibrational motion of molecular chains is transferred to heat energy. Beyond this range, the polymer often presents a low damping property. Therefore, the application for polymers is often restricted to a narrow temperature range associated with glass transition. Generally, low elastic modulus of the polymers also restricts their engineering applications to some extent. Thus, properly extending the damping range and enhancing the modulus of polymers are meaningful for the practical applications. Several methods, such as mechanical blending [5], interpenetrating polymer networks (IPNs) [6, 7] and forming organic hybrid composites composed of polar polymers and functional organic small molecules [8], have been proposed and investigated to improve the damping ability of the polymers. In comparison, the last method of forming intermolecular hydrogen bonds to improve the damping properties has presented a higher cost-effective prospect. The DZ (N,N′-dicyclohexyl-benzothiazole-2-sulfonamide), AO-80 (3,9-bis{1,1-dimethyl-2[β(3-tert-butyl-4-hydroxy-5-methylphenyl)propionyloxy]ethyl}-2,4,8,10-tetraoxaspiro [5,5]-undecane), AO-60 (tetrakis[methylene-3-(3-5-di-tert-butyl-4-hydroxy phenyl) propionyloxy] methane), etc. hindered phenol have been selected as the functional organic small molecules to be added in the chlorinated polyethylene matrix [8–10], and an apparent improvement of the damping factor of the polymer matrix has been observed. However, little information on the effect of organic hybridizing on the damping property of polyurethane is available. In this article, the ultraviolet absorbent with functional organic small molecules has been added into the thermoplastic polyurethanes (TPUs) to investigate the effect of organic hybridizing on the microstructure and damping property of the TPU.

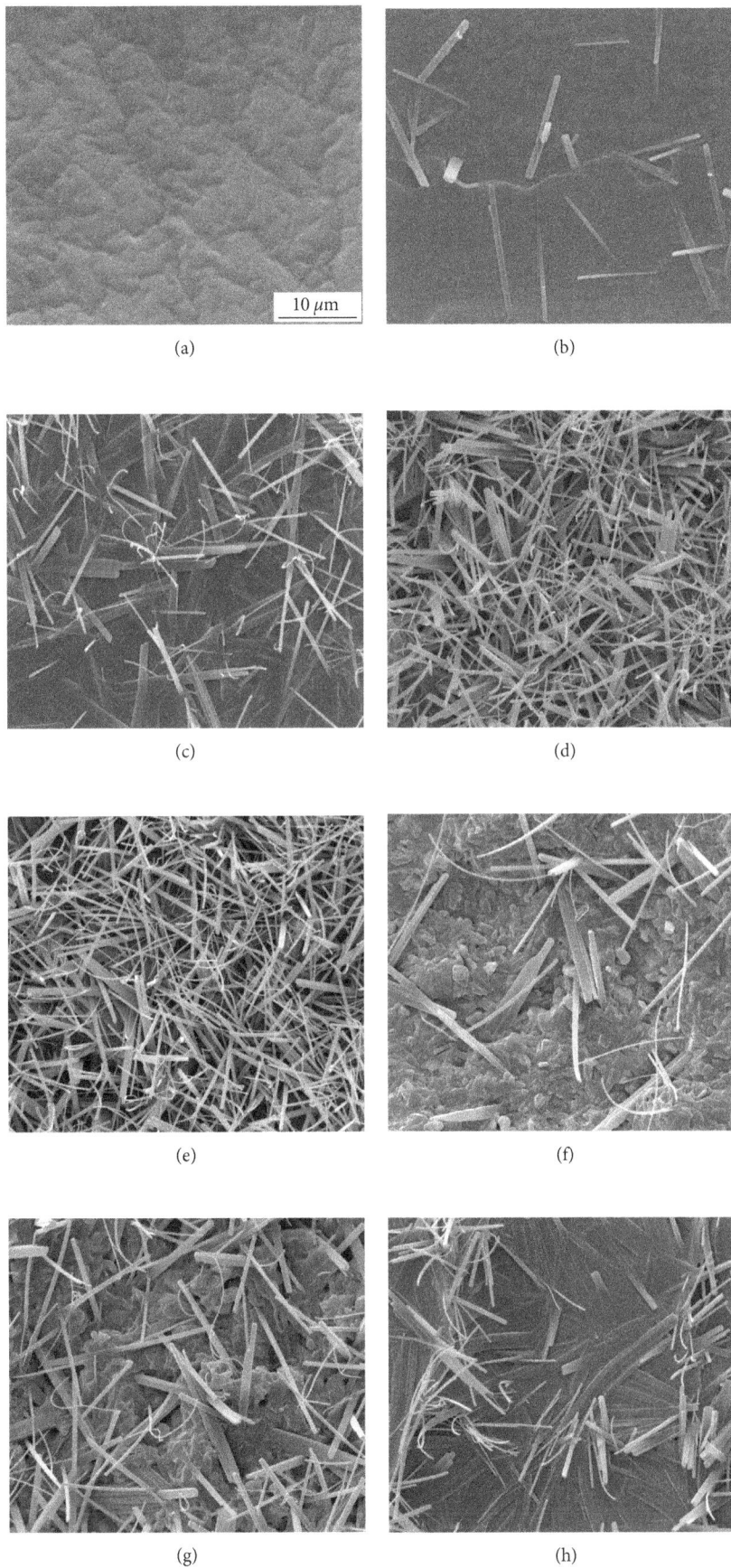

FIGURE 1: SEM images of fracture surfaces of (a) TPU and 328-composites with different mass ratios of 328: (b) 10%, (c) 20%, (d) 25%, (e) 30%, (f) 40%, (g) 50%, and (h) 60%.

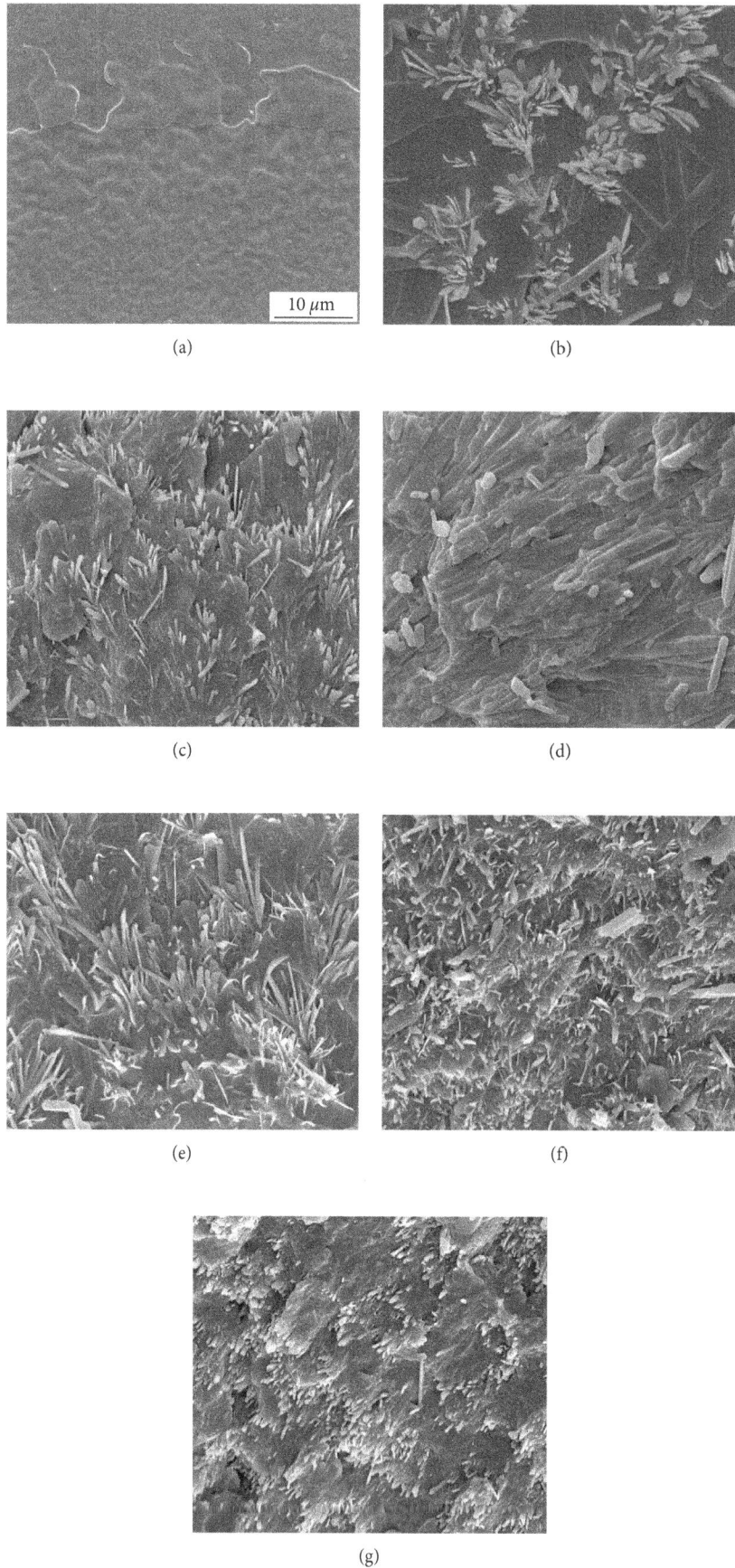

FIGURE 2: SEM images of fracture surfaces of 329-composites with different mass ratios of 329: (a) 10%, (b) 20%, (c) 25%, (d) 30%, (e) 40%, (f) 50%, and (g) 60%.

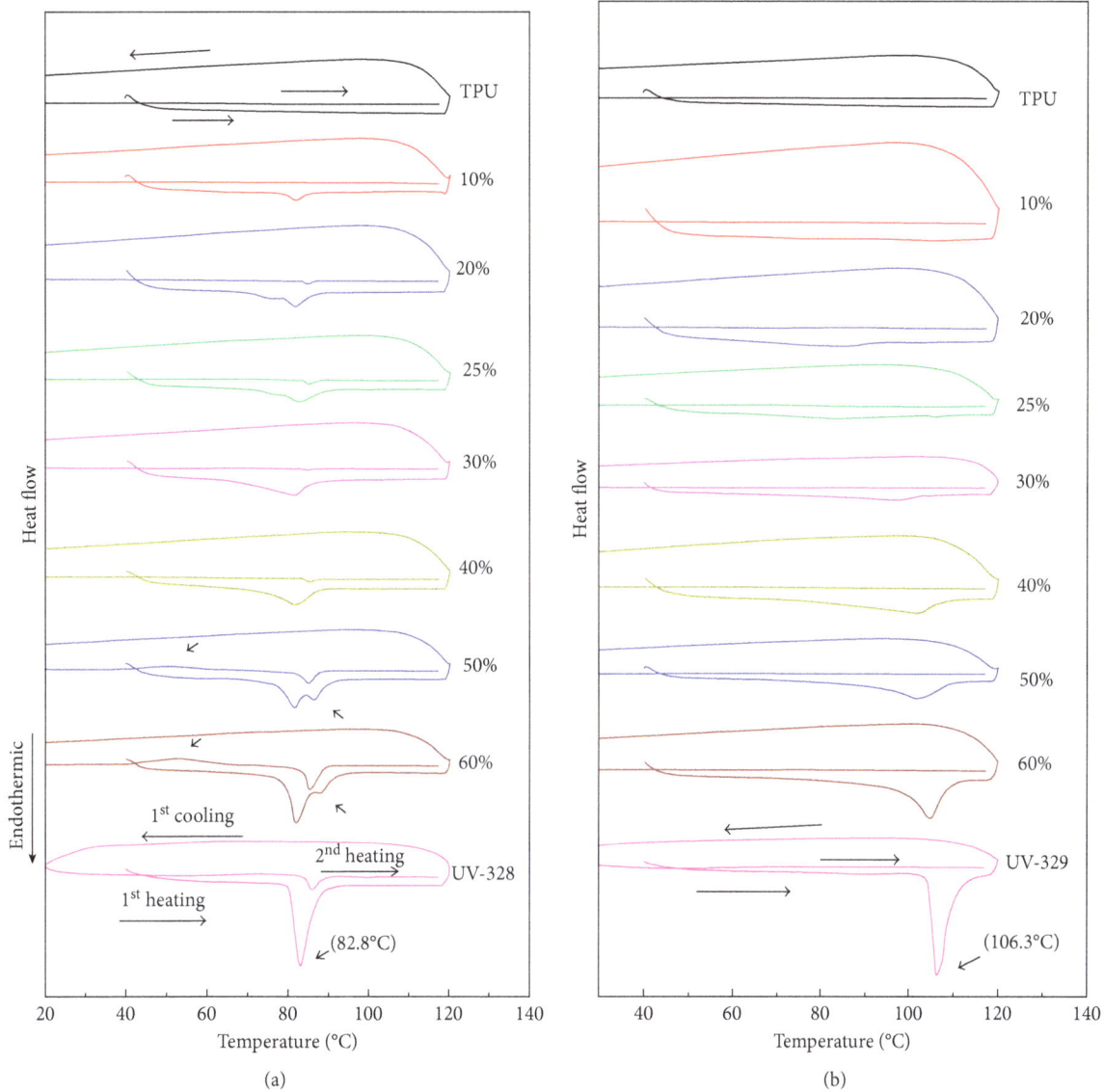

Figure 3: DSC curves of (a) 328-composites and (b) 329-composites with different mass ratios of UV-absorbents.

2. Experimental Details

The ultraviolet (UV) absorbents 2-(2'-hydroxy-3',5'-di-tert-amylphenyl) benzotriazole UV-328 ($C_{22}H_{29}N_3O$) and 2-(2'-hydroxy-5'-tert-octylphenyl) benzotriazole UV-329 ($C_{20}H_{25}N_3O$) (powders supplied by Nippon Kasei Chemical Co., Ltd., Japan) were selected as the functional organic small molecules to be added into the WHT-8185 polyether thermoplastic polyurethane (TPU, supplied by Yantai wanhua polyurethanes Co., Ltd., China), respectively, with a mass ratio of 10–60%. The melting point of UV-328 and UV-329 is at 80–83°C and 101–106°C, respectively. Both UV-328 and UV-329 belong to benzotriazole-type chemicals, and the difference between them is the different substituent adjacent to the hydroxyl group. The typical parameters of WHT-8185 obtained from the manufacturer are as follows: glass transition temperature: ~−45°C, tensile strength: 26 MPa, elongation: 500%, hardness: 85 (shore A), and processing temperature: 180–195°C. Firstly, the TPU particles were dissolved in the solvent mixture with a volume ratio of 1 : 1 of N, N-dimethylformamide (DMF) and tetrahydrofuran (THF) and heated in a vacuum oven at 100°C for 40 min to accelerate dissolution. Then, the UV-absorbents were dissolved into the above mixture and ultrasonically dispersed at 35°C for 0.5 h. Finally, the mixture solution was poured into a mold and then dried in a vacuum oven to fully evaporate the dissolvent of DMF and THF to obtain the UV-absorbent/TPU composite thin films. For simplicity, thereafter, the composites containing UV-328 and UV-329 absorbents were denoted as 328-composites and 329-composites, respectively.

The damping curves were recorded by a dynamic mechanical analyzer (DMA$^+$450) under a tension mode, in which the frequency was 10 Hz, the vibration amplitude was 5 μm, and the heating rate was 3°C min^{-1}. The dimension of samples for DMA testing was ~30 × 8 × 0.3 mm^3. DSC

(a)

(b)

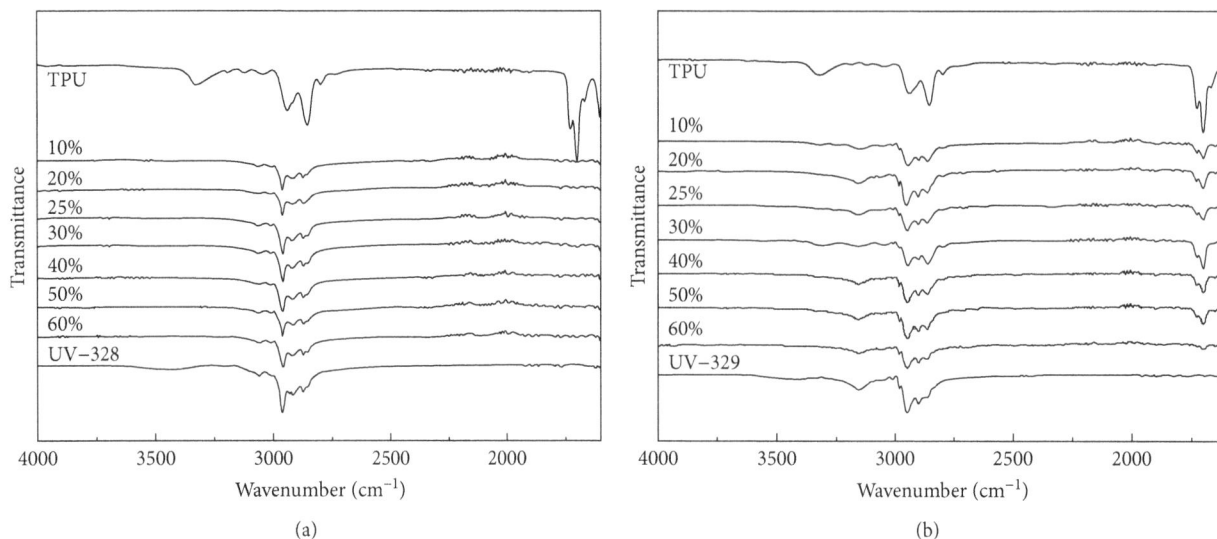

FIGURE 4: FTIR spectra of (a) 328-composites and (b) 329-composites with different mass ratios of UV-absorbents.

measurements were performed in a TA-Q200 apparatus with a heating/cooling rate of 20°C min⁻¹. FTIR test was performed in a Perkin-Elmer Spectrum-100 fourier transform infrared spectrometer. IR spectrum was obtained by scanning 32 times at the resolution of 4 cm⁻¹ within the wavenumber range of 450–4000 cm⁻¹. IR spectrum of the UV-absorbent powders was obtained using the KBr pallet method, and IR spectrum of UV-absorbent/TPU composite films was obtained using the attenuated total reflection (ATR) method. Microstructure of the samples was observed using FEI Quanta200 scanning electron microscope (SEM).

3. Results and Discussion

The TPU composites exhibit a very large plasticity at room temperature and are difficult to be fractured, so the composites are bent to fracture under liquid nitrogen to observe microstructure of the samples. Figure 1 shows fracture surface of the composites with different mass ratios of 328 UV-absorbents. All the micrographs have the same magnification to that of Figure 1(a). It can be seen that the pure TPU presents a flat fracture surface with a single phase, as shown in Figure 1(a). After adding 328 for 10% (Figure 1(b)), the composite is found to be composed of two phases, including a flat matrix and some short fibers that should be 328 absorbents precipitated from the solution mixture after removing the solvents. With increasing 328 from 20% to 30% (Figures 1(c)–1(e)), it is seen that the fiber volume has been greatly increased and the matrix was nearly fully covered and cannot be seen on the surface. However, continue increasing 328 to 40%–60% (Figures 1(f)–1(h)), it seems that the 328 fiber volume is decreased and the TPU matrix can be found, which could be caused by the fact that the distribution of 328 fibers is not homogeneous in the composites and these composites mainly fractured on the areas with few 328 fibers.

For the 329-composites, as shown in Figure 2, it is different from that of the 328-composites. For the 10%

composite (Figure 2(a)), there is no 329 phases observed on the surface. For the composites with 20% 329, the petaloid 329 phases can be found, as shown in Figure 2(b). Continue increasing 329 from 25% to 60% (Figures 2(c)–2(g)), the fracture surface becomes rougher gradually as compared to the pure TPU, but there is no apparent two-phase separation like 328-composites observed on the surface and the 329 phases seem to be mostly embedded in the TPU matrix. The above microstructure results demonstrate that the UV-329 should have a better compatibility with the TPU matrix than the UV-328.

Figures 3(a) and 3(b) show the DSC results of the 328-composites and 329-composites, respectively. All the samples are subjected to a thermal cycle of heating, cooling, and reheating. As shown in Figure 3(a), it can be seen that there is no any thermal phenomenon observed for the pure TPU at this thermal cycle. For the pure 328, an endothermic peak centered at around 82.8°C during first heating process can be found, which should stand for the melting of the UV-328. Upon cooling, there is no thermal event observed, but one much weaker endothermic peak than that of the first heating at similar temperature is observed during the second heating. The occurrence of the weak endothermic peak means the melting of some crystalline 328 phases that should recrystallize during the cooling process although the thermal event associated with it is not detected on cooling. For the composites with 328 of >10%, these two endothermic peaks associated with the melting of 328 phases are also observed, and these two peaks become stronger gradually with increasing 328 content. In addition, it is noted that, for the 50% and 60% composites, except for the abovementioned two endothermic peaks, an extra endothermic peak during first heating and an exothermic peak on the second heating are observed, as indicated by arrows. From Figure 1, it can be seen that the 328 fibers mainly have two forms in the composites, one is "free form" that has almost no interaction with the TPU matrix (Figures 1(d) and 1(e)), and the other is "embedding form" that has some interfacial bonding with the matrix

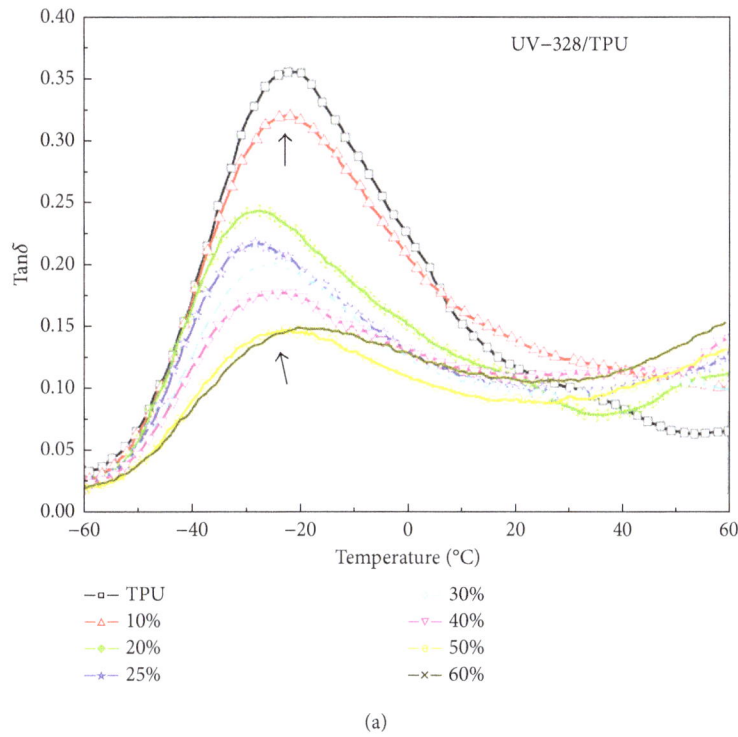

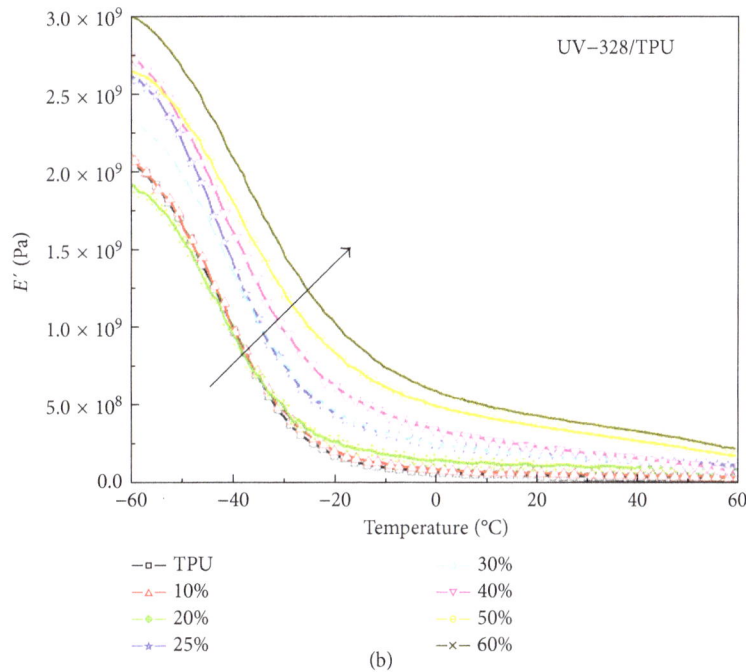

FIGURE 5: Temperature dependence of (a) tanδ and (b) E' for the 328-composites.

(Figures 1(c) and 1(h)). Therefore, during the first heating, the second endothermic peak indicated by the arrow should be related to the melting of the UV fiber with "embedding form" and the exothermic peak indicated by the arrow on the second heating means the debonding process between the "embedding form" UV fiber and the TPU matrix.

For the 329-composites, it is a little different from that of the 328-composites, as shown in Figure 3(b). Only one endothermic peak centered at 106.3°C associated with

melting of 329 phases is observed during the first heating process and no endothermic peak is found on the second heating, meaning that the 329 phase is also in the crystalline state and no recrystallization happened on subsequent cooling. In addition, it is found that, unlike 328-composite, the melting peak on the first heating process for the 329-composite with 329 content of <30% cannot be found. The 329-composite exhibits a relatively wide melting peak with low latent heat as compared to the 328-composite.

(a)

(b)

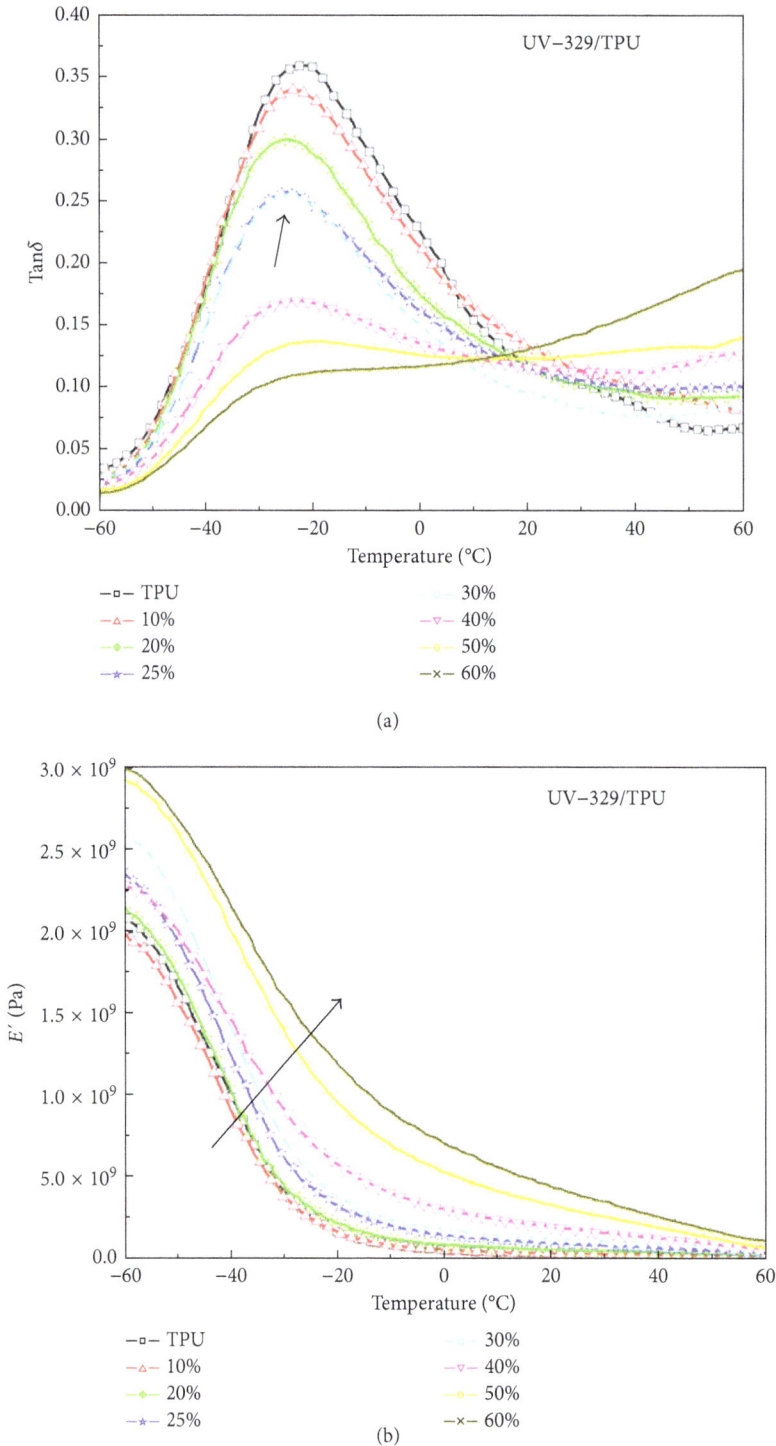

FIGURE 6: Temperature dependence of (a) tanδ and (b) E' for the 329-composites.

Therefore, the absence of melting peaks should be caused by the low melting latent heat of the 329-composite with 329 content of <30%. The above results demonstrate that the 328/329 UV absorbents in the as-prepared composites are mainly in the crystalline state at room temperature.

Figures 4(a) and 4(b) show the infrared spectra of the 328-composites and 329-composites, respectively. It can be seen that, for 328-composites, they are mainly reflected by

the absorption peaks that belong to the UV-328, as shown in Figure 4(a). With increasing the content of 328, there is no apparent change of the spectra. This should be caused by the fact that a large amount of 328 phases formed on the sample surface that covered the TPU matrix (Figure 1) and resulted in the disappearance of the absorption peaks for the matrix. In contrast, for the 329-composites, they are mainly composed of the absorption peaks of 329 and TPU matrix, which is also in agreement with the SEM results shown in Figure 2, in which the 329 phases is well embedded in the TPU matrix. However, the intermolecular hydrogen bonding observed in the organic hybrid materials [8–10] was not detected in the present composites. The undetected intermolecular hydrogen bonding for both composites could result from two reasons: one is the crystalline UV absorbents that are not favorable for the formation of hydrogen bonding with TPU matrix, and the other is the less functional group in these UV absorbents.

Figure 5(a) shows the temperature dependence curve of the damping factor tanδ for the 328-composites. It is seen that the composites exhibit a similar tanδ curve to that of the pure TPU; that is, the tanδ firstly increased with increasing temperature and then decreased and reached the maximum at the glass transition temperature ($\sim-20°C$) of the matrix. It is clear that the tanδ maximum is gradually reduced with increasing the 328 content for the composites. However, it is also noted that the tanδ seems to be gradually increased to some extent with increasing 328 content when the temperature is higher than 30°C when the TPU matrix is in a full elastomeric state. Therefore, it is suggested that the damping mechanism is different for these two temperature ranges. The energy absorption for the present composites should come from two ways, the first is the intrinsic damping of the TPU matrix caused by molecular movements, and the other is the interfacial friction between the matrix and 328 phases [11]. During the glass transition process (−60 to 30°C), the former way plays a predominant role; therefore, the tanδ was gradually decreased with increasing 328 that diluted the TPU matrix. After glass transition, the energy harvesting coming from TPU has been minimum and the interfacial friction between the matrix and 328 phases has become predominant, thus resulting in the increase of tanδ when the matrix is in the elastomeric state.

Figure 5(b) shows the storage modulus (E') curves with variation of temperature for the 328-composites. It can be seen that the E' exhibits a sudden drop during glass transition from the glassy state to the elastomeric state. The E' of composites has similar changing trend to that of the TPU. It is noted that the E' has been increasingly enhanced with increasing 328 content, which should be caused by the reinforcing effect of 328 phases on the TPU matrix. The 328 phases are in the crystalline state and harder than the TPU matrix, therefore resulting in the reinforcement of the matrix.

Figures 6(a) and 6(b) show the tanδ curves and E' curves with variation of temperature for the 329-composites, respectively. It can be seen that they are similar to the results obtained in Figure 5 for the 328-composites. The tanδ maximum of the composites is gradually reduced with increasing 329 content during glass transition range, whereas the tanδ is enhanced when the matrix is at a full elastomeric state (>20°C). It is noted that, for the 329-composites with 60% 329, the tanδ at temperature of >20°C is even higher than that at the glass transition temperature range. For the E', similar to the 328-composites, the E' of the 329-composites is gradually enhanced due to the reinforcing effect of the crystalline 329 phases on the TPU matrix. Based on the

above results, it is concluded that the addition of 328 and 329 phases reduced the damping capacity at the glass transition range of the composites due to the dilution of TPU matrix but increased the damping property when the TPU matrix is at elastomeric state because of the interfacial frictions between the matrix and UV-absorbents. Meanwhile, the addition of UV-absorbents enhanced the TPU matrix and increased the elastic modulus of the composites.

4. Conclusions

The microstructure and damping property of polyurethane composites with different contents of UV-absorbents (UV-328 and UV-329) are investigated in this article. It is found that the UV-328 absorbents formed fiber-shaped phases in the TPU matrix and exhibited a very weak compatibility with the matrix, whereas the UV-329 absorbents are embedded in the matrix and form a relatively better compatibility with the matrix. Both 328 and 329 UV-absorbents are in the crystalline state in the composites at room temperature. The tanδ for the composites at glass transition range is gradually reduced with increasing content of UV-absorbents due to the dilution of matrix by the UV-absorbents. But the tanδ for the composites is increasingly enhanced with increasing UV-absorbents when the matrix is at a full elastomeric state, which is attributed to the interfacial frictions between the UV-absorbents and the matrix. The storage modulus of the composites is gradually enhanced with increasing content of UV-absorbents due to the reinforcing effect of the crystalline UV-absorbents on the TPU matrix.

Conflicts of Interest

The authors declare that they have no conflicts of interest.

References

[1] A. Eceiza, M. D. Martin, K. de la Caba et al., "Thermoplastic polyurethane elastomers based on polycarbonate diols with different soft segment molecular weight and chemical structure: mechanical and thermal properties," *Polymer Engineering and Science*, vol. 48, no. 2, pp. 297–306, 2008.

[2] B. C. Chun, T. K. Cho, M. H. Chong, and Y. C. Chung, "Structure–property relationship of shape memory polyurethane cross-linked by a polyethyleneglycol spacer between polyurethane chains," *Journal of Materials Science*, vol. 42, no. 21, pp. 9045–9056, 2007.

[3] W. L. Li, J. L. Liu, C. W. Hao, K. Jiang, D. F. Xu, and D. J. Wang, "Interaction of thermoplastic polyurethane with polyamide 1212 and its influence on the thermal and mechanical properties of TPU/PA1212 blends," *Polymer Engineering and Science*, vol. 48, no. 2, pp. 249–256, 2008.

[4] M. Nakamura, Y. Aoki, G. Enna, K. Oguro, and H. Wada, "Polyurethane damping material," *Journal of Elastomers and Plastics*, vol. 47, no. 6, pp. 515–522, 2015.

[5] N. Yamada, S. Shoji, H. Sasaki et al., "Developments of high performance vibration absorber from poly(vinyl chloride)/chlorinated polyethylene/epoxidized natural rubber blend," *Journal of Applied Polymer Science*, vol. 71, no. 6, pp. 855–863, 1999.

[6] X. Yu, G. Gao, J. Wang, F. Li, and X. Tang, "Damping materials based on polyurethane/polyacrylate IPNs: dynamic mechanical spectroscopy, mechanical properties and multiphase morphology," *Polymer International*, vol. 48, no. 9, pp. 805–810, 1999.

[7] C. L. Qin, W. M. Cai, J. Cai, D. Y. Tang, J. S. Zhang, and M. Qin, "Damping properties and morphology of polyurethane/vinyl ester resin interpenetrating polymer network," *Materials Chemistry and Physics*, vol. 85, no. 2-3, pp. 402–409, 2004.

[8] C. F. Wu and S. Akiyama, "Enhancement of damping performance of polymers by functional small molecules," *Chinese Journal of Polymer Science*, vol. 20, pp. 119–127, 2002.

[9] C. F. Wu, Y. Otani, N. Namiki, H. Emi, K. Nitta, and S. Kubota, "Dynamic properties of an organic hybrid of chlorinated polyethylene and hindered phenol compound," *Journal of Applied Polymer Science*, vol. 82, no. 7, pp. 1788–1793, 2001.

[10] C. Zhang, P. Wang, C. Ma, and M. Sumita, "Damping properties of chlorinated polyethylene-based hybrids: effect of organic additives," *Journal of Applied Polymer Science*, vol. 100, no. 4, pp. 3307–3311, 2006.

[11] J. Suhr, N. Koratkar, P. Keblinski, and P. Ajayan, "Viscoelasticity in carbon nanotube composites," *Nature Materials*, vol. 4, no. 2, pp. 134–137, 2005.

19

Preparation and Sealing Performance of a New Coal Dust Polymer Composite Sealing Material

Bo Li[ID],[1,2,3,4] Junxiang Zhang,[4,5] Jianping Wei[ID],[1,2] and Qiang Zhang[1,2]

[1]State Key Laboratory Cultivation Base for Gas Geology and Gas Control, Jiaozuo 454000, China
[2]The Collaborative Innovation Center of Coal Safety Production of Henan Province, Jiaozuo 454000, China
[3]Hebei State Key Laboratory of Mine Disaster Prevention, North China Institute of Science and Technology, Beijing 101601, China
[4]State and Local Joint Engineering Laboratory for Gas Drainage & Ground Control of Deep Mines, Henan Polytechnic University, Jiaozuo 454003, China
[5]School of Energy Science and Engineering, Henan Polytechnic University, Jiaozuo 454003, China

Correspondence should be addressed to Bo Li; anquanlibo@163.com

Academic Editor: Peter Majewski

In this study, a new coal-dust polymer composite material (CP) was fabricated to improve the borehole gas drainage effect and address the limitations of traditional sealing materials. This material used coal dust generated during underground drilling construction as filler, amino resin as binder, and also some other additives. Based on the orthogonal test scheme, 16 groups of experiments were conducted to investigate the effects of coal-dust content, water-material ratio, and cross-linking agent content on viscosity, 3-day (3-d) compressive strength, and gel time of the CP. The influences of the various factors were studied using a range of analytical methods to obtain the optimal proportion of the CP. Moreover, the mechanical property and microstructure of the CP were also analyzed to evaluate the sealing performance of this material. Results showed that this material exhibited obvious plastic characteristic and could provide an effective support for the borehole; it can adapt to the deformation of the borehole to some extent and achieve a long-term sealing effect. The scanning electron microscopy (SEM) results show a good correlation between the resin molecules and coal dust inside the material, which were conducive to improve the sealing performance of the CP material.

1. Introduction

Coal bed methane (CBM) is an unconventional gas associated with coal and is largely adsorbed in coal seams and surrounding rocks. The CBM reserve in China is $3.68 \times 10^{12} \, m^3$, which ranks third in the world and is equal to the total amount of reserves of natural gas resources. Therefore, exploiting CBM is considerably significant, given that CBM is a clean and efficient energy [1, 2]. The calorific value of $1 \, m^3$ of pure methane can reach 35.9 MJ, which is equivalent to 1.21 kg of standard coal, and pure methane hardly generates any exhaust gas after combustion [3, 4]. However, CBM is also a major hazard source in coal mines. CBM pressure and content in the coal seam dramatically increase with increasing mining depth, which easily leads to a large number of gas disasters. Borehole gas drainage is the most important technical means to control gas disaster. This approach can rapidly eliminate the danger of coal and gas outburst but also can increase CBM utilization and prevent environment pollution. In 2016, the underground CBM drainage quantity in China was $1.28 \times 10^{11} \, m^3$, but the CBM utilization amounting was only $4.8 \times 10^{10} \, m^3$, with the CBM utilization rate of 37.5%, which means that $8 \times 10^{10} \, m^3$ of CBM had escaped into the atmosphere. This phenomenon does not only result in a significant waste of energy but also exacerbates the greenhouse effect. The low rate of CBM utilization is mainly due to the poor sealing effect of gas drainage boreholes; approximately 65% of the coal face boreholes' prepumping gas concentrations in China are below 30%, which completely illustrates the poor sealing quality of current boreholes [5, 6].

Various sealing material properties directly influence the sealing quality of boreholes. At present, sealing materials can be categorized into inorganic and organic. Inorganic sealing materials are mainly composed of cement mortar, cement-water glass, cement-fly ash, and other cement-based materials. Although these materials possess the advantages of low cost and a wide range of sources, there also exist some disadvantages, such as easy precipitation and bleeding during solidification and with volume shrinkage above 15%, which tend to limit the use of the cement-based materials [7, 8]. Beyond this amount, because of poor adhesion of the cement-based material, this material easily separates from the borehole wall, resulting in the formation of leakage channels that seriously affect the borehole gas drainage effect [9–12]. In recent years, many studies on sealing materials have been conducted. Polyurethane, as a representative of organic sealing materials, has been widely used. Due to its fast curing speed, large expansion rate, and excellent adhesion, polyurethane has achieved significant advances in sealing boreholes [6, 13, 14]. However, its fast foaming speed and high viscosity make it difficult to diffuse adequately around the borehole, which fails to achieve the ideal sealing effect; therefore, it results in a number of defects in the practical application [15–19]. Moreover, Zhou has proposed "mucus sealing" based on the idea of "solid seal liquid and liquid seal gas," but the high-pressure mucus easily leaks through the fractures around the borehole; so supplementing this mucus repeatedly is necessary, which will lead to a high cost of sealing borehole [20, 21]. In view of the limitations of traditional sealing materials, developing a low-cost and superior-performance sealing material is necessary.

Considering the high adsorption and low permeability characteristics of coal seams in China, massive boreholes must be drilled in the coal seam to relieve CBM pressure and drain CBM to ensure underground safety during production [22, 23]. The number of underground drilling in China has exceeded 1.5×10^8 m every year [24]; thereby, large amounts of coal dust from drilling constructions are discharged into the work site. The accumulated coal dust not only takes up the limited workspace and increases mine ventilation resistance but also pollutes the work environment, which significantly threatens the health of workers in the locale. If coal dust is transported to the ground, it will increase the production cost of coal mines [25–27]. In view of this, this study developed a new CP material, which used coal dust as filler and is supplemented by amino resin as binder and also some other additives. The CP can be locally obtained to reduce the cost and promote an efficient and rational utilization of coal resources; it is a new way to promote direct utilization of coal resources without burning. In this study, an orthogonal test scheme was designed, with three factors and four levels for each factor, based on the proportion conditions of the CP. The effects of coal-dust content, water-material ratio, and cross-linking agent content on viscosity, compressive strength, and gel time of the CP were studied to obtain the optimal proportion. Then, the sealing performance

Figure 1: The sample of the prepared CP material.

of the CP was evaluated using mechanical analysis and SEM. The results in this study can provide a theoretical basis for the promotion and application of the CP material.

2. Preparation of the CP

2.1. Material Compositions. The CP material used the coal dust, which was generated from the underground drilling construction as filler, and was supplemented by amino resin as binder, and added some other additives, such as cross-linking, foaming, toughening, and coupling agents. The sample of the prepared CP material is shown in Figure 1.

Considering that the basic structural unit of the CP is composed of aromatic hydrocarbons with various condensation degrees and bridge bonds, the coal molecule has a specific rigidity, which can be added into the polymer as the rigid particles. Meanwhile, the pore structure characteristic of the coal mass and abundant groups on the margin of the coal molecule are easier to modify than other mineral fillers, which are favorable for improving the compatibility and dispersivity between the coal mass and the resin matrix [17]. On the contrary, coal dust is used as the dispersed phase to fill in the resin matrix; therefore, the size of the coal-dust particles should not be too large or too small to ensure interior integrity of the composite material. Otherwise, several internal defects or agglomeration may occur. Based on extensive researches, the size of coal dust less than 1 mm can meet the actual operation requirements.

The amino resin used in the material is a linear polymer, as the matrix of the composites; the resin plays the role of connecting coal particles and transferring stress. The active groups, such as carbonyl, amino, and subamino groups, on the resin molecules have strong binding characteristics that can enhance the cohesion between the coal particles and resin matrix. Moreover, the resin molecules contain many hydrophilic groups, such as ether bonds and hydroxyl groups, which can be associated with water molecules, thus forming a stable dispersion system. -CH_2OH, -NH-, and other active groups in the resin molecules allow the polycondensation reaction to form a three-dimensional network structure in the presence of a cross-linking agent. Therefore, the coal-dust particles can be tightly wrapped in the resin matrix with a specific mechanical intensity.

The foaming agent is a modified carbonate, which can react with hydrogen ions to produce CO_2 as the expansion source that leads to volume expansion of the material. The toughening agent is a type of linear polymer compound that reacts with resin molecules under an acidic condition and embeds itself into the macromolecular chain of the resin to enhance the toughness and aging resistance of the material. The coupling agent contains a bifunctional group, which is composed of hydrolytic and nonhydrolytic groups; on the one hand, the condensation products of the hydrolytic group can react with hydroxyl (-OH) and carboxyl (-COOH) on the coal surface to produce a single-molecular-layer chemisorption; on the other hand, the nonhydrolytic group can result in a physical entanglement or chemical reaction with the resin molecules, which creates an interface phase between the dispersed (coal dust) and matrix (resin) phases, thereby improving the coupling process of the allogenic material.

2.2. Design of Orthogonal Test. This study utilized the orthogonal test to investigate the effect of the preparation parameters on the properties of the CP and to obtain the optimal proportion of the material. By conducting a number of experiments, the various contents of coal dust and water-material ratio that significantly influence the properties of the CP were examined. Meanwhile, the sealing material should possess a reasonable gel time to meet the requirements of the grouting process, which is mainly influenced by the addition amount of the cross-linking agent. Therefore, three influence factors were selected that characterize the CP, namely, coal-dust content (A), water-material ratio (B), and cross-linking agent content (C). Each factor comprised four levels. In addition, the viscosity, 3-d compressive strength, and gel time of the CP were used as indicators in the test. A total of 16 group experiments were designed using the three factors and the four levels of the $L_{16}(4^5)$ scheme. Tables 1 and 2 show the values of the factors and levels and the scheme of the orthogonal test.

2.3. Experimental Methods. The viscosity of the CP was tested using an NJD-8S rotary viscometer. The prepared grout of the CP was placed in a 500 ml beaker, and then the rotor of the viscometer was immerged to a specified depth of the grout. And then the viscosity of the grout was measured at 20°C and 80% relative humidity.

The compressive strength of the composite materials was tested using the RMT-150B electrohydraulic servo test machine. The displacement control method was utilized in the experiments; the specimens of the CP were loaded with the speed of 0.02 mm/s until damage. In the test, all samples of the CP material were curing in standard conditions for 3 days, and then the samples were processed into standard size with $\Phi 50\,\text{mm} \times H\,100\,\text{mm}$ and tested the compressive strength.

The inverted cup method was adopted to measure the gel time of the CP. We first placed the prepared grout of the CP

TABLE 1: Values of factors and levels.

Level	Coal-dust content (%) (A)	Water-material ratio (B)	Cross-linking agent content (%) (C)
		Factor	
1	50	1.0:1	1.5
2	75	1.2:1	2.0
3	100	1.5:1	2.5
4	125	1.9:1	3.0

TABLE 2: Scheme of the orthogonal test.

Test group	Coal-dust content (%) (A)	Water-material ratio (B)	Cross-linking agent content (%) (C)
1	(A_1) 50	(B_1) 1.0:1	(C_1) 1.5
2	50	(B_2) 1.2:1	(C_2) 2.0
3	50	(B_3) 1.5:1	(C_3) 2.5
4	50	(B_4) 1.9:1	(C_4) 3.0
5	(A_2) 75	1.0:1	2.0
6	75	1.2:1	1.5
7	75	1.5:1	3.0
8	75	1.9:1	2.5
9	(A_3) 100	1.0:1	2.5
10	100	1.2:1	3.0
11	100	1.5:1	1.5
12	100	1.9:1	2.0
13	(A_4) 125	1.0:1	3.0
14	125	1.2:1	2.5
15	125	1.5:1	2.0
16	125	1.9:1	1.5

in a 200 ml beaker A and started to record time at the same time. We then poured the grout from beaker A into beaker B and repeated the above operations at every 1 minute, until the grout no longer flowed. The time of this process was the gel time.

3. Analysis of Orthogonal Test Results

The viscosity, 3-d compressive strength, and gel time of the 16 different group tests were tested based on the scheme of the orthogonal test; the results are listed in Table 3. According to the results, the direct analysis method was used to analyze the significance of the three factors by calculating the range value of each factor, which reflects the effect of the factor on the indicators under different levels. The order of importance of the factors can be determined on the basis of the range value, and the optimal level of each factor can be determined by combining it with the effective curves. The orthogonal test result can provide quantitative basis for the optimum proportion of the CP.

3.1. Effect of Preparation Parameters on the Viscosity of the CP. Viscosity is an important index for determining the pumpability of the sealing material. It indicates the internal friction resistance when fluid flows. A lower viscosity of the material entails better fluidity and a larger grout diffusion range in the stratum, which are conducive for sealing massive fractures around boreholes. Table 4 shows the range results of the sample viscosity as influenced by several

TABLE 3: Experimental results of orthogonal test.

Test group	Viscosity (mPa·s)	3-d compressive strength (MPa)	Gel time (min)
1	1035	4.45	67
2	513	4.05	39
3	364	2.82	19
4	299	1.60	23
5	1484	5.02	20
6	774	4.57	88
7	566	3.25	12
8	411	2.35	38
9	2337	5.98	9
10	1339	5.17	7
11	849	3.23	119
12	569	2.48	83
13	4150	4.12	11
14	2769	4.58	27
15	1988	3.15	83
16	1299	2.05	113

TABLE 4: Range analysis of the influencing factors of viscosity.

Level	Coal-dust content (%) (A)	Water-material ratio (B)	Cross-linking agent content (%) (C)
1	552.75	2251.50	989.25
2	808.75	1348.75	1138.50
3	1237.50	941.75	1470.25
4	2551.50	644.50	1588.50
Range	1998.75	1607.00	599.25

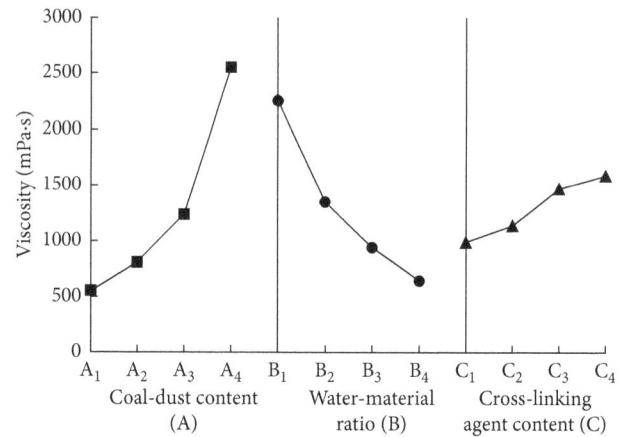

FIGURE 2: Effective curves of the factors affecting viscosity.

factors. The range values of the three factors, that is, coal-dust content, water-material ratio, and cross-linking agent content, are 1998.75, 1607.00, and 599.25, respectively. Thus, the order of the importance of the factors to the viscosity of the CP is coal-dust content > water-material ratio > cross-linking agent content. The results show that coal-dust content plays a significant role in controlling the viscosity.

The effective curves of the sample viscosity were constructed based on the data in Table 4, as shown in Figure 2, to analyze intuitively the effect of the various factors on viscosity. The effective curves show that the viscosity increases with the content of coal dust and cross-linking agent increasing, and decreases with the increase of the water-material ratio. Due to the viscosity of the material grout mainly depending on the force between the molecules in the material, as the coal-dust content increases, the friction between the coal particles will increase and lead to a sharp rise in the viscosity; as shown in Figure 2, when coal-dust content exceeds 100% (A$_3$), the viscosity increases sharply, thus the content of coal dust should not be more than 100%. However, the increase in the water-material ratio can widen the distance between the particles, the friction resistance between the particles will be reduced, and the viscosity of the grout can be improved. Figure 2 shows that when the water-material ratio reaches 1.2 : 1 (B$_2$), the viscosity of the material grout is lower with a better fluidity. As for the cross-linking agent, it can promote the resin molecules to allow the polycondensation reaction to take place, and therefore the increase in the content of the cross-linking agent will accelerate this reaction; as a result, the prepolymer is condensed to form resin particles in the grout, and then the particles are aggregated and deposited, which result in the increase of the viscosity of the material grout. Therefore, the optimal proportion for decreasing the viscosity of the CP is A$_1$B$_4$C$_1$ from the orthogonal test.

3.2. Effect of Preparation Parameters on the 3-d Compressive Strength of the CP. For sealing materials, obtaining the specific early strength is necessary to achieve a timely support to the borehole and prevent the formation of leakage channels that resulted from the instability failure of the borehole. The 3-d compressive strength was used as an indicator in the orthogonal test to analyze the effect of each factor on the strength of the CP. Table 5 lists the range of the various factors from the orthogonal test results. The water-material ratio achieved the maximum range of 2.77, followed by the coal-dust content of 0.99 and the cross-linking agent content of 0.40. The results indicate that the water-material ratio is the most important factor that influences the early strength of the CP.

Figure 3 shows the effective curves of the factors that influence the 3-d compressive strength. It can be seen that the 3-d compressive strength decreases with the increase in the water-material ratio; this is because that the excess water molecules tend to attach to the surface of the coal particles, which results in the reduction of the bonding area between the resin matrix and coal particles and the internal cohesion of the material; in addition to that, during the curing of the material, when water molecules on the surface of the coal particles evaporate, it leads to the generation of a large amount of microfissures between the resin matrix and coal dust, which seriously affect the strength of the material; thus, the water-material ratio should be controlled at 1.2 : 1~1.5 : 1 (B$_2$~B$_3$) by using the above analysis of the viscosity that ensures the fluidity of the material. Meanwhile, the influence of coal-dust and cross-linking agent content on the 3-d compressive strength of the material is relatively similar. The material strength increases first and then decreases with an

TABLE 5: Range analysis of the influencing factors of 3-d compressive strength.

Level	Coal-dust content (%) (A)	Water-material ratio (B)	Cross-linking agent content (%) (C)
1	3.23	4.89	3.58
2	3.80	4.59	3.68
3	4.22	3.11	3.93
4	3.48	2.12	3.54
Range	0.99	2.77	0.40

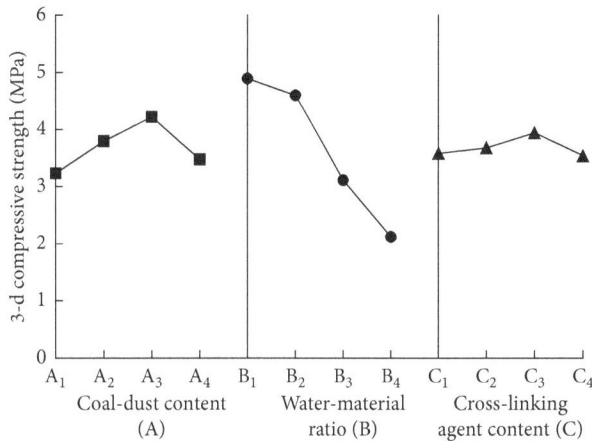

FIGURE 3: Effective curves of the factors affecting 3-d compressive strength.

increase in their content. Coal dusts form the skeleton system of the composites due to the presence of rigid particles in the resin matrix; therefore, increasing the content of coal dust in a specific range can improve the density of the material, so the compactness of the material can also be increased as the mechanical strength. By contrast, if the coal-dust content is too much, then the distance of coal particles will be decreased; therefore, the resin matrix cannot easily fill the gaps among the coal particles to transfer stress and, as a consequence, internal defects will be formed in the material that reduce the mechanical strength of the material. Figure 3 reveals that when the coal-dust content reaches 100% (A_3), the CP obtains the highest 3-d compressive strength of 4.22 MPa. For the cross-linking agent, if the content is too small, the cross-linking density of the resin matrix will be reduced. Turning the resin molecules into the stable macromolecular structure is difficult, thus resulting in the decrease in the material strength. On the contrary, if too much of the cross-linking agent is added, the cross-linking rate between the resin molecules will be too high. The resin matrix will produce a greater internal stress, thereby decreasing the mechanical strength of the CP. As shown in Figure 2, the optimal proportion for increasing the 3-d compressive strength of the CP is $A_3B_2C_3$.

3.3. Effect of Preparation Parameters on the Gel Time of the CP.
Gel time is the controllability index of the sealing material, which directly determines the operation design of

sealing technology. The appropriate gel time of the material can ensure the complete diffusion of the material gout around the borehole to achieve the ideal sealing effect. The gel time of 16 different samples are tested at a temperature of $(20 \pm 2)°C$, and the results are listed in Table 6, which show that the range value of coal-dust content, water-material ratio, and cross-linking agent content is 21.50, 37.50, and 83.50. Thus, the cross-linking agent content has a significant effect on the gel time of the material. The water-material ratio achieves an intermediate significance, whereas the coal-dust content possesses a slight significance.

The effective curves of the various factors that influence the gel time are constructed based on the data in Figure 4. Figure 4 indicates that the cross-linking agent content is the most significant factor that influences the gel time. With the increase in the content of the cross-linking agent, the gel time continuously decreased with a nonlinear change trend due to the production of a specific amount of free radicals in the grout by the cross-linking agent, which can induce the occurrence of resin molecules during the polycondensation reaction. More content of the cross-linking agent will accelerate the cross-linking rate of the reaction, which results in the shortened gel time of the material. As shown in the figure, when the content is up to 2.5 %, it yields a favorable promotion effect. However, if the content of cross-linking agent continues to increase, the gel time is shortened, and the drop gradient is also decreased, considering that the gel time of sealing material is generally approximately 30–60 min in practical application. Therefore, the content of the cross-linking agent should be controlled at 2.0%~2.5% (C_2~C_3). In addition, the increase in the water content in the grout can reduce the collision probability of the resin molecules. Therefore, as shown in Figure 3, the gel time increases with the increase in the water-material ratio. Meanwhile, the excess moisture can absorb the reaction temperature of the grout system, which reduces the reaction rate of the resin molecules. Furthermore, the increase in the coal-dust content can also prevent the collision of the resin molecules, which reduces the cross-linking rate and prolongs the gel time. Consequently, the optimal proportion of the appropriate gel time of the CP is $A_2B_2C_2$.

3.4. Optimal Proportion of the Composite Sealing Material.
The optimal proportion for decreasing the viscosity is $A_1B_4C_1$, for increasing the 3-d compressive strength is $A_3B_2C_3$, and for the appropriate gel time of the CP is $A_2B_2C_2$, which were obtained by analyzing the experiment results of the orthogonal test. In order to acquire the optimal proportion of the CP, three types of proportion conditions were prepared for the sample materials, and the indicators were also tested. The results are shown in Table 7.

Table 7 shows that, for $A_1B_4C_1$, although this sample has the lowest viscosity, but the 3-d compressive strength is low, it cannot provide the effective support and protection for the borehole, and the grout of this sample material is easily leaked through the fractures around the borehole that resulted from the longer gel time. As for $A_3B_2C_3$, due to the high viscosity, its grout cannot easily penetrate into the

TABLE 6: Range analysis of the influencing factors of gel time.

Level	Coal-dust content (%) (A)	Water-material ratio (B)	Cross-linking agent content (%) (C)
1	552.75	2251.50	989.25
2	808.75	1348.75	1138.50
3	1237.50	941.75	1470.25
4	2551.50	644.50	1588.50
Range	1998.75	1607.00	599.25

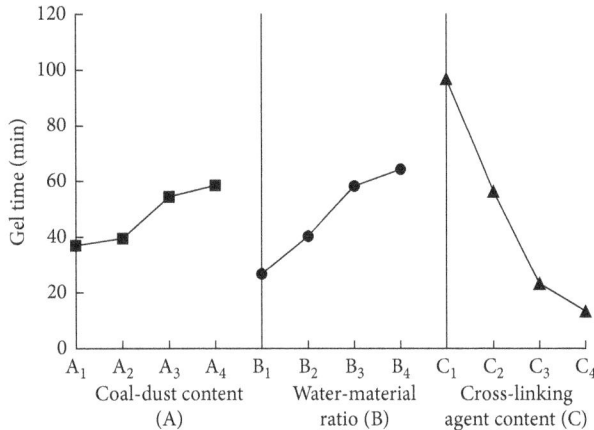

FIGURE 4: Effective curves of the factors affecting gel time.

fractures to achieve a sealing effect, so sample $A_3B_2C_3$ is not the ideal result. However, sample $A_2B_2C_2$ has a lower viscosity, which allows the material grout to completely diffuse around the borehole, and this sample also has a higher compressive strength and a more suitable gel time, which can achieve the technical requirements of sealing. Accordingly, the best optimal proportion of the material properties of the CP is $A_2B_2C_2$.

4. Analysis of Sealing Performance of the CP

4.1. Analysis of Mechanical Properties. The mechanical strength of the sealing material is a significant factor that influences the sealing effect. It does not only need a higher early strength to appropriately support the borehole timely, but it should also possess the plastic characteristic to adapt to the deformation of the borehole. In accordance with the orthogonal test results, the optimal proportion is $A_2B_2C_2$. The samples were cured at a temperature of $(20 \pm 2)°C$, with a humidity of 90%. For 28 curing days, the samples were processed into the standard specimens with $\Phi 50\,mm \times H$ 100 mm, which were used to examine the mechanical properties of CP.

4.1.1. Analysis of the Stress Strain Relation Curves of the CP. The stress-strain relation curve is an important method to examine the deformation property of the material. Figure 5 shows the results of the uniaxial compressive strength test of the three specimens.

As shown in this figure, the changing trends of the curves are basically similar. On the basis of the curves' characteristics in the chart, the failure process of the specimens comprises five stages, that is, compaction, elastic, plastic, softening, and residual strength stages. With the gradual increase of the load, all the specimens first enter the compaction stage; the stress-strain relation curves in this range are concave upside-down, then the pores and holes inside the specimen are gradually closed, which results in a slight strain change of the specimens. The specimens enter the elastic stage as the load continues to increase, the stress represents the linear relationship with the strain, and the fractures have appeared in the material in this stage. As specimens enter the plastic stage with the further increase in load, the stress varied nonlinearly with strain, which results in the deformation of the unrecoverable plastic of the material; this reveals that the fractures in the material have gradually extended and coalesced and eventually evolved into the macrocracks. When the stress on the specimens reaches the peak, the softening stage occurs, the deformation type of the specimens is transformed from compaction to expansion, and the internal structures of the material are placed along the crack surfaces; the strain of the specimens raises sharply with the development of the macrocracks; however, the mechanical strength declines. When the specimens are broken, the stress drops rapidly to a relatively stable level due to friction between the fracture planes in the material. Therefore, the specimen still experiences a specific carrying capacity.

4.1.2. Analysis of the Failure Forms of the CP. The failure form of the specimen is a significant characteristic to describe the failure mechanism of the material. Figure 6 shows the three specimens after the uniaxial compression failure. As shown in the figure, all the three specimens present a significant characteristic of plastic failure. There is a clear shear slip surface presented in each specimen. During uniaxial compression, the deformation of the specimens is axial compression and radial expansion, whereas the radial deformation in the middle part of the specimen is larger than that of both sides. This is due to the dislocation of the resin colloidal particles in the material. When the radial strain value of the specimen exceeds the ultimate tensile strain, the cracks are gradually generated on the surface of the material and gradually extended to the interior, until the specimen is damaged.

The mechanical properties of the CP indicate that this material possesses a typical plastic characteristic, and it can provide effective supports for the borehole and can adapt to the borehole deformation to some extent. Thus, it can avoid the instant damage caused by the stress changes, which is conducive to improve the stability of the borehole to achieve a long-term drainage effect.

4.2. Microstructure Analysis of the CP. The microstructure of the CP was observed using a JSM-6390/LV SEM, which can magnify 5–300000 times with resolution of 3.0 nm and an acceleration voltage of 0.5–30 kV. Figure 7 shows the SEM images of the CP at 300 times magnification.

TABLE 7: Experimental results of the three kinds of materials.

Test group	Viscosity (mPa·s)	3-d compressive strength (MPa)	Gel time (min)
$A_1B_4C_1$	234	1.32	86
$A_3B_2C_3$	1201	5.33	23
$A_2B_2C_2$	781	4.71	47

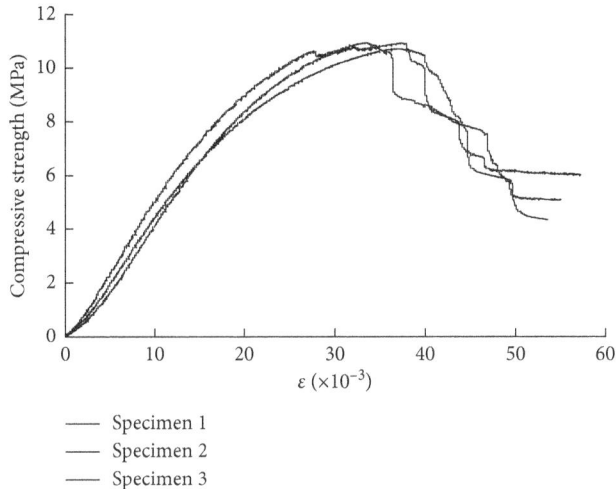

FIGURE 6: The failure type of the specimen.

— Specimen 1
— Specimen 2
— Specimen 3

FIGURE 5: Stress-strain relation curves of the specimens.

Figure 7(a) shows that the structure of the CP is dense and that several isolated pores are distributed in the material, which are produced by a foaming agent to promote the volume expansion of the material. It can be seen that the coal-dust particles are dispersed uniformly in the material and closely embedded in the resin matrix, which form the carrier of the composites, thereby improving the mechanical strength of the material.

Figure 7(b) shows the scanning result of the resin matrix at 2500 times magnification. This figure clearly shows that a great number of colloidal particles are distributed in the resin matrix. The matrix skeleton and internal pores are intersected and intertwined into a dense network structure due to the unique structure of the resin. When the CP is in the loading process, the resin matrix first yields and produces plastic deformation. Then, the colloidal particles are dislocated to allow the deformation of the material until failure. Therefore, the plastic capacity characteristic of the resin plays a significant role in transferring stress and dissipating energy to the composite materials. However, a modulus difference exists between coal mass and resin matrix. The interfaces are relatively weaker than the surfaces of the CP, which easily yield stress concentration under load. Therefore, the binding ability between them directly influences the overall performance of the material.

Figure 7(c) shows the result of the interface structure of the CP. This figure shows a better combination effect on the interface. No obvious gaps are found between the coal particle and the resin matrix. In addition, the adhesion layer of the resin can be observed on the surface of the coal particles due to a large number of active groups present in

the resin molecule, such as the carbonyl group (-C=O-), amino group (-NH$_2$), and imino group (-NH-). These groups have a fine adhesion attached on the surface of the coal dust through physical reaction or chemical grafting to form a unified unit. Therefore, the silicon-oxygen (Si-O) and magnesium-oxygen (Mg-O) bonds in the coal molecules can create an association with the hydrogen atoms of the hydroxymethyl (-CH$_2$OH) in the resin, which improves the stability of the interface between the coal dust and resin matrix.

5. Discussion of Sealing Performance of the CP

The results show that the CP has a superior sealing performance that can satisfy the technical requirements for sealing boreholes. As shown in Figure 8, the sealing mechanisms of CP can be generalized as follows.

5.1. Preferable Fluidity. The optimal proportion of the CP is $A_2B_2C_2$, with a low viscosity of 781 MPa·s, using the orthogonal test. This material can meet the technical requirements of low-pressure grouting and can completely diffuse in the coal seam. The solid-phase particles of the coal dust in the material can fill the holes and macrofractures. The liquid-phase of the resin matrix can seal the microfractures around the borehole. This combination can further improve the sealing effect of the drainage borehole.

5.2. Adapt to the Deformation of Borehole. In the CP, coal dust is used as filler, and amino is chosen as binder. The coal dust continuously distributed in the material forms the skeleton system of the CP. The resin matrix disperses and transfers stress. The CP, with the characteristic of plastic strengthening, can adapt to the deformation of boreholes that resulted from the stress change in the surrounding rocks, which ensures the stability of gas drainage borehole as well as a long-term drainage effect.

5.3. Stable Combination with Coal Mass. SEM results show a good correlation between the resin molecules and coal dust; the resin matrix plays a role of transferring stress, and the coal dust forms the carrier to bear external stress changes. Accordingly, combination of them improves the sealing effect and achieves a superior drainage effect.

FIGURE 7: Microstructure of the composite material. (a) The CP material, (b) resin matrix, and (c) interface structure.

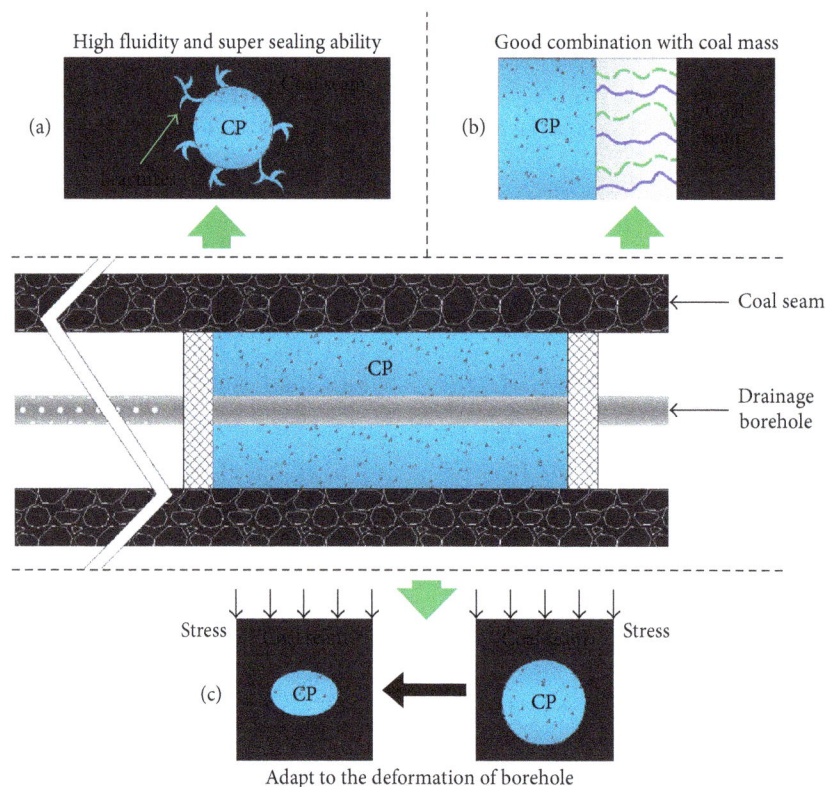

FIGURE 8: Sealing performance of CP material.

6. Conclusions

(1) An orthogonal test scheme of 16 groups of experiments was designed to determine the optimal coal-dust content, water-material ratio, and cross-linking agent and the four levels used for each factor as well as the effects of viscosity, 3-d compressive strength, and gel time on the properties of the CP.

(2) The range analysis of the results of orthogonal test showed that the significant factors that influence the viscosity followed the order coal-dust content > water-material ratio > cross-linking agent content. The order of importance of the three factors based on their influence on the 3-d compressive strength is water-

material ratio > coal-dust content > cross-linking agent content. The rank of importance in terms of gel time is cross-linking agent content > water-material ratio > coal-dust content. The optimal proportion of the CP is found to be $A_2B_2C_2$ by comparing and analyzing the selected materials.

(3) Mechanical property test showed that the CP possessed a high compressive strength with a significant plastic characteristic. Thus, the CP can provide the support to boreholes and adapt to deformation to a certain extent. And SEM results reveal that a good correlation between the resin molecules and coal dust. In general, these results demonstrated that the CP has a superior sealing performance.

Conflicts of Interest

The authors declare that they have no conflicts of interest.

Acknowledgments

This study was supported by Program for National Natural Science Foundation of China (nos. 51774120 and 51604096), Hebei State Key Laboratory of Mine Disaster Prevention (KJZH2017K08), and Henan Research Program of Application Foundation and Advanced Technology (162300410031) and funded by the Research Fund of State and Local Joint Engineering Laboratory for Gas Drainage & Ground Control of Deep Mines (Henan Polytechnic University) (G201609), Key Research Project of Higher Education Institution of Henan Province in 2015 (no. 15A440001), and Dr. Funds of Henan Polytechnic University (no. B2015-05).

References

[1] M. Qingmin, Z. Zhonglin, and P. Fangwei, "Development trends of natural gas resources and gas-fired power generations in China," *Environmental Progress and Sustainable Energy*, vol. 35, no. 3, pp. 853–858, 2016.

[2] C. Peng, C. Zou, T. Zhou et al., "Factors affecting coalbed methane (CBM) well productivity in the Shizhuangnan block of southern Qinshui basin, North China: investigation by geophysical log, experiment and production data," *Fuel*, vol. 191, pp. 427–441, 2017.

[3] G. J. Millar, S. J. Couperthwaite, and C. D. Moodliar, "Strategies for the management and treatment of coal seam gas associated water," *Renewable and Sustainable Energy Reviews*, vol. 57, pp. 669–691, 2016.

[4] Y. P. Cheng, J. Dong, W. Li, M. Y. Chen, and K. Liu, "Effect of negative pressure on coalbed methane extraction and application in the utilization of methane resource," *Journal of the China Coal Society*, vol. 42, no. 6, pp. 1466–1474, 2017.

[5] C. Zhai, X. Xiang, Q. Zou, X. Yu, and Y. Xu, "Influence factors analysis of a flexible gel sealing material for coal-bed methane drainage boreholes," *Environmental Earth Sciences*, vol. 75, p. 385, 2016.

[6] N. G. Hua, L. B. Quan, Z. Cheng et al., "Microscopic properties of drilling sealing materials and their influence on the sealing performance of boreholes," *Journal of University of Science and Technology Beijing*, vol. 35, no. 5, pp. 572–579, 2013.

[7] H. Samouh, E. Rozière, and A. Loukili, "Influence of formwork duration on the shrinkage, microstructure, and durability of cement-based materials," *Nature*, vol. 185, no. 4711, pp. 439–441, 2015.

[8] P. Chaunsali and P. Mondal, "Influence of mineral admixtures on early-age behavior of calcium sulfoaluminate cement," *ACI Materials Journal*, vol. 112, no. 1, pp. 1–6, 2015.

[9] X. Chen, Y. Zhang, W. Sun, X. Ge, and R. Zhao, "Research on expansion property of inorganic hole sealing material of coal mine," *Materials Research Innovations*, vol. 19, no. 8, pp. S8-142–S8-144, 2015.

[10] M. I. Childers, M. T. Nguyen, K. A. Rod et al., "Polymer-cement composites with self-healing ability for geothermal and fossil energy applications," *Chemistry of Materials*, vol. 29, no. 11, pp. 4708–4718, 2017.

[11] S. W. Lee, G. T. Chae, M. Jo, and T. Kim, "Comparison of portland cement (KS and API class G) on cement carbonation for carbon storage," *Journal of Materials in Civil Engineering*, vol. 27, no. 1, article 04014105, 2015.

[12] Z. Cheng, Y. Xu, N. Guanhua, L. Min, and H. Zhiyong, "Microscopic properties and sealing performance of new gas drainage drilling sealing material," *International Journal of Mining Science and Technology*, vol. 23, no. 4, pp. 475–480, 2013.

[13] Y. Jiang, T. C. Zhu, and T. X. Wang, "Research and application of pressure measurement technology by combination sealing of capsule and polyurethane," *Journal of Safety Science and Technology*, vol. 35, no. 5, pp. 80–84, 2015.

[14] Z. Q. Feng and H. P. Kang, "Technology research of chemical grouting for cracked coalrock mass and demonstration project," *Journal of Yangtze River Scientific Research Institute*, vol. 26, no. 7, pp. 60–65, 2013.

[15] Y. Wei, A. M. Asce, F. Wang, X. Gao, and Y. Zhong, "Microstructure and fatigue performance of polyurethane grout materials under compression," *Journal of Materials in Civil Engineering*, vol. 29, no. 9, article 04017101, 2017.

[16] P. K. R. Vennapusa, Y. Zhang, and D. J. White, "Comparison of pavement slab stabilization using cementitious grout and injected polyurethane foam," *Journal of Performance of Constructed Facilities*, vol. 30, no. 6, article 04016056, 2016.

[17] M. Genedy, U. F. Kandil, E. N. Matteo, J. Stormont, and M. M. R. Taha, "A new polymer nanocomposite repair material for restoring wellbore seal integrity," *International Journal of Greenhouse Gas Control*, vol. 58, pp. 290–298, 2017.

[18] J. Scucka, P. Martinec, and K. Soucek, "Polyurethane grouted gravel type geomaterials—a model study on relations between material structure and physical–mechanical properties," *Geotechnical Testing Journal*, vol. 38, no. 2, article 20140100, 2015.

[19] H. Liu, W. Hui, C. Zhuang, Y. Liu, and Y. Jian, "Preparation of EA/PU IPN grouting material and its performances research," *Journal of Central South University*, vol. 44, no. 8, pp. 3129–3136, 2013.

[20] Z. Cheng, X. W. Xiang, Y. U. Xu, P. Shen, N. I. Guan-Hua, and L. I. Min, "Sealing performance of flexible gel sealing material of gas drainage borehole," *Journal of China University of Mining and Technology*, vol. 42, no. 6, pp. 982–988, 2013.

[21] F. B. Zhou, Y. N. Sun, H. J. Li, and G. F. Yu, "Research on the theoretical model and engineering technology of the coal seam gas drainage hole sealing," *Journal of China University of Mining and Technology*, vol. 45, no. 3, pp. 433–439, 2016.

[22] T. Xia, F. Zhou, J. Liu, and F. Gao, "Evaluation of the predrained coal seam gas quality," *Fuel*, vol. 130, no. 130, pp. 296–305, 2014.

[23] Y. S. Kang, L. Z. Sun, B. Zhang, J. Y. Gu, and D. L. Mao, "Discussion on classification of coalbed reservoir permeability in china," *Journal of the China Coal Society*, vol. 42, pp. 186–194, 2017.

[24] Y. L. Wang, W. B. Song, Y. N. Sun, X. X. Zhai, and Z. F. Wang, "Clogging mechanical model and its application in gas extraction borehole," *Journal of Chongqing University*, vol. 37, no. 9, pp. 119–127, 2014.

[25] Y. Chen, L. Y. Wang, and S. Q. Zhu, "Surface modification effects of microorganism from lignite on fine coal," *Advanced Materials Research*, vol. 868, no. 9, pp. 423–428, 2013.

[26] J. Lu, S. Jiang, J. Tao, and J. Hu, "Analysis of dust to evaluate the incidence of pneumoconiosis in Huainan coal mines," *Analytical Letters*, vol. 49, no. 11, pp. 1783–1793, 2016.

[27] Q. Huang and R. Honaker, "Optimized reagent dosage effect on rock dust to enhance rock dust dispersion and explosion mitigation in underground coal mines," *Powder Technology*, vol. 301, pp. 1193–1200, 2016.

20

Preparation of Poly(L,L-Lactide) Microparticles via Pickering Emulsions Using Chitin Nanocrystals

T. S. Demina ⓘ,[1,2] **Yu. S. Sotnikova,**[1,3] **A. V. Istomin,**[1] **Ch. Grandfils,**[4]
T. A. Akopova ⓘ,[1] **and A. N. Zelenetskii**[1]

[1]*Enikolopov Institute of Synthetic Polymer Materials, Russian Academy of Sciences, 70 Profsoyuznaya St., 117393 Moscow, Russia*
[2]*Institute for Regenerative Medicine, Sechenov University, 8-2 Trubetskaya St., Moscow 119991, Russia*
[3]*Moscow Aviation Institute (National Research University), 4 Volokolamskoe Shosse, 125993 Moscow, Russia*
[4]*Interfaculty Research Centre on Biomaterials (CEIB), Chemistry Institute, University of Liège, B6C Allée du 6 Août 11,*
 B-4000 Liege (Sart-Tilman), Belgium

Correspondence should be addressed to T. S. Demina; detans@gmail.com

Academic Editor: Renal Backov

The aim of the present study was to investigate an ability of chitin nanocrystals to be used as stabilizing components for fabrication of poly(L,L-lactide) microparticles via the Pickering oil/water emulsion solvent evaporation technique. The anisometric chitin nanocrystals were extracted from two different samples of crab shell chitin via acetic hydrolysis and analyzed using atomic force microscopy, dynamic light scattering, and FTIR spectroscopy. The extracted nanocrystals showed no difference in the chemical structure but possessed different morphology and aspect ratios as a function of raw chitin used. The effect of chitin nanocrystals characteristics and concentration in the aqueous phase on the total yield, size distribution, and shape and surface morphology of the prepared polylactide microparticles was evaluated.

1. Introduction

Polylactide micro- and nanoparticles are widely used as drug delivery systems and cell microcarriers [1, 2]. To control the cell attachment, biocompatibility, and drug kinetic release, the particles could be coated with functional polysaccharides during or after a preparation procedure [3–6]. As an alternative, the surface of the particles could be modified using especially synthesized amphiphilic copolymers as functional macromolecular emulsifiers [7]. Another approach to fabrication of core-shell microparticles is based on the application of solid nanoparticles as stabilizing components, that is, preparation via the Pickering emulsions.

The Pickering emulsions, that is, emulsions stabilized by solid nanoparticles instead of molecular emulsifiers, provide a number of benefits, such as high resistance to coalescence and low amount of nanoparticles needed to stabilize an interface [8–10]. The Pickering emulsions were successfully used for preparation of a so-called magnetic polymer microspheres consisted of polystyrene core and Fe_3O_4 nanoparticle shells [11]. A range of polymeric nanoparticles made of polymer brushes or polymer-grafted particles of various natures could be also used for the Pickering emulsions, which allows one to synthetize new nanocomposites dedicated to specific applications [12–18]. Usage of biocompatible nanoparticles, such as hydroxyapatite, SiO_2, proteins, lignin, flavonoids, and various polysaccharides, for the Pickering emulsion stabilization is especially interesting for food or biomedical applications [19–22]. Polysaccharides could be transformed into nanodimensional forms using controlled aggregation of macromolecules [23, 24] or extraction of nanocrystals via hydrolysis of amorphous regions of raw polysaccharides [25–27]. Nanocrystals extracted from cellulose, chitin, and starch are highly crystalline rigid nanoparticles, which morphology and size could significantly vary as a function of raw polysaccharide used [26]. The most widely used cellulose and chitin nanocrystals have a rod-like anisometric

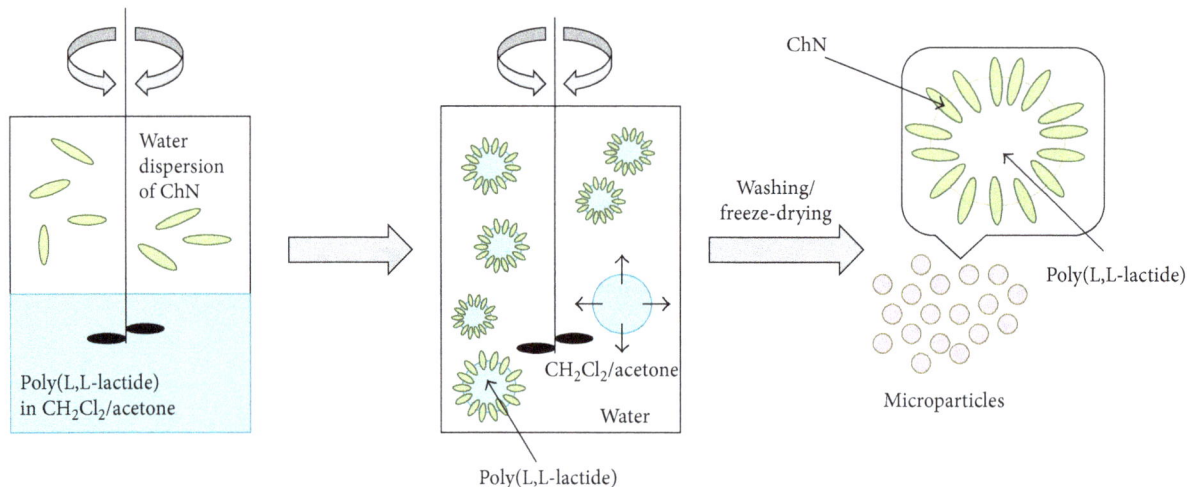

Figure 1: Scheme of PLLA-ChN core-shell microparticles preparation via Pickering oil/water emulsion solvent evaporation technique.

morphology with a length and diameter varied from several nanometers to micrometers. They are proposed as reinforcing fillers, templates for preparation of mesoporous materials, for fabrication of semiconductor and smart materials, including mechanically adaptive and self-healing ones. The use of polysaccharide nanocrystals as stabilizing agents in the Pickering emulsions was previously described mainly for food industry [28–31]. Efficiency of nanocrystals as stabilizing components for preparation of polymeric microparticles via the emulsion solvent evaporation technique was studied mostly on cellulose nanocrystals [32, 33].

This work was aimed at evaluating the effect of chitin source on characteristics of the obtained chitin nanocrystals and on a possibility to use them as stabilizing agents for fabrication of core-shell poly(L,L-lactide)/chitin microparticles via the Pickering emulsion solvent evaporation technique.

2. Experimental

2.1. Materials. Crab shell chitins were purchased from "Kombio" (Russia) and "Xiamen Fine Chemical" (China) and were used as raw materials for preparation of chitin nanocrystals, which were subsequently marked as ChN-K and ChN-X, respectively. The ash contents of raw chitins were 0.16 wt.% and 1.76 wt.% for "Kombio" and "Xiamen Fine Chemical," respectively. The nanocrystals were obtained by acidic hydrolysis of the chitin samples according to the procedure described in [34, 35]. Poly(L,L-lactide) (PLLA) with an average molecular weight of 160 kDa was purchased from Sigma-Aldrich and used as a core material for the microparticle fabrication. Polyvinyl alcohol (PVA) marked as Mowiol 10–98 (Germany) was used as received.

2.2. Atomic Force Microscopy (AFM). The size and morphology of the prepared ChN were evaluated using atomic force microscopy NtegraPrima (NT-MDT, Russia) in tapping mode. The samples for AFM analysis were prepared as follows: 0.1 wt.% aqueous dispersions were dropped on

a cover glass and dried in a dust-free chamber [35]. The AFM images were analyzed using Image Analysis 2.0 (NT-MDT, Russia) software.

2.3. Dynamic Light Scattering. The mean size and size distribution of ChN in aqueous dispersions at concentrations of 0.1, 0.5 and 1 wt.% were evaluated by dynamic light scattering (DLS) using a Zetatrac particle size analyzer (Microtrac, Inc., USA) with the aid of Microtrac application software program (V.10.5.3).

2.4. FTIR. The chitin nanocrystals in a form of films cast from 1 wt.% ChN aqueous dispersions were characterized by FTIR using a PerkinElmer Spectrum 100 FTIR spectrometer (PerkinElmer Inc., Wellesley, MA) equipped with universal ATR accessory with 8 scans at $4 \, \text{cm}^{-1}$ resolution.

2.5. Preparation of Poly(L,L-Lactide) Microparticles Stabilized with Chitin Nanocrystals. PLLA-ChN core-shell microparticles were prepared via Pickering emulsion solvent evaporation technique by mixing of 6 wt.% solutions of PLLA in CH_2Cl_2 : acetone (9 : 1 v/v) with 0.1–1 wt.% aqueous dispersions of ChN. The scheme of the microparticle fabrication is shown in Figure 1. Briefly, the oil phase, that is PLLA solution, was rapidly added to ChN dispersion to achieve oil/water phase ratio as 1/9 v/v. Microparticles stabilized with polyvinyl alcohol (PVA) solution (2.5 wt.%) as emulsifier in aqueous phase were prepared as model samples. The mixing of the phases was carried out using a four-blade stirrer at a rotation speed of 700 rpm for 3 hours. The emulsions were kept in a water bath through the mixing at 15°C during the first 15 min, and, afterwards, the temperature was increased up to 30°C. After the evaporation of the organic solvents, the obtained PLLA-ChN core-shell microparticles were collected, washed several times with deionized water and fractionated using sieves with apertures of 400, 315, 200, and 100 μm, then, freeze-dried and weighted.

(a)

(b)

(c)

(d)

(e)

(f)

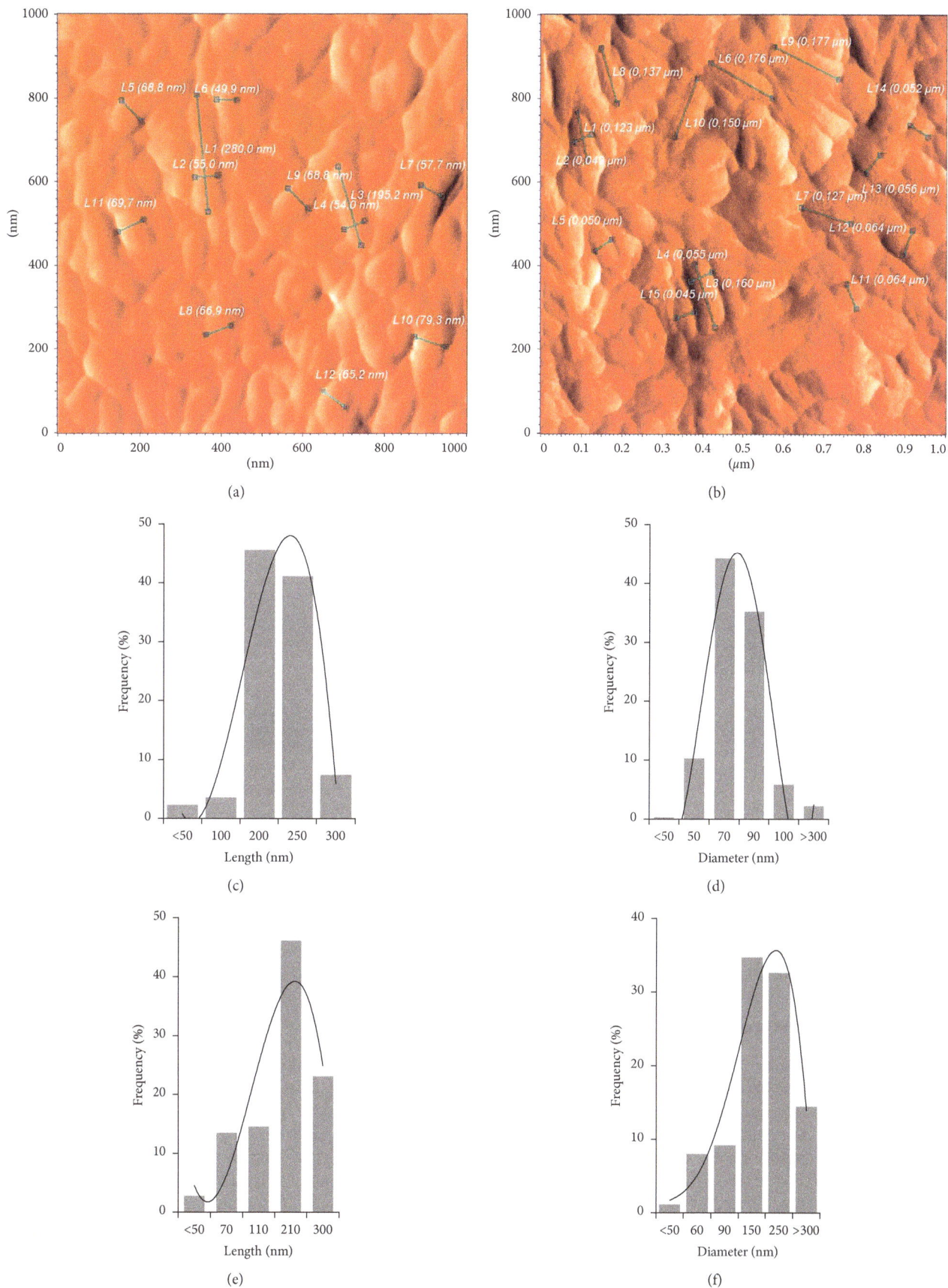

FIGURE 2: AFM images (a, b) and histograms of length (c, e) and diameter (d, f) distributions of ChN-X (a, c, and d) and ChN-K (b, e, and f).

(a)

(b)

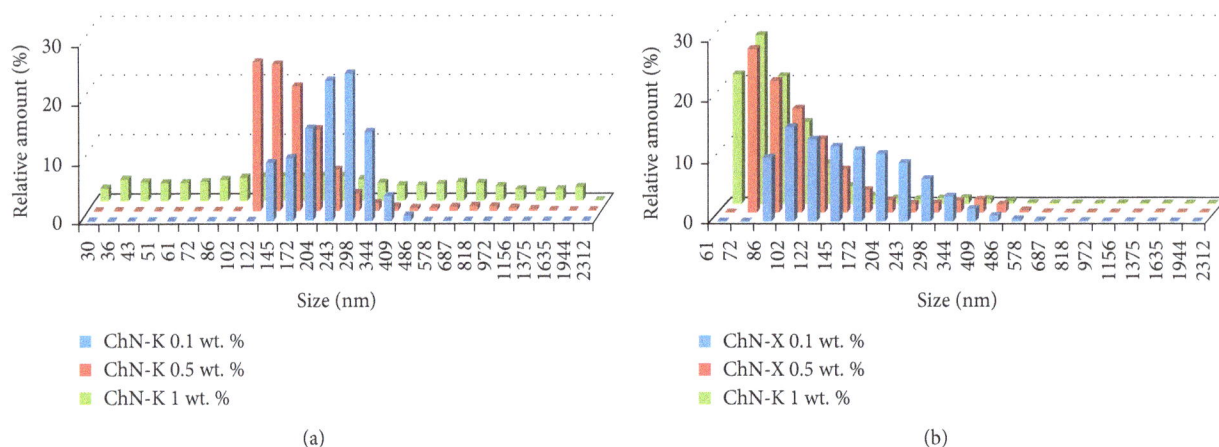

FIGURE 3: Number-weighted size distribution of ChN-K (a) and ChN-X (b) in their aqueous dispersions at various concentrations.

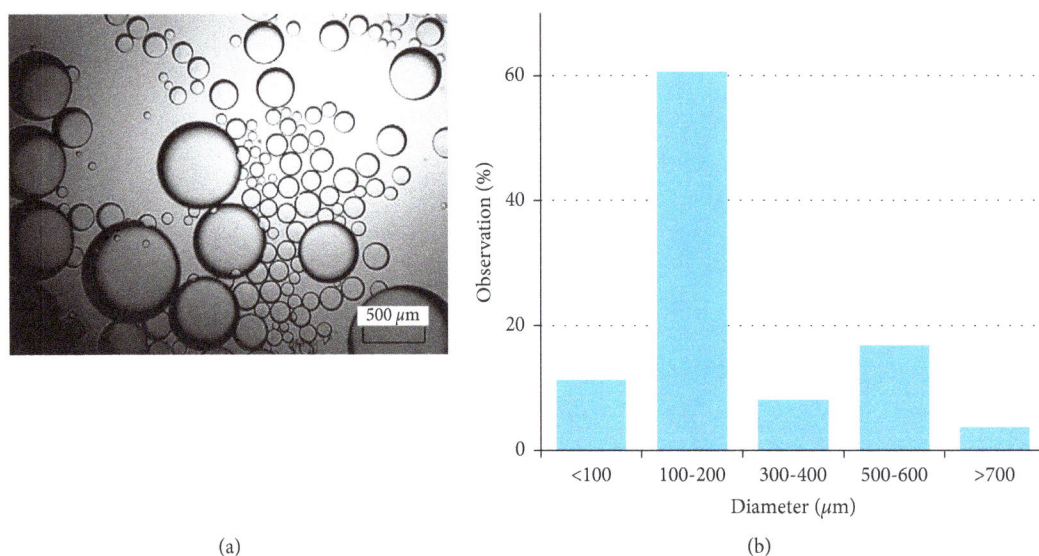

(a)

(b)

FIGURE 4: Optical micrographs of CH_2Cl_2/ChN-K (1 wt.%) emulsion (a) and oil droplet size distribution histogram (b).

2.6. Scanning Electron Microscopy. The surface morphology of the fabricated microparticles was studied by scanning electron microscopy (SEM) using PhenomProX (Phenom-World, Netherlands) at 10–15 kV.

3. Results and Discussions

3.1. Chitin Nanocrystals Characterization. According to AFM, both types of the chitin nanocrystals had anisometric morphology that was strongly dependent on the chitin source (Figure 2). As could be seen from the calculation of the length and diameter of the nanocrystals, both types of samples had a rather wide size distribution, which was more pronounced for ChN-K nanocrystals. The mean lengths of the nanocrystals were 181 ± 63 nm and 131 ± 94 nm; diameters of 65 ± 14 nm and 93 ± 42 nm for ChN-X and ChN-K, respectively. The calculation of the nanocrystals dimensions showed that the ChN-X had more anisometric morphology than ChN-K: aspect ratios were of 2.8 and 1.4, respectively. Thus, in spite of both raw chitins were from

crab shells, the structure of the polysaccharide could significantly vary as a function of animal source (species, parts, etc.) as well as conditions and quality of demineralization and deproteinization stages [36]. An X-ray analysis showed that a crystallinity of the extracted nanocrystals was higher than that of corresponding raw polysaccharides, while the positions of characteristic peaks were unchanged [35].

FTIR spectra of the extracted ChN showed no significant differences between the samples and contained a full set of typical characteristic bands of chitin, such as C-O stretching vibrations of pyranose cycle at 1070 cm^{-1}, asymmetric bridge oxygen stretching at 1155 cm^{-1}, the vibration of a C=O group bonded to hydroxyl at 1619 cm^{-1}, CH bending and symmetric CH_3 deformations at 1376 cm^{-1} as well as a set of amide bands: Amide I at 1654 cm^{-1}, Amide II at 1553 cm^{-1} and Amide III at 1309 cm^{-1} [37].

Since the fabricated ChN were intended to be used as stabilizing components for oil/water Pickering emulsions, their water dispersions were analyzed by DLS. As could be seen in Figure 3 the ChN-X dispersions were less sensitive to

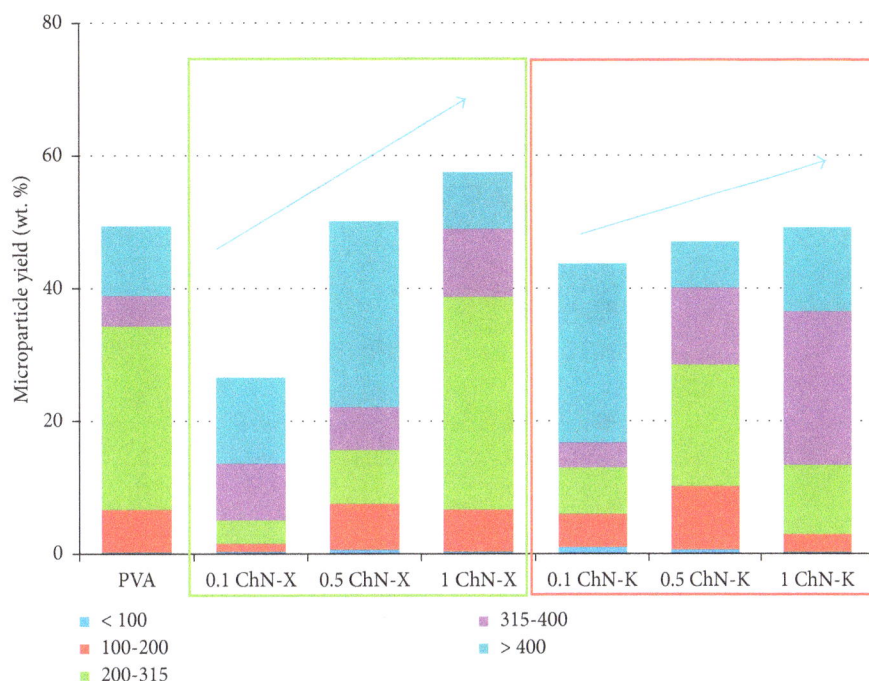

FIGURE 5: The total yield and microparticle size distribution as a function of origin and concentration of chitin nanocrystals in the aqueous phase.

the concentration change, while ChN-K showed a wide-range size distribution at 1 wt.% concentration.

Thus, the extraction of ChN allowed to prepare nano-dimensional forms of chitins having chemical structure of native polysaccharides and anisometric morphology, which was varied in a range of 1.4–2.8 as a function of chitin source.

3.2. Microparticle Fabrication.

As could be seen in Figure 4, the ChN could effectively stabilize a simple CH_2Cl_2/water (1/9 v/v) emulsion. The size distribution of oil droplets depends on mixing parameters and effectiveness of oil/water interface stabilization, which is controlled by the ChN hydrophilic-hydrophobic balance and concentration in the aqueous phase.

Total yield, size distribution, and shape and surface morphology of the fabricated PLLA microparticles are additionally regulated by an arrangement of nanocrystals at the interface during the evaporation of the organic solvent and decreasing of the oil/water surface area. As could be seen in Figure 5, the total yield of the PLLA microparticles stabilized with ChN increased with increase in nanocrystals concentration in the aqueous phase and reached up to 58 wt.% in the case of 1 wt.% of ChN-X. This yield was higher than that obtained using PVA as the emulsifier in spite of lower amount of ChN in comparison with PVA concentration used: 1 and 2.5 wt.%, respectively.

The comparative stabilization effectiveness of 0.1 wt.% ChN-K was higher than that of 0.1 wt.% ChN-X, but the impact of nanocrystals concentration in the aqueous phase was more pronounced for ChN-X as compared with microparticles stabilized by ChN-K. Increase of the ChN-X content in the aqueous phase led to a clear increase in stabilization effectiveness, that is, increase of total yield and

decrease of mean microparticle size, while this dependence was not so obvious in the case of ChN-K. This difference could be caused by the narrower size distribution of ChN-K nanocrystals and their better stability in the aqueous phase at the concentrations used (Figure 3).

SEM observations of all ChN-stabilized microparticles showed that they had more rough surface morphology than that of the ones stabilized using PVA (Figure 6). However, the shape of the fabricated PLLA-ChN microparticles was strongly varied as a function of origin and concentration of ChN. The microparticles stabilized with ChN-X and the ones stabilized with ChN-K at the lowest concentration (0.1 wt.%) had the spherical form, while the microparticles obtained using ChN-K at higher concentrations (0.5 and 1 wt.%) had an irregular shape.

It could be assumed that the ChN particles were attached so firmly at the oil/water interface that their rearrangement during the solvent evaporation process was impaired. The excess of ChN entrapped at the interface was not so critical in the case of usage of low nanocrystals concentration (0.1 wt. %), and the microparticles were able to remain in the spherical form (Figures 6(b) and 6(c)), while a presence of significant excess of ChN at the higher concentration in the aqueous phase could lead to the development of "raisin" effect. The difference in the shape of ChN-K- and ChN-X-stabilized microparticles could be caused by different aspect ratios of the nanocrystals: 1.4 and 2.8, respectively. The ChN-X entrapped at the interface could turn its orientation to the interface from lengthwise to crosswise, thus, to decrease the covered surface area. From another point of view, the higher stability of ChN-X size at various concentrations (see DLS data) could be a reason of such behavior as well.

(a)

(b)

(c)

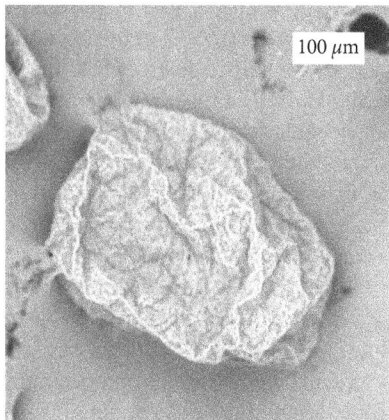

(d)

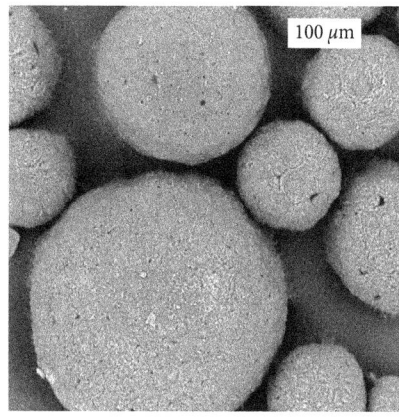

(e)

FIGURE 6: Continued.

FIGURE 6: SEM images of the microparticles stabilized with PVA (a), ChN-K (b, d, f), and ChN-X (c, e, g) at concentrations of 0.1 (b, c), 0.5 (d, e), and 1 wt.% (f, g) in the aqueous phase.

As a summary, chitin nanocrystals could effectively serve as a stabilizing component of oil/water emulsions during the preparation of polylactide-ChN core-shell microparticles, which total yield, size distribution, and shape were strongly depended on concentration, size, and size distribution of the nanocrystals used. These microparticles could be used as biodegradable cell microcarriers with improved cell affinity.

4. Conclusions

A promising approach to prepare poly(L,L-lactide) microparticles in a course of solvent evaporation from the Pickering oil/water emulsions stabilized with chitin nanocrystals has been discussed in terms of nanocrystals characteristics and processing parameters. Chitin nanocrystals extracted from two commercially available crab shell chitins showed different morphology and anisometric aspect ratios as well as an ability to be used as stabilizing agents in an aqueous phase during the preparation of polylactide microparticles via the solvent evaporation technique. The nanocrystals with higher aspect ratio allowed us to achieve a higher total yield of the fabricated spherical microparticles than that in a case of classical emulsifier application, that is, polyvinyl alcohol. The increase of nanocrystals concentration in an aqueous phase led to an increase in total yield and to decrease of mean microparticles size. The use of chitin nanocrystals with two fold lower aspect ratio at higher concentration in an aqueous phase caused the formation of microparticles with an irregular shape that could be explained by a lower ability of the nanocrystals to be rearranged at an interface during the solvent evaporation.

Conflicts of Interest

The authors declare that they have no conflicts of interest.

Acknowledgments

The authors are grateful to Dr. P. S. Timashev and his colleagues from the Institute of Photonic Technologies of "Crystallography and Photonics" of the Russian Academy of Sciences and Institute for Regenerative Medicine of Sechenov University for the access to the equipment.

References

[1] B. Tyler, D. Gullotti, A. Mangraviti, T. Utsuki, and H. Brem, "Polylactic acid (PLA) controlled delivery carriers for biomedical applications," *Advanced Drug Delivery Reviews*, vol. 107, pp. 163–175, 2016.

[2] R. Dorati, A. DeTrizio, I. Genta et al., "An experimental design approach to the preparation of pegylated polylactide-co-glicolide gentamicin loaded microparticles for local antibiotic delivery," *Materials Science and Engineering: C*, vol. 58, pp. 909–917, 2016.

[3] Y. Hong, C. Gao, Y. Xie, Y. Gong, and J. Shen, "Collagen-coated polylactide microspheres as chondrocyte microcarriers," *Biomaterials*, vol. 26, no. 32, pp. 6305–6313, 2005.

[4] J. Wu, T. Kong, K. W. K. Yeung et al., "Fabrication and characterization of monodisperse PLGA–alginate core–shell microspheres with monodisperse size and homogeneous shells for controlled drug release," *Acta Biomaterialia*, vol. 9, no. 7, pp. 7410–7419, 2013.

[5] L. Lao, H. Tan, Y. Wang, and C. Gao, "Chitosan modified poly (L-lactide) microspheres as cell microcarriers for cartilage tissue engineering," *Colloids and Surfaces B: Biointerfaces*, vol. 66, no. 2, pp. 218–225, 2008.

[6] A. Privalova, E. Markvicheva, Ch. Sevrin et al., "Biodegradable polyester-based microcarriers with modified surface tailored for tissue engineering," *Journal of Biomedical Materials Research Part A*, vol. 103, no. 3, pp. 939–948, 2015.

[7] T. S. Demina, T. A. Akopova, L. V. Vladimirov, A. N. Zelenetskii, E. A. Markvicheva, and Ch. Grandfils, "Polylactide-based microspheres prepared using solid-state copolymerized chitosan and D,L-lactide," *Materials Science and Engineering: C*, vol. 59, pp. 333–338, 2016.

[8] Y. Chevalier and M.-A. Bolzinger, "Emulsions stabilized with solid nanoparticles: Pickering emulsions," *Colloids and Surfaces A: Physicochemical and Engineering Aspects*, vol. 439, pp. 23–34, 2013.

[9] J. Wu and G.-H. Ma, "Recent studies of Pickering emulsions: particles make the difference," *Small*, vol. 12, no. 34, pp. 4633–4648, 2016.

[10] C. Linke and S. Drusch, "Pickering emulsions in foods–opportunities and limitations," *Critical Reviews in Food Science and Nutrition*, pp. 1–15, 2017, In press.

[11] C. Wang, C. Zhang, Y. Li, Y. Chen, and Z. Tong, "Facile fabrication of nanocomposite microspheres with polymer cores and magnetic shells by Pickering suspension polymerization," *Reactive & Functional Polymers*, vol. 69, no. 10, pp. 750–754, 2009.

[12] J. O. Zoppe, R. A. Venditti, and O. J. Rojas, "Pickering emulsions stabilized by cellulose nanocrystals grafted with thermo-responsive polymer brushes," *Journal of Colloid and Interface Science*, vol. 369, no. 1, pp. 202–209, 2012.

[13] T. Saigal, H. Dong, K. Matyjaszewski, and R. D. Tilton, "Pickering emulsions stabilized by nanoparticles with thermally responsive grafted polymer brushes," *Langmuir*, vol. 26, no. 19, pp. 15200–15209, 2010.

[14] N. Popadyuk, A. Popadyuk, I. Tarnavchyk et al., "Synthesis of covalently cross-linked colloidosomes from peroxidized Pickering emulsions," *Coatings*, vol. 6, pp. 1–14, 2015.

[15] H. Guo, D. Yang, M. Yang, Y. Gao, Y. Liu, and H. Li, "Dual responsive Pickering emulsions stabilized by constructed core crosslinked polymer nanoparticles via reversible covalent bonds," *Soft Matter*, vol. 12, no. 48, pp. 9683–9691, 2016.

[16] K. S. Silmore, C. Gupta, and N. R. Washburn, "Tunable Pickering emulsions with polymer-grafted lignin nanoparticles," *Journal of Colloid and Interface Science*, vol. 466, pp. 91–100, 2016.

[17] J. Lu, W. Zhou, J. Chen, Y. Jin, K. B. Walters, and S. Ding, "Pickering emulsions stabilized by palygorskite particles grafted with pH-responsive polymer brushes," *RSC Advances*, vol. 5, no. 13, pp. 9416–9424, 2015.

[18] Y. Zhu, J. Sun, C. Yi, W. Wei, and X. Liu, "One-step formation of multiple Pickering emulsions stabilized by self-assembled poly(dodecyl acrylate-co-acrylic acid) nanoparticles," *Soft Matter*, vol. 12, no. 36, pp. 7577–7584, 2016.

[19] Z. Wei, C. Wang, H. Liu, S. Zou, and Z. Tong, "Facile fabrication of biocompatible PLGA drug-carrying microspheres by O/W Pickering emulsions," *Colloids and Surfaces B: Biointerfaces*, vol. 91, pp. 97–105, 2012.

[20] F. Ye, M. Miao, B. Jiang et al., "Elucidation of stabilizing oil-in-water Pickering emulsion with different modified maize starch-based nanoparticles," *Food Chemistry*, vol. 229, pp. 152–158, 2017.

[21] S. Ge, L. Xiong, M. Li et al., "Characterizations of Pickering emulsions stabilized by starch nanoparticles: influence of starch variety and particle size," *Food Chemistry*, vol. 234, pp. 339–347, 2017.

[22] T. Zeng, Z.-L. Wu, J.-Y. Zhu et al., "Development of antioxidant Pickering high internal phase emulsions (HIPEs) stabilized by protein/polysaccharide hybrid particles as potential alternative for PHOs," *Food Chemistry*, vol. 231, pp. 122–130, 2017.

[23] X.-Y. Wang and M.-C. Heuzey, "Chitosan-based conventional and Pickering emulsions with long-term stability," *Langmuir*, vol. 32, no. 4, pp. 929–936, 2016.

[24] W. W. Mwangi, K.-W. Ho, B.-T. Tey, and E.-S. Chan, "Effect of environmental factors on the physical stability of Pickering emulsions stabilized by chitosan particles," *Food Hydrocolloids*, vol. 60, pp. 543–550, 2016.

[25] A. M. Salaberria, J. Labidi, and S. C. M. Fernandes, "Different routes to turn chitin into stunning nano-objects," *European Polymer Journal*, vol. 68, pp. 503–515, 2015.

[26] J. Huang, P. R. Chang, and N. Lin, *Polysaccharide-Based Nanocrystals: Chemistry and Applications*, Willey, Hoboken, NJ, USA, 2015.

[27] T. Abitbol, A. Rivkin, Y. Cao et al., "Nanocellulose, a tiny fiber with huge applications," *Current in Opinion on Biotechnology*, vol. 39, pp. 76–88, 2016.

[28] X. Zhai, D. Lin, D. Liu, and X. Yang, "Emulsions stabilized by nanofibers from bacterial cellulose: new potential food-grade Pickering emulsions," *Food Research International*, vol. 103, pp. 12–20, 2018.

[29] M. V. Tzoumaki, T. Moschakis, V. Kiosseoglou, and C. G. Biliaderis, "Oil-in-water emulsions stabilized by chitin nanocrystal particles," *Food Hydrocolloids*, vol. 25, no. 6, pp. 1521–1529, 2011.

[30] M. V. Tzoumaki, T. Moschakis, E. Scholten, and C. G. Biliaderis, "In vitro lipid digestion of chitin nanocrystal stabilized o/w emulsions," *Food & Function*, vol. 4, no. 1, pp. 121–129, 2013.

[31] T. Angkuratipakorn, A. Sriprai, S. Trantrawong, W. Chaiyasit, and J. Singkhonrat, "Fabrication and characterization of rice bran oil-in-water Pickering emulsion stabilized by cellulose nanocrystals," *Colloids and Surfaces A: Physicochemical and Engineering Aspects*, vol. 522, pp. 310–319, 2017.

[32] I. Kalashnikova, H. Bizot, B. Cathala, and I. Capron, "New Pickering emulsions stabilized by bacterial cellulose nanocrystals," *Langmuir*, vol. 27, no. 12, pp. 7471–7479, 2011.

[33] N. Rescignano, E. Fortunati, I. Armentano et al., "Use of alginate, chitosan and cellulose nanocrystals as emulsion stabilizers in the synthesis of biodegradable polymeric nanoparticles," *Journal of Colloid and Interface Science*, vol. 445, pp. 31–39, 2015.

[34] A. V. Istomin, T. S. Demina, E. N. Subcheva, T. A. Akopova, and A. N. Zelenetskii, "Nanocrystalline cellulose from flax stalks: preparation, structure, and use," *Fibre Chemistry*, vol. 48, no. 3, pp. 199–201, 2016.

[35] Yu. S. Sotnikova, T. S. Demina, A. V. Istomin et al., "Materials based on guar and hydroxypropylguar filled with nanocrystalline polysaccharides," *Fibre Chemistry*, vol. 49, no. 3, pp. 188–194, 2017.

[36] I. Younes and M. Rinaudo, "Chitin and chitosan preparation from marine sources. Structure, properties and applications," *Marine Drugs*, vol. 13, no. 3, pp. 1133–1174, 2015.

[37] M. L. Duarte, M. C. Ferreira, M. R. Marvao, and J. Rocha Duarte, "An optimised method to determine the degree of acetylation of chitin and chitosan by FTIR spectroscopy," *International Journal of Biological Macromolecules*, vol. 31, no. 1–3, pp. 1–8, 2002.

The Synthesis and Characterization of Poly(methyl methacrylate-tourmaline acrylate)

Yingmo Hu, Xubo Chen, and Yunhua Li

Beijing Key Laboratory of Materials Utilization of Nonmetallic Minerals and Solid Wastes, National Laboratory of Mineral Materials, School of Materials Science and Technology, China University of Geosciences, Beijing 100083, China

Correspondence should be addressed to Yingmo Hu; huyingmo@cugb.edu.cn

Academic Editor: Jainagesh A. Sekhar

In order to synthesize the tourmaline-containing functional copolymer, the polymerizable organic tourmaline acrylate was prepared by the surface modification on tourmaline powder and then copolymerization with methyl methacrylate to get poly(methyl methacrylate-tourmaline acrylate). The synthetic processes were optimized, and the structures of the as-prepared samples were characterized by IR, SEM, and X-ray fluorescence analysis. The characterization results indicated that the tourmaline has been introduced into the copolymer by means of surface modification with acryloyl chloride and following copolymerization with methyl methacrylate. The results from the negative ions measurement revealed that the amounts of released negative ions from both the modified tourmaline and its copolymer are much larger than that of tourmaline.

1. Introduction

Tourmaline is one of the typical cyclosilicate minerals; the general chemical formula of the tourmaline group can be expressed as $(Na,Ca)(Mg,Fe,Mn,Li,Al)_3Al_6(Si_6O_{18})(BO_3)_3(OH,F)_4$ and divided into schorl, dravite, and elbaite depending on different isomorphism states [1]. The complex crystal structures of tourmaline lead to the specific functional characteristics, such as permanently spontaneous polarization [2], thermoelectricity [3], piezoelectricity [4], far-infrared radiation [5], and negative ion releasing [6]. Tourmaline has been paid broad attention as a novel functional mineral material with lots of potential merits. Many researchers have done much work to explore the functional tourmaline composites and their applications.

Tourmaline/titanium dioxide/nylon-6 composite materials show good photocatalytic properties on photodegradation of organic pollutant and have potential application in water treatment [7, 8], and tourmaline/zinc composites have also been applied as catalytic materials in heat exchangers for mitigation of calcium carbonate ($CaCO_3$) fouling or scaling by physical water treatment [9]. Wu et al. [10] used the hydroxyl silicone-oil to modify tourmaline and then composited it with polyurethane to get superhydrophobic film materials

that can release negative air ions. The surface organic modification on tourmaline powder with span-60 was studied [11]; the modified tourmaline powder approached 100% activation index and showed an excellent hydrophobicity and then composited with polypropylene to get well-dispersed tourmaline/PP composite. The amounts of released negative ions over both modified tourmaline and its composite were greater than that of unmodified tourmaline. Tijing et al. [12] have fabricated tourmaline/polyurethane hybrid mat by electrospinning the homogeneous dispersion of superhydrophilic tourmaline in polyurethane and showed that it possesses high antibacterial activity against *Escherichia coli* and enterococci bacteria. Zheng and Wang [13] applied polyvinyl alcohol semi-IPN poly(acrylic acid)/tourmaline composite to remove heavy metals in waste water.

In order to synthesize the tourmaline-containing functional polymer, the surface modification on tourmaline powder with acryloyl chloride was made in this work to synthesize the polymerizable tourmaline acrylate, and then it was copolymerized with methyl methacrylate to get the tourmaline-containing copolymer. The preparation process was investigated in detail by optimizing the experimental parameters of contact angles and turbidities of modified tourmaline. The structures were characterized by IR,

SEM, and X-ray fluorescence analysis, and the amounts of released negative ions on both modified tourmaline and its copolymer have been measured quantificationally.

2. Experimental and Materials

2.1. Materials. Tourmaline powder (TM, d_{50} = 1.75 μm, d_{97} = 5.23 μm) was obtained from Yanxin Mineral Co. Ltd., Hebei, China; N,N-dimethylformamide (DMF, CP, Xilong Chemical Co., Ltd., Shantou, China) was dehydrated by anhydrous sodium sulfate; methyl methacrylate (CP, Tianjin Fuchen Chemical Reagents Factory, China) was newly distilled before being used; acryloyl chloride (CP, Huateng Technology Co., Ltd., Beijing, China), sulfuric acid (CP, Beijing Chemical Works, Beijing, China), anhydrous alcohol (CP, Beijing Chemical Works, Beijing, China), and liquid paraffin (CP, Tianjin Fuchen Chemical Works, Tianjin, China) were used as original.

2.2. The Preparation of Tourmaline Acrylate. In a 100 mL four-neck flask equipped with condenser, funnel, thermometer, and electromagnetic stirrer, 5 grams of tourmaline powder was dispersed in 15 mL dried DMF and heated to 40°C with sufficient stirring. 0.1 mL of sulfuric acid was added to a flask, and 3.0 mL acryloyl chloride was dropped into the flask with funnel. Then, the mixture was stirred for 3 h at 40°C. The samples were separated by filtration with vacuum and washed for three times with anhydrous ethanol and finally dried to get the tourmaline acrylate.

2.3. The Synthesis of Poly(methyl methacrylate-tourmaline acrylate). Anhydrous dimethylformamide (DMF, 70 mL) was poured into a 250 mL four-neck flask with condenser, funnel, thermometer, and electromagnetic stirrer, then nitrogen was added into DMF for 30 min, and then the flask was heated to 80°C. The 50 mL of newly distilled methyl methacrylate (MMA) solution 0.2 g benzoperoxide (BPO) was added dropwise. And tourmaline acrylate was added into the flask when the viscosity of reaction system began to increase and continued reacting for 8 hr at 80°C to get poly(methyl methacrylate-tourmaline acrylate) copolymer solution, and the tourmaline-containing fibers were prepared via electrospinning.

2.4. The Determination of Turbidity. 1.0 gram of modified tourmaline powder was added into a 100 mL beaker with 50 mL liquid paraffin. The mixture was stirred for 5 min and allowed to stand for 24 h. Then, the supernatant liquid was collected and its turbidity was tested by SGZ-2 nephelometer.

2.5. The Measurement of Contact Angle. The sample was pressed into a circle using the presser, and then the distilled water was dropped onto the surface of the sample to evaluate the contact angle with DSA100M optical instruments; each value was averaged over three tests.

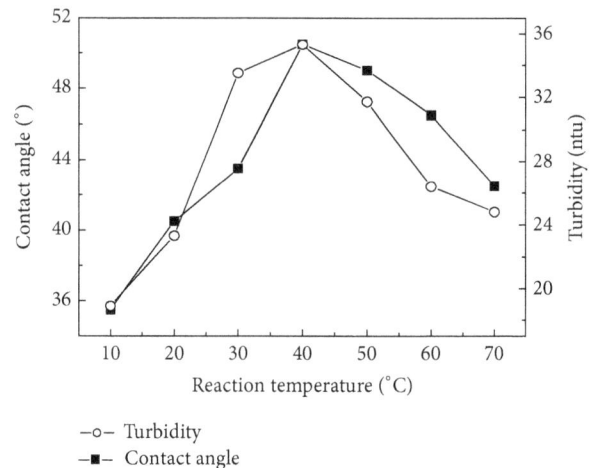

−o− Turbidity
−■− Contact angle

FIGURE 1: Effect of reaction temperature.

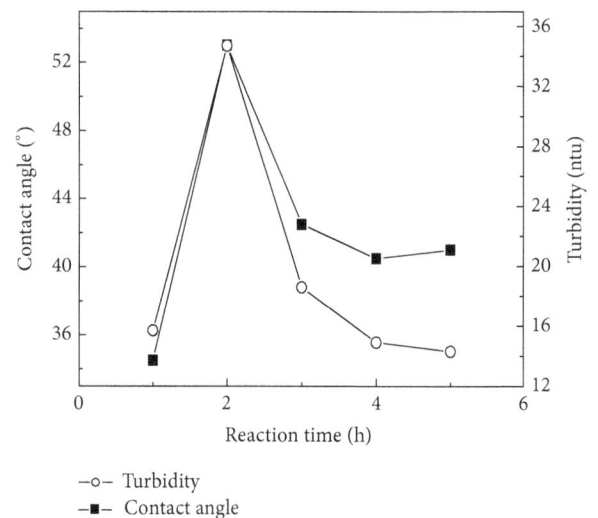

−o− Turbidity
−■− Contact angle

FIGURE 2: Effect of reaction time.

3. Results and Discussion

3.1. The Optimization of Acryloyl Chloride Modified Tourmaline. The modified effects of tourmaline powder with acryloyl chloride under various experimental conditions were studied by focusing on the experimental parameters of turbidity in liquid paraffin and contact angles of product.

3.1.1. Effect of the Reaction Temperature. The curves of contact angle and turbidity of modified tourmaline at different reaction temperature were shown in Figure 1. Both contact angle and turbidity of modified tourmaline increased with rising of the reaction temperature below 40°C and then reduced with the persistent rising of reaction temperature, reaching the maximum at 40°C.

3.1.2. Effect of Reaction Time. Figure 2 revealed the influence of reaction time on the contact angle and turbidity of modified tourmaline. Figure 2 showed the same trend for

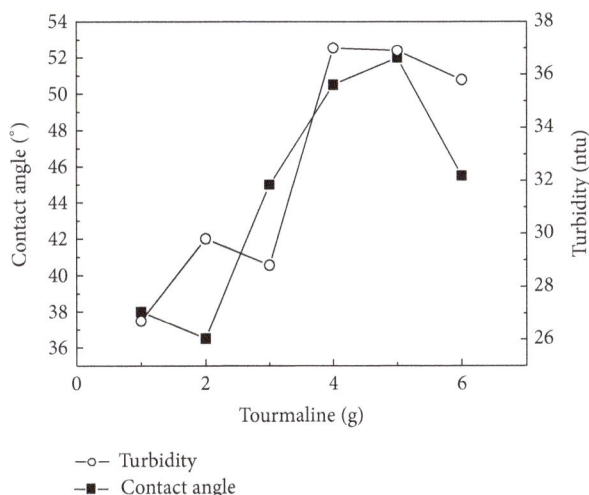

FIGURE 3: Effect of the amount of tourmaline.

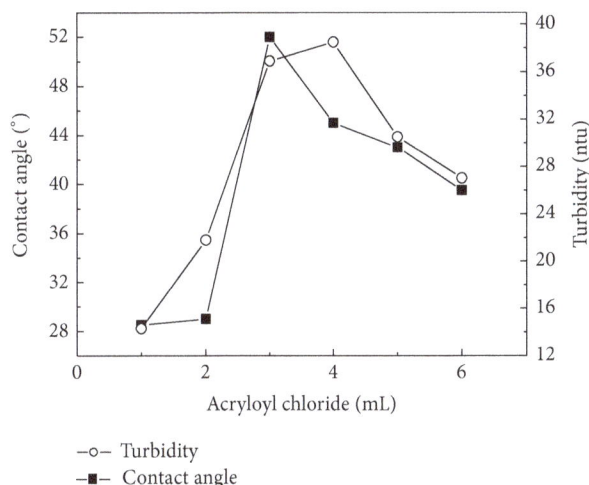

FIGURE 4: Effect of acryloyl chloride.

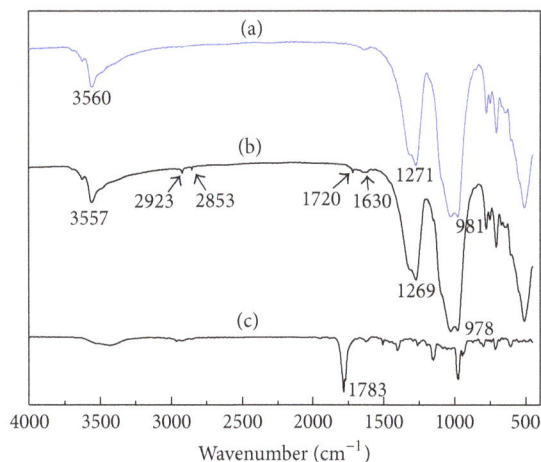

FIGURE 5: FT-IR spectra of tourmaline (a), modified tourmaline (b), and acryloyl chloride (c).

SCHEME 1

contact angle and turbidity of modified tourmaline, which increased sharply at first and then decreased with prolonging the reaction time. The optimal reaction time was 2 h.

3.1.3. Effect of the Dosage of Tourmaline.

By fixing the dosage of solvent (DMF) as 50 mL, the effect of the amount of tourmaline on the contact angle and turbidity of modified tourmaline were shown in Figure 3. The contact angle and turbidity of modified tourmaline tended to increase with the increase of tourmaline and then decrease when the amount of tourmaline was over 5 g. Accordingly, the optimal amount of tourmaline was 5 g in this work.

3.1.4. Effect of Acryloyl Chloride.

The effect of acryloyl chloride on turbidity and contact angle of modified tourmaline was exhibited in Figure 4. The turbidity and contact angle of modified tourmaline increased speedily with the increase of acryloyl chloride amount at the beginning and then reduced with further increasing acryloyl chloride when the dosage of acryloyl chloride was over 3 mL, which is the optimum amount of acryloyl chloride.

3.2. Structural Characterization of Tourmaline Acrylate and Its Copolymer

3.2.1. IR Analysis of Tourmaline Acrylate.

The IR spectra of unmodified tourmaline and modified tourmaline were shown in Figure 5. In Figure 5(a), unmodified tourmaline [14] has absorption bands at $3558\,cm^{-1}$ (–OH group), $1272\,cm^{-1}$ (B–O group), and $975\,cm^{-1}$ (Si–O group), and the IR spectrum of modified tourmaline (Figure 5(b)) shows absorption bands of methyl ($2923\,cm^{-1}$), methylene ($2853\,cm^{-1}$), carbonyl ($1720\,cm^{-1}$), and double bond ($1630\,cm^{-1}$) of acryloyl group. After modification, the shape of absorption bands of tourmaline was almost unchanged, but the bands all showed red shift compared with that of the unmodified tourmaline owing to the electronic effect of acryloyl group. This indicated that acryloyl group has been introduced on the surface of tourmaline to produce the polymerizable tourmaline acrylate via the reaction of acryloyl chloride with tourmaline powder (Scheme 1).

3.2.2. The IR Spectra of Poly(methyl methacrylate-tourmaline acrylate).

The IR spectra of poly(methyl methacrylate-tourmaline acrylate) were shown in Figure 6. In Figure 6, besides the characteristic absorption peaks of poly(methyl methacrylate) at $3000\,cm^{-1}$ and $2954\,cm^{-1}$ (CH_3 and CH_2), $1731\,cm^{-1}$ (C=O), $1485\,cm^{-1}$ and $1447\,cm^{-1}$ (C–O), and $1190\,cm^{-1}$ and $1151\,cm^{-1}$ (C–O–C) [15], the absorption bands of tourmaline were found clearly at $3436\,cm^{-1}$ (–OH group), $1273\,cm^{-1}$ (B–O bands), and $986\,cm^{-1}$ (Si–O bands) [14],

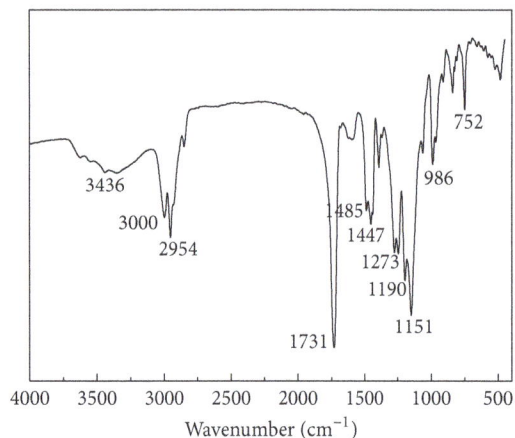

FIGURE 6: The IR spectrum of poly(methyl methacrylate-tourmaline acrylate).

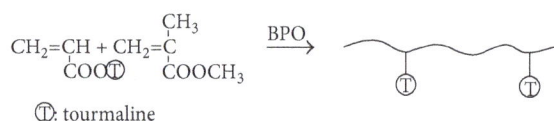

: tourmaline

SCHEME 2

which correspond to those of poly(methyl methacrylate-tourmaline acrylate). This demonstrated that tourmaline was successfully introduced into copolymer by copolymerization of tourmaline acrylate with methyl methacrylate (Scheme 2).

3.2.3. Stabilization of Copolymer.

Figure 7 shows the storage stability of copolymers. The poly(methyl methacrylate-tourmaline acrylate) displayed excellent stability (Figure 7(b)) after storage for 30 days. There is no sedimentation of tourmaline observed, and only a little solvent separated out in the top layer due to the dissolubility of copolymer in the process of longtime storage. In contrast, obvious tourmaline sediment appeared in less than a week for the poly(methyl methacrylate-tourmaline) obtained by the same procedure (Figure 7(a)). These observations also indicated that tourmaline was introduced into the copolymer to be poly(methyl methacrylate-tourmaline acrylate) via the preparation of tourmaline acrylate and copolymerization.

3.2.4. X-Ray Fluorescence (XRF) Analysis of Copolymer.

In order to confirm the existence of tourmaline in copolymer, the XRF analysis was shown in Table 1. The results indicated that Si, Al, Na, Mg, Fe, Ca, and other elements were detected in copolymer, which is in accordance with the characteristic of tourmaline [14]. So the tourmaline was suitable to be incorporated into copolymer to get tourmaline-containing copolymer.

3.3. SEM Images of Tourmaline-Containing Fibers.

Figures 8(a) and 8(b) show the SEM images of electrospinning fibers of poly(methyl methacrylate-tourmaline acrylate) and poly(methyl methacrylate-tourmaline), respectively.

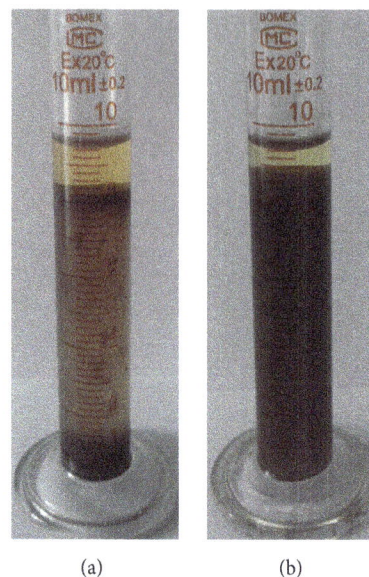

(a)   (b)

FIGURE 7: The storage stability of copolymer.

TABLE 1: The X-ray fluorescence analysis of copolymer.

Components	Mass percent (%)
SiO_2	1.88
Al_2O_3	0.431
Na_2O	0.399
MgO	0.297
Fe_2O_3	0.295
CaO	0.158
TiO_2	0.005
MnO	0.0036
K_2O	0.0359

TABLE 2: The amount of negative ions released.

Sample	The amount of negative ions (ions/cm^3)
Tourmaline	300
Modified tourmaline	500
Copolymer	400

Figure 8(a) indicated that the particle size of tourmaline was much smaller and well-distributed in copolymer, while tourmaline showed agglomeration obviously in poly(methyl methacrylate-tourmaline) (Figure 8(b)) and the fibers were brittle. This is because unmodified tourmaline could not be well dispersed and is easy to agglomerate, which would badly influence the performance of its composites. After modification with acryloyl chloride, the tourmaline was introduced evenly into copolymer by copolymerization to get tourmaline-containing copolymer with well distribution and good performance.

3.4. The Measurement of the Negative Ions.

The amount of negative ions released from the tourmaline, modified tourmaline, and tourmaline-containing copolymer was listed in Table 2. From the data in Table 2, the amounts of released

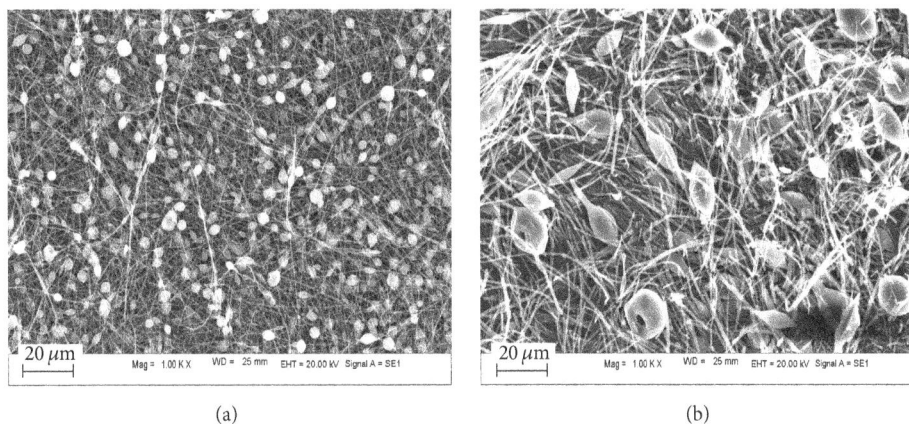

(a)							(b)

FIGURE 8: SEM images of electrospinning fibers of poly(methyl methacrylate-tourmaline acrylate) (a) and poly(methyl methacrylate-tourmaline) (b).

negative ions of both modified tourmaline and tourmaline-containing copolymer were greater obviously than that of tourmaline. This was because the surface energy and aggregation of the modified tourmaline particles were reduced, and the surface area of tourmaline particle increased, which resulted in an increased amount of released negative ions [11]. The negative ions amount from copolymer was a little lower than that from modified tourmaline because partial tourmaline was coated. But it is greater than that of tourmaline.

4. Summary

As a summary of this work, there are a few informative points: (1) the tourmaline was modified with acryloyl chloride to get polymerizable tourmaline acrylate; (2) the tourmaline-containing poly(methyl methacrylate-tourmaline acrylate) was obtained by means of copolymerizing the tourmaline acrylate and methyl methacrylate; (3) the copolymer has an excellent storage stability; (4) the negative ions measurement indicated that the amount of released negative ions of both modified tourmaline and its copolymer is larger than that of unmodified tourmaline.

This work provided more extensive channels for development of novel functional materials and high-worth comprehensive utilization of tourmaline.

Competing Interests

The authors declare that they have no competing interests.

Acknowledgments

This work is supported by the National Natural Science Foundation of China (no. 51372233).

References

[1] L. Zhang, Y. Wu, and B. Xiao, "Research on the composition, structure and deep processing technology of tourmaline," *Multipurpose Utilization of Mineral Resources*, vol. 4, pp. 30–34, 2009.

[2] T. Kubo, "Interface activity of water given rise by tourmaline," *Solid State Physics*, vol. 24, pp. 1055–1060, 1989.

[3] J. Wei, Y. Y. Liu, and K. Y. Zhang, "Study on application features of tourmaline," *Non-Metallic Mines*, vol. 26, pp. 34–36, 2003.

[4] X. F. Zhang, R. H. Wu, and Y. Dong, "Research on surface electricity propriety of tourmaline powder," *China Non-Metallic Minerals Industry*, vol. 14, pp. 73–76, 2005.

[5] X. Yang, Y. M. Hu, and J. H. Zhu, "Surface modification and characterization of tourmaline powder by span 60," *Multipurpose Utilization of Mineral Resources*, vol. 1, pp. 14–17, 2011.

[6] D. J. Henry and M. K. de Brodtkorb, "Mineral chemistry and chemical zoning in tourmalines, Pampa del Tamboreo, San Luis, Argentina," *Journal of South American Earth Sciences*, vol. 28, no. 2, pp. 132–141, 2009.

[7] S.-J. Kang, L. D. Tijing, B.-S. Hwang, Z. Jiang, H. Y. Kim, and C. S. Kim, "Fabrication and photocatalytic activity of electrospun nylon-6 nanofibers containing tourmaline and titanium dioxide nanoparticles," *Ceramics International*, vol. 39, no. 6, pp. 7143–7148, 2013.

[8] H.-Y. Xu, Z. Zheng, and G.-J. Mao, "Enhanced photocatalytic discoloration of acid fuchsine wastewater by TiO$_2$/schorl composite catalyst," *Journal of Hazardous Materials*, vol. 175, no. 1–3, pp. 658–665, 2010.

[9] L. D. Tijing, M.-H. Yu, C.-H. Kim et al., "Mitigation of scaling in heat exchangers by physical water treatment using zinc and tourmaline," *Applied Thermal Engineering*, vol. 31, no. 11-12, pp. 2025–2031, 2011.

[10] Z. F. Wu, H. Wang, M. Xue et al., "Facile preparation of superhydrophobic surfaces with enhanced releasing negative air ions by a simple spraying method," *Composites Science and Technology*, vol. 94, pp. 111–116, 2014.

[11] Y. M. Hu and X. Yang, "The surface organic modification of tourmaline powder by span-60 and its composite," *Applied Surface Science*, vol. 258, no. 19, pp. 7540–7545, 2012.

[12] L. D. Tijing, M. T. G. Ruelo, A. Amarjargal et al., "Antibacterial and superhydrophilic electrospun polyurethane nanocomposite fibers containing tourmaline nanoparticles," *Chemical Engineering Journal*, vol. 197, pp. 41–48, 2012.

[13] Y. Zheng and A. Q. Wang, "Removal of heavy metals using polyvinyl alcohol semi-IPN poly(acrylic acid)/tourmaline composite optimized with response surface methodology," *Chemical Engineering Journal*, vol. 162, no. 1, pp. 186–193, 2010.

[14] P. S. R. Prasad and D. S. Sarma, "Study of structural disorder in natural tourmalines by infrared spectroscopy," *Gondwana Research*, vol. 8, no. 2, pp. 265–270, 2005.

[15] J. M. Zhang, X. G. Sun, and C. H. Wang, "Synthesis and kinetics study of TiO_2/P(MMA-BA-MAA) composite particles by emulsion polymerization," *CIESC Journal*, vol. 60, no. 10, pp. 2640–2649, 2009.

Permissions

List of Contributors

Woshington S. Brito and André L. Mileo Ferraioli Silva
Pará Federal University, 66075-110 Belém, PA, Brazil

Rozineide A. A. Boca Santa and Humberto Gracher Riella
Chemical Engineering Department, Santa Catarina Federal University, 88040-900 Florianópolis, SC, Brazil

Kristoff Svensson, José Antônio da Silva Souza and Herbert Pöllmann
Department of Geosciences and Geography, Martin Luther University of Halle-Wittenberg, Halle 06108, Germany

M. Moreno, G. M. Alonzo-Medina and A. I. Oliva-Avilés
División de Ingeniería y Ciencias Exactas, Universidad Anáhuac Mayab, Carretera Mérida-Progreso Km. 15.5 AP 96 Cordemex, 97310 Mérida, YUC, Mexico

A. I. Oliva
Centro de Investigación y de Estudios Avanzados del IPN, Unidad Mérida, Departamento de Física Aplicada, AP 73 Cordemex, 97310 Mérida, YUC, Mexico

Lingfeng Zhang
School of Civil Engineering, Southeast University, Nanjing, China

Weiqing Liu
School of Civil Engineering, Southeast University, Nanjing, China
Advanced Engineering Composites Research Center, Nanjing Tech University, Nanjing, China

Guoqing Sun and Lu Wang
College of Civil Engineering, Nanjing Tech University, Nanjing, China

Lingzhi Li
College of Civil Engineering, Tongji University, Shanghai, China

D. Kioupis, S. Tsivilis and G. Kakali
National Technical University of Athens, School of Chemical Engineering, 9 Heroon Polytechniou St., 15773 Athens, Greece

Ch. Kavakakis
Hellenic Open University, School of Science and Technology, 18 Parodos Aristotelous St., 26335 Patras, Greece

Shujahadeen B. Aziz
Advanced Polymeric Materials Research Laboratory, School of Science-Department of Physics, Faculty of Science and Science Education, University of Sulaimani, Sulaimani, Kurdistan Region, Iraq

Sujal Bhattacharjee and Dilpreet S. Bajwa
Department of Mechanical Engineering, North Dakota State University, Fargo, ND 58108-6050, USA

Muhammad Bilal Khan, Rahim Jan, Amir Habib and Ahmad Nawaz Khan
School of Chemical and Materials Engineering (SCME), National University of Sciences and Technology (NUST), H-12 Campus, Islamabad 44000, Pakistan

J. Ramesh Babu and K. Vijaya Kumar
Department of Physics, K L University, Vaddeswarram, 522 502 Guntur, India

K. Ravindhranath
Department of Chemistry, K L University, Vaddeswarram, 522 502 Guntur, India

Dayong Yang, Zhenping Wan, Peijie Xu and Longsheng Lu
School of Mechanical and Automotive Engineering, South China University of Technology, Guangzhou 510640, China

Richa Saxena
Ultrasonic and Dielectric Laboratory, Department of Physics, H.N.B. Garhwal University, Srinagar, Uttarakhand, India
IFTM University, Lodhipur Rajput, Delhi Road, Moradabad 244001, India

S. C. Bhatt
IFTM University, Lodhipur Rajput, Delhi Road, Moradabad 244001, India

Wei Shang and Zhongxian Liu
Key Laboratory of Protection and Retrofitting for Civil Building and Construction of Tianjin, Tianjin Chengjian University, Tianjin 300384, China

Xiaojing Yang
School of Materials Science and Engineering, Hebei University of Technology, Tianjin 300130, China

Xinhua Ji
Department of Mechanics, Tianjin University, Tianjin 300072, China

D. N. Gómez-Balbuena
CIATEQ A.C., Av. Del Retablo No. 150, Col. Constituyentes Fovissste, 76150 Santiago de Querétaro, QRO, Mexico
Instituto Tecnológico Superior de Huichapan, Domicilio conocido sn Col. El Saucillo, 42411 Huichapan, HGO, Mexico

T. López-Lara, J. B. Hernandez-Zaragoza, J. Horta-Rangel and E. Rojas-Gonzalez
División de Estudios de Posgrado, Facultad de Ingeniería, Universidad Autónoma de Querétaro, Cerro de las Campanas S/N, Col. Niños Héroes, 76010 Santiago de Querétaro, QRO, Mexico

R. G. Ortiz-Mena
Instituto Tecnológico Superior de Huichapan, Domicilio conocido sn Col. El Saucillo, 42411 Huichapan, HGO, Mexico

M. G. Navarro-Rojero
CIATEQ A.C., Av. Del Retablo No. 150, Col. Constituyentes Fovissste, 76150 Santiago de Querétaro, QRO, Mexico

R. Salgado-Delgado
División de Estudios de Posgrado e Investigación, Instituto Tecnológico de Zacatepec, Calzada Tecnológico No. 27, Col. Centro, 62780 Zacatepec, MOR, Mexico

V. M. Castano
Centro de Física Aplicada y Tecnología Avanzada, Universidad Nacional Autónoma de México, Boulevard Juriquilla 3001, 76230 Santiago de Querétaro, QRO, Mexico

Abu Shaid, Lijing Wang, Rajiv Padhye and Amit Jadhav
School of Fashion and Textiles, RMIT University, Melbourne, VIC, Australia

N. Moonprasith, C. Kongkaew, S. Loykulnant and K. Suchiva
Natural Rubber Focus Unit, National Metal and Materials Technology Center, 114 Thailand Science Park, Phahonyothin Road, Khlong Nueng, Khlong Luang, PathumThani 12120, Thailand

A. Poonsrisawat and V. Champreda
Enzyme Technology Laboratory, National Center for Genetic Engineering and Biotechnology, 113 Thailand Science Park, Phahonyothin Road, Khlong Nueng, Khlong Luang, PathumThani 12120, Thailand

Hamad F. Alharbi, Monis Luqman, Ehab El-Danaf and Nabeel H. Alharthi
Mechanical Engineering Department, College of Engineering, King Saud University, 11421 Riyadh, Saudi Arabia

Mengting Li, Jun Wang and Weiqing Liu
College of Civil Engineering, Nanjing Tech University, Nanjing 211816, China

Ruifeng Liang and Hota Ganga Rao
Department of Civil and Environmental Engineering, West Virginia University, Morgantown, WV, USA

Yang Li
Jiangsu Wei Xin Engineering Consultants Ltd., Nanjing, China

Victor Stroganov
Kazan State University of Architecture and Engineering, Zelenaya 1, 420043 Kazan, Russia

Oleg Stoyanov and Ilya Stroganov
Kazan National Research Technology University, Karl Marx 68, 420015 Kazan, Russia

Eduard Kraus
SKZ-German Plastic Center, Friedrich-Bergius-Ring 22, 97076 Wuerzburg, Germany

Jiang Chang
Institute of Materials Processing and Intelligent Manufacturing, College of Materials Science and Chemical Engineering, Harbin Engineering University, Harbin 150001, China
College of Light Industry and Textile Engineering, Qiqihar University, Qiqihar 161006, China

Bing Tian, Li Li and Yufeng Zheng
Institute of Materials Processing and Intelligent Manufacturing, College of Materials Science and Chemical Engineering, Harbin Engineering University, Harbin 150001, China

Bo Li
State Key Laboratory Cultivation Base for Gas Geology and Gas Control, Jiaozuo 454000, China
The Collaborative Innovation Center of Coal Safety Production of Henan Province, Jiaozuo 454000, China
Hebei State Key Laboratory of Mine Disaster Prevention, North China Institute of Science and Technology, Beijing 101601, China
State and Local Joint Engineering Laboratory for Gas Drainage & Ground Control of Deep Mines, Henan Polytechnic University, Jiaozuo 454003, China

Junxiang Zhang
State and Local Joint Engineering Laboratory for Gas Drainage & Ground Control of Deep Mines, Henan Polytechnic University, Jiaozuo 454003, China
School of Energy Science and Engineering, Henan Polytechnic University, Jiaozuo 454003, China

Jianping Wei and Qiang Zhang
State Key Laboratory Cultivation Base for Gas Geology and Gas Control, Jiaozuo 454000, China
The Collaborative Innovation Center of Coal Safety Production of Henan Province, Jiaozuo 454000, China

T. S. Demina
Enikolopov Institute of Synthetic Polymer Materials, Russian Academy of Sciences, 70 Profsoyuznaya St., 117393 Moscow, Russia
Institute for Regenerative Medicine, Sechenov University, 8-2 Trubetskaya St., Moscow 119991, Russia

Yu. S. Sotnikova
Enikolopov Institute of Synthetic Polymer Materials, Russian Academy of Sciences, 70 Profsoyuznaya St., 117393 Moscow, Russia

Moscow Aviation Institute (National Research University), 4 Volokolamskoe Shosse, 125993 Moscow, Russia

A. V. Istomin, T. A. Akopova and A. N. Zelenetskii
Enikolopov Institute of Synthetic Polymer Materials, Russian Academy of Sciences, 70 Profsoyuznaya St., 117393 Moscow, Russia

Ch. Grandfils
Interfaculty Research Centre on Biomaterials (CEIB), Chemistry Institute, University of Liège, B6C Allée du 6 Aôut 11, B-4000 Liege (Sart-Tilman), Belgium

Yingmo Hu, Xubo Chen and Yunhua Li
Beijing Key Laboratory of Materials Utilization of Nonmetallic Minerals and Solid Wastes, National Laboratory of Mineral Materials,
School of Materials Science and Technology, China University of Geosciences, Beijing 100083, China

Index

A

Absorbance Spectrum, 11

Acoustic Impedance, 89-92

Adiabatic Compressibility, 89-91

Aerogel Particle, 113, 116

Alternating Direction Implicit, 17, 20, 33

Ambient Temperature, 1-2, 4, 23

Anisotropic Plastic, 126-127

Arrhenius Equation, 45, 50-52, 55, 146

Atomic Force Microscopy, 187-188

B

Bayer Process, 1-4

C

Carbon Fiber-reinforced Polymer, 18

Chemical Bath Deposition, 7, 9, 15-16

Chitin Nanocrystals, 187-188, 190-191, 193

Chitosan, 45, 47, 49-52, 55, 67, 71, 79, 193-194

Coal Bed Methane, 177

Coal Dust Polymer, 177

Coefficient of Thermal Expansion, 57-58, 60-62, 64

Crack Initiation, 81-84, 86

D

Deproteinized Natural Rubber, 119, 121, 123

Differential Scanning Calorimetry, 56, 58, 141

E

Electric Force, 112, 117

Electrical Discharge Machining, 81, 88

Electrochemical Cells, 72-74, 78

Electrospinning, 112-114, 116-118, 196, 198-199

Electrospraying, 112-113, 115, 117

Epoxy Oligomers, 152-154, 157, 163, 165-166

Ethylenediaminetetraacetic Acid, 119-120

F

Fiber-reinforced Polymer, 17-18, 32, 151

Fly Ash, 1-5, 34-36, 39, 43-44, 110, 178

Fourier Transform Infrared Spectroscopy, 56, 58

Fracture Mechanical Properties, 94-96, 98

Fracture Morphology, 94-95, 98-99

G

Gauss-seidel Iterative Approach, 17

Gel Permeation Chromatography, 56, 58

Geopolymer Synthesis, 1-3, 34

Geopolymerization Reaction, 1, 3-4

Geopolymers, 1-6, 34-38, 40-41, 43-44

Glass Fiber, 17, 138-139, 151

H

Heat Deflection Temperature, 8, 56, 58-61

Hydroxyurethane Polymers, 152, 154

I

Intermolecular Free Length, 89-90, 92

Ion Transport, 45, 49-52, 79

L

Light Ash, 1-2

M

Melt Flow Index, 58, 60-62, 64

Methyl Methacrylate, 79, 100, 195-199

Molybdenum Disulphide, 66

Multiwall Carbon Nanotubes, 7-8

N

Nanocomposite Polymer Electrolyte, 79

Needleless Electrospinning, 112-113, 117

O

Orthogonal Cutting, 81-84, 87-88

P

Polyimide, 8, 67, 70, 79

Polymer Composites, 10, 15, 18, 32-33, 44, 66-67, 71, 81-83, 86-88, 111, 138, 151

Polymer Electrolyte Films, 72-73, 78

Polymer Nanocomposites, 67

Polymeric Substrate, 7-8, 12-13

Polymethyl Methacrylate, 94

Polystyrene, 66-67, 70, 187

Polysulfone, 7-8, 12

Polyurethane, 67, 70, 113, 160, 166, 168, 171, 175-176, 178, 185, 195, 199

Polyvinyl Alcohol, 89-93, 113, 188, 193, 195, 199

Q

Quarry Waste, 102-106, 109-110

S

Sand Replacement, 102-103

Scanning Electron Microscopy, 4, 48, 50, 55-56, 66, 177, 190

Sealing Performance, 177-178, 182-185
Sodium Hydroxide, 1-2, 4, 154
Strain Paths, 126-127
Sustained Loads, 138-139, 143-145, 148-151

T
Tetrahydrofuran, 58, 67
Texture Evolution, 126-128, 130, 137
Thermogravimetric Analysis, 56, 58
Thermomechanical Model, 17-18, 20, 25
Thin Films Deposited, 7-8, 12, 14
Tourmaline, 195-199

Tourmaline Acrylate, 195-199
Transmission Electron Microscopy, 55, 67

U
Ultimate Tensile Strength, 66-67, 69, 144-146
Ultrasonic Velocities, 89
Ultraviolet Absorbents, 168

W
Wet-dry Cycles, 138-139, 150

X
X-ray Diffraction, 35, 47, 67, 102

www.ingramcontent.com/pod-product-compliance
Lightning Source LLC
Chambersburg PA
CBHW080703200326
41458CB00013B/4951